Problem Books in Mathematics

Series Editor

Peter Winkler, Department of Mathematics, Dartmouth College, Hanover, USA

Books in this series are devoted exclusively to problems - challenging, difficult, but accessible problems. They are intended to help at all levels - in college, in graduate school, and in the profession. Arthur Engels "Problem-Solving Strategies" is good for elementary students and Richard Guys "Unsolved Problems in Number Theory" is the classical advanced prototype. The series also features a number of successful titles that prepare students for problem-solving competitions.

Corneliu Mănescu-Avram

Selection Tests in Algebra for Mathematical Olympiads

Springer

Corneliu Mănescu-Avram
Ploiesti, Romania

ISSN 0941-3502 ISSN 2197-8506 (electronic)
Problem Books in Mathematics
ISBN 978-3-032-01031-5 ISBN 978-3-032-01032-2 (eBook)
https://doi.org/10.1007/978-3-032-01032-2

Mathematics Subject Classification: 26C99; 97H99; 97U40; 39-01; 26D05

This Springer imprint is published by the registered company Springer Nature Switzerland AG
The registered company address is: Gewerbestrasse 11, 6330 Cham, Switzerland

Competing Interests The author has no competing interests to declare that are relevant to the content of this manuscript.

About the Book

This book contains some selection tests proposed to the national IMO teams in many countries. Most solutions are original or adapted by the author. The material is organized chronologically, in order to have an image of the evolution of the phenomenon. The index allows the direct access of a theme.

The book is a possible preparation material for the interested students or other mathematics lovers. Further readings are indicated in the references.

Contents

About the Author

Corneliu Mănescu-Avram is graduated from the Faculty of Mathematics, University of Bucharest, Romania. In 2010, he obtained the QTS certification. He is an active problem-solver and proposer in many mathematical journals, such as *Mathematical Reflections, Crux Mathematicorum, Pi Mu Epsilon Journal, Pentagon Kappa Mu Epsilon Journal, Mathematical Excalibur, School Science and Mathematics, Mathematical Recreations* (in Romanian), and *Mathematical Review from Timişoara* (in Romanian).

He was a computer programmer and a high school teacher of mathematics, computer science, and astronomy. At present, he is retired, but remains active.

About the Authors

Chapter 1
Introduction

1.1 Inequalities

Bergström's inequality. For any real numbers a_1, a_2, ..., a_n and positive real numbers b_1, b_2, ..., b_n, it holds

$$\frac{a_1^2}{b_1} + \frac{a_2^2}{b_2} + \ldots + \frac{a_n^2}{b_n} \geq \frac{(a_1 + a_2 + \ldots + a_n)^2}{b_1 + b_2 + \ldots + b_n}.$$

Equality holds only when

$$\frac{a_1}{b_1} = \frac{a_2}{b_2} = \ldots = \frac{a_n}{b_n}.$$

Example. Let a_1, a_2, ..., a_n be nonnegative real numbers such that

$$a_1 + a_2 + \ldots + a_n = \sqrt{n}.$$

Prove that

$$\frac{\sqrt{n}}{n+1} \leq \sum_{i=1}^{n} \frac{a_i}{a_i^2 + 1} \leq \frac{n\sqrt{n}}{n+1}.$$

Solution. From $a_i^2 \leq n$, it follows that

$$\sum_{i=1}^{n} \frac{a_i}{a_i^2 + 1} \geq \sum_{i=1}^{n} \frac{a_i}{n+1} = \frac{\sqrt{n}}{n+1},$$

C. Mănescu-Avram, *Selection Tests in Algebra for Mathematical Olympiads*, Problem Books in Mathematics, https://doi.org/10.1007/978-3-032-01032-2_1

with equality only when $n - 1$ numbers are 0. For the second inequality, we will use the AM–GM inequality, then the Bergström's inequality:

$$\sum_{i=1}^{n} \frac{a_i}{a_i^2+1} = \sum_{i=1}^{n} \frac{a_i}{a_i^2+\frac{1}{n}+\frac{n-1}{n}} \leq \sum_{i=1}^{n} \frac{a_i}{2a_i \cdot \frac{1}{\sqrt{n}}+\frac{n-1}{n}} =$$

$$= \sum_{i=1}^{n} \frac{a_i}{\frac{a_i}{\sqrt{n}}+\frac{a_i}{\sqrt{n}}+\frac{1}{n}+\ldots+\frac{1}{n}} \leq \frac{1}{(n=1)^2}\left(2\sum_{i=1}^{n}\sqrt{n}+(n-1)\sum_{i=1}^{n} na_i\right) =$$

$$= \frac{2n\sqrt{n}+n(n-1)\sqrt{n}}{(n+1)^2} = \frac{n\sqrt{n}}{n+1},$$

with equality only when $a_1 = a_2 = \ldots = a_n$.

Cauchy–Schwarz inequality. If $a_1, a_2, \ldots, a_n$ and $b_1, b_2, \ldots, b_n$ are reals, then

$$\left(\sum_{i=1}^{n} a_i b_i\right)^2 \leq \left(\sum_{i=1}^{n} a_i\right)\left(\sum_{i=1}^{n} b_i\right),$$

with equality if and only if there exists $\lambda \in \mathbb{R}$ such that $a_i = \lambda b_i$, $1 \leq i \leq n$.

Examples

1. For positive real numbers a, b, c, the following inequality holds:

$$\frac{a+b+c}{abc} + \sum_{\text{cyc}} \frac{1}{a^2+bc} \geq 3\sum_{\text{cyc}} \frac{1}{a(b+c)}.$$

Solution. By the Cauchy–Schwarz inequality, we have

$$\frac{1}{(a+b)^2} = \frac{1}{\left(a \cdot 1 + \sqrt{bc} \cdot \sqrt{\frac{b}{c}}\right)^2} \geq \frac{1}{(a^2+bc)\left(1+\frac{b}{c}\right)} = \frac{c}{(a^2+bc)(b+c)}.$$

Analogously,

$$\frac{1}{(a+c)^2} \geq \frac{b}{(a^2+bc)(b+c)}.$$

Adding these two inequalities, we get

$$\frac{1}{(a+b)^2} + \frac{1}{(a+c)^2} \geq \frac{1}{a^2+bc},$$

whence

$$\frac{a^3}{(a+b)^2}+\frac{a^3}{(a+c)^2}\geq\frac{a^3}{a^2+bc}.$$

Write two similar inequalities and by addition, we obtain

$$\sum_{cyc}\frac{b^3+c^3}{(b+c)^2}\geq\sum_{cyc}\frac{a^3}{a^2+bc}\Leftrightarrow\sum_{cyc}\frac{(b+c)^2-3bc}{b+c}\geq\sum_{cyc}\frac{a(a^2+bc)-abc}{a^2+bc}\Leftrightarrow$$

$$\Leftrightarrow\sum a+abc\sum_{cyc}\frac{1}{a^2+bc}\geq 3\sum_{cyc}\frac{bc}{b+c}.$$

Dividing by abc, we get the proposed inequality. Equality holds only when $a=b=c$.

2. For positive real numbers a, b, c, the following inequality holds:

$$\sum_{\text{cyc}}\frac{1}{a^2+ab+b^2}\geq\frac{1}{\sum_{\text{cyc}}ab}+\frac{4}{\sum_{\text{cyc}}a^2+\sum_{\text{cyc}}ab}.$$

Solution. Since

$$\frac{a^2+ab+b^2+(a+b+c)c}{a^2+ab+b^2}=1+\frac{(a+b+c)c}{a^2+ab+b^2},$$

$$\frac{a^2+b^2+c^2+ab+bc+ca}{ab+bc+ca}=\frac{(a+b+c)^2}{ab+bc+ca}-1,$$

we can write the proposed inequality as

$$\frac{(a+b+c)c}{a^2+ab+b^2}+\frac{(a+b+c)a}{b^2+bc+c^2}+\frac{(a+b+c)b}{c^2+ca+a^2}\geq\frac{(a+b+c)^2}{ab+bc+ca}\Leftrightarrow$$

$$\Leftrightarrow\frac{c}{a^2+ab+b^2}+\frac{a}{b^2+bc+c^2}+\frac{b}{c^2+ca+a^2}\geq\frac{a+b+c}{ab+bc+ca}.$$

The last inequality follows from the Cauchy–Schwarz inequality:

$$\frac{c}{a^2+ab+b^2}+\frac{a}{b^2+bc+c^2}+\frac{b}{c^2+ca+a^2}=$$

$$=\frac{c^2}{a^2c+abc+b^2c}+\frac{a^2}{ab^2+abc+ac^2}+\frac{b^2}{bc^2+abc+a^2b}\geq$$

$$\geq\frac{(a+b+c)^2}{a^2b+ab^2+b^2c+bc^2+a^2c+ac^2+3abc}=$$

$$= \frac{(a+b+c)^2}{(a+b+c)(ab+bc+ca)} = \frac{a+b+c}{ab+bc+ca}.$$

3. For positive real numbers x, y, z, prove the inequality

$$(x+y)\sqrt{(z+x)(z+y)} + (y+z)\sqrt{(x+y)(x+z)} + (z+x)\sqrt{(y+z)(y+x)} \geq$$
$$\geq 4(xy+yz+zx).$$

Solution. Denote $t = \sqrt{xy+yz+zx}$. By the Cauchy–Schwarz inequality,

$$\sum_{cyc}(x+y)\sqrt{(z+x)(z+y)} = \sum_{cyc}\sqrt{(z+x)(x+y)(z+y)(x+y)} =$$

$$= \sum_{cyc}\sqrt{(x^2+t^2)(y^2+t^2)} \geq \sum_{cyc}\sqrt{(xy+t^2)^2} = t^2+3t^2 = 4(xy+yz+zx).$$

A stronger version of this inequality was done by *Titu Zvonaru*:

$$\sum_{cyc}(x+y)\sqrt{(z+x)(z+y)} \geq 4(xy+yz+zx) + \frac{xy+yz+zx}{3(x^2+y^2+z^2)} \times$$

$$\times \left[(x-y)^2+(y-z)^2+(z-x)^2\right].(\textit{Mathematical Excalibur, Volume } 17, no.4)$$

Chebyshev's inequality. Let $a_1 \leq a_2 \leq \ldots \leq a_n$ and $b_1 \leq b_2 \leq \ldots \leq b_n$ be two nondecreasing sequences of real numbers. Then

$$\frac{a_1b_1+a_2b_2+\ldots+a_nb_n}{n} \geq \frac{a_1+a_2+\ldots+a_n}{n} \cdot \frac{b_1+b_2+\ldots+b_n}{n} \geq$$
$$\geq \frac{a_1b_n+a_2b_{n-1}+\ldots+a_nb_1}{n}.$$

Examples

1. Prove the inequality for positive real numbers a, b, c:

$$\frac{b+c}{a^2}+\frac{c+a}{b^2}+\frac{a+b}{c^2} \geq \frac{18}{a+b+c}.$$

Solution. If $a \leq b \leq c$, then

$$\frac{1}{b^2}+\frac{1}{c^2} \leq \frac{1}{c^2}+\frac{1}{a^2} \leq \frac{1}{a^2}+\frac{1}{b^2}.$$

By the Chebyshev's inequality,

$$3\left[a\left(\frac{1}{b^2}+\frac{1}{c^2}\right)+b\left(\frac{1}{c^2}+\frac{1}{a^2}\right)+c\left(\frac{1}{a^2}+\frac{1}{b^2}\right)\right]\geq 2(a+b+c)\left(\frac{1}{a^2}+\frac{1}{b^2}+\frac{1}{c^2}\right)\geq$$

$$\geq\frac{2}{3}(a+b+c)\left(\frac{1}{a}+\frac{1}{b}+\frac{1}{c}\right)^2\geq\frac{2}{3}\cdot 9\left(\frac{1}{a}+\frac{1}{b}+\frac{1}{c}\right)\geq\frac{54}{a+b+c}.$$

2. If a, b, c are positive real numbers and $a+b+c=1$, then

$$\frac{-a+b+c}{\sqrt{a}+\sqrt{b}}+\frac{a-b+c}{\sqrt{b}+\sqrt{c}}+\frac{a+b-c}{\sqrt{c}+\sqrt{a}}\geq\frac{3}{2(\sqrt{a}+\sqrt{b}+\sqrt{c})}.$$

Solution. From

$$\sum_{\text{cyc}}\frac{a-b}{\sqrt{a}+\sqrt{b}}=\sum_{\text{cyc}}\left(\sqrt{a}-\sqrt{b}\right)=0,$$

it follows that

$$\sum_{\text{cyc}}\frac{-a+b+c}{\sqrt{a}+\sqrt{b}}=\sum_{\text{cyc}}\frac{c}{\sqrt{a}+\sqrt{b}}.$$

The sequences

$$(a,b,c),\left(\frac{1}{\sqrt{b}+\sqrt{c}},\frac{1}{\sqrt{c}+\sqrt{a}},\frac{1}{\sqrt{a}+\sqrt{b}}\right)$$

have the same ordering. By the Chebyshev's inequality,

$$\sum_{\text{cyc}}\frac{c}{\sqrt{a}+\sqrt{b}}\geq\frac{1}{3}\left(\sum_{\text{cyc}}a\right)\left(\sum_{\text{cyc}}\frac{1}{\sqrt{a}+\sqrt{b}}\right)=\frac{1}{3}\sum_{\text{cyc}}\frac{1}{\sqrt{a}+\sqrt{b}}.$$

Now, the desired result follows from the Bergström's inequality.

3. Let a, b, c be positive real numbers such that $abc=1$. Show that

$$\frac{1}{a^3(b+c)}+\frac{1}{b^3(c+a)}+\frac{1}{c^3(a+b)}\geq\frac{3}{2}.(\mathit{IMO},1995)$$

Solution. For

$$x=\frac{1}{a},y=\frac{1}{b},z=\frac{1}{c},$$

we have $xyz=1$ and the inequality becomes

$$\frac{x^2}{y+z}+\frac{y^2}{z+x}+\frac{z^2}{x+y}\geq\frac{3}{2}.$$

It is easy to see that the triples

$$(x,y,z),\left(\frac{x}{y+z},\frac{y}{z+x},\frac{z}{x+y}\right)$$

have the same ordering. By the Chebyshev's inequality,

$$\frac{x^2}{y+z}+\frac{y^2}{z+x}+\frac{z^2}{x+y}\geq\frac{1}{3}(x+y+z)\left(\frac{x}{y+z}+\frac{y}{z+x}+\frac{z}{x+y}\right).$$

Denote $y + z = u$, $z + x = v$, $x + y = w$. Then

$$x=\frac{-u+v+w}{2},y=\frac{u-v+w}{2},z=\frac{u+v-w}{2},$$

and we obtain

$$\frac{x}{y+z}+\frac{y}{z+x}+\frac{z}{x+y}=\frac{1}{2}\left(\frac{-u+v+w}{u}+\frac{u-v+w}{v}+\frac{u+v-w}{w}\right)=$$

$$=\frac{1}{2}\left[\left(\frac{u}{v}+\frac{v}{u}\right)+\left(\frac{v}{w}+\frac{w}{v}\right)+\left(\frac{w}{u}+\frac{u}{w}\right)-3\right]\geq\frac{1}{2}(2+2+2-3)=\frac{3}{2}.$$

Back to the above inequality,

$$\frac{x^2}{y+z}+\frac{y^2}{z+x}+\frac{z^2}{x+y}\geq\frac{x+y+z}{3}\cdot\frac{3}{2}\geq\sqrt[3]{xyz}\cdot\frac{3}{2}=\frac{3}{2},$$

with equality only when $x = y = z$.

Hölder's inequality. Let $a_1, \ldots, a_n$; $b_1, \ldots, b_n$; $\ldots$; $z_1, \ldots, z_n$ be sequences of non-negative real numbers and let $\lambda_a, \lambda_b, \ldots, \lambda_z$ be positive reals with sum 1. Then

$$(a_1+\ldots+a_n)^{\lambda_a}(b_1+\ldots+b_n)^{\lambda_b}\ldots(z_1+\ldots+z_n)^{\lambda_z}\geq$$
$$\geq a_1^{\lambda_a}b_1^{\lambda_b}\ldots z_1^{\lambda_z}+\ldots+a_n^{\lambda_a}b_n^{\lambda_b}\ldots z_n^{\lambda_z}.$$

Example. Let $n \geq 2$ be a positive integer and $a_1, a_2, \ldots, a_n$ be positive real numbers such that $a_1a_2\ldots a_n = 1$. Prove that

$$\prod_{i=1}^{n}\left(a_i^n+1\right)\geq\frac{1}{n^n}\left(\sum_{i=1}^{n}\left(a_i+\frac{1}{a_i}\right)\right)^n.$$

Solution. By the Hölder's inequality,

$$\left(a_1^n+1\right)\dots\left(a_{i-1}^n+1\right)\left(1+a_i^n\right)\left(a_{i+1}^n+1\right)\dots\left(a_n^n+1\right)\geq$$

$$\geq\left(a_1\dots a_{i-1}\cdot 1\cdot a_{i+1}\dots a_n+1\cdot\dots\cdot 1\cdot a_i\cdot 1\cdot\dots\cdot 1\right)^n=\left(\frac{1}{a_i}+a_i\right)^n.$$

Summing all inequalities

$$\sqrt[n]{\prod_{i=1}^{n}\left(a_i^n+1\right)}\geq a_i+\frac{1}{a_i}$$

from 1 to n, we obtain an equivalent inequality to the proposed.

Jensen's inequality. Let I be an interval of real numbers and $f\colon I\to\mathbb{R}$ be a convex function. Then for any $x_1, \dots, x_n \in I$ and any nonnegative real numbers $a_1, \dots, a_n$,

$$\frac{a_1 f(x_1)+\dots+a_n f(x_n)}{a_1+\dots+a_n}\geq f\left(\frac{a_1x_1+\dots+a_nx_n}{a_1+\dots+a_n}\right).$$

If f is concave, then the inequality is reversed.

Example. Prove that if $a, b, c \in (0, 1)$, then

$$\sqrt{a-a^2}+\sqrt{2b-b^2}+\sqrt{3c-c^2}\leq\sqrt{6(a+b+c)-(a+b+c)^2}.$$

Solution. By the substitution $a=x$, $b=2y$, $c=3z$, the inequality becomes

$$\sqrt{x-x^2}+2\sqrt{y-y^2}+3\sqrt{z-z^2}\leq 6\sqrt{\frac{x+2y+3z}{6}-\left(\frac{x+2y+3z}{6}\right)^2}.$$

Apply the Jensen's inequality to the concave function $f(x)=\sqrt{x-x^2}$.

Karamata's inequality. Let $f\colon I\to\mathbb{R}$ be a convex function and suppose that the sequence $x_1, \dots, x_n$ majorizes the sequence $y_1, \dots, y_n$, where $x_i, y_i \in I$ for all i. Then $f(x_1)+\dots+f(x_n)\geq f(y_1)+\dots+f(y_n)$. Majorization means that

$$x_1\geq x_2\geq\dots\geq x_n, y_1\geq y_2\geq\dots\geq y_n,$$

$$x_1+\dots+x_i\geq y_1+\dots+y_i, 1\leq i\leq n-1, x_1+\dots+x_n=y_1+\dots+y_n.$$

If f is a strictly convex function, then equality holds if and only if $x_i=y_i$ for all i.

Example. Find the maximum of $a^{12}+b^{12}+c^{12}$ if $a, b, c \in [-1, 1]$ and $a+b+c=-\frac{1}{2}$.(*China*, 1997)

Solution. The continuous function $f(x) = x^{12}$ is convex on $[-1,1]$. If $a+b+c=-\frac{1}{2}, 1\geq a\geq b\geq c\geq -1$, then $\left(1, -\frac{1}{2}, -1\right)$ majorizes (a,b,c), so

$$a^{12}+b^{12}+c^{12}=f(a)+f(b)+f(c)\leq f(1)+f\left(-\frac{1}{2}\right)+f(-1)=2+\frac{1}{2^{12}},$$

with equality when $a=1, b=-\frac{1}{2}, c=-1$.

Maclaurin's inequalities. Let x_1, …, x_n be nonnegative real numbers. Define the averages S_k by

$$S_k=\frac{\sum_{1\leq i_1<i_2<\ldots<i_k\leq n}x_{i_1}x_{i_2}\ldots x_{i_k}}{\binom{n}{k}}, 1\leq k\leq n.$$

Then $S_1\geq\sqrt{S_2}\geq\sqrt[3]{S_3}\geq\ldots\geq\sqrt[n]{S_n}$, with equality if and only if all x_i are equal.

Means inequalities. If x_1, …, x_n are nonnegative real numbers and a_1, …, a_n are nonnegative real numbers with a positive sum, then the function

$$f(r)=\left(\frac{a_1x_1^r+\ldots+a_nx_n^r}{a_1+\ldots+a_n}\right)^{\frac{1}{r}}$$

(weighted power mean) is nondecreasing, with the convention that $f(0)$ is the weighted geometric mean. The function f is strictly increasing, unless all x_i are equal, except possibly for $r\leq 0$, where if some x_i is zero, then f is identically zero. In particular, $f(2)\geq f(1)\geq f(0)\geq f(-1)$ give

$$QM\geq AM\geq GM\geq HM,$$

that is, the inequalities between the quadratic, arithmetic, geometric, and harmonic means, respectively.

Examples

1. Prove the inequality for positive real numbers a, b, c:

$$(a+b)(b+c)(c+a)\geq 8abc+\frac{[(a-b)(b-c)(c-a)]^2}{(a+b)(b+c)(c+a)}.$$

Solution. Denote $r=abc$, $u=ab^2+bc^2+ca^2$, $v=a^2b+b^2c+c^2a$. Then

$$(a+b)(b+c)(c+a)=2r+u+v, (a-b)(b-c)(c-a)=u-v.$$

Eliminating denominators, we obtain

$$(2r+u+v)^2 \geq 8r(2r+u+v)+(u-v)^2,$$

which is equivalent to $(u-r)(v-r) \geq 4r^2$. By the AM–GM inequality, we have $u \geq 3r$, $v \geq 3r$, therefore $(u-r)(v-r) \geq 2r \cdot 2r = 4r^2$.

2. If $x, y, z \geq 0$ and $x+y+z=1$, then

$$\sqrt{x(y+z)}+\sqrt{y(z+x)}+\sqrt{z(x+y)} \leq \sqrt{2}.$$

Solution. By the GM–AM inequality,

$$\text{LHS} = \frac{\sqrt{2x(y+z)}+\sqrt{2y(z+x)}+\sqrt{2z(x+y)}}{\sqrt{2}} \leq$$

$$\leq \frac{(2x+y+z)+(2y+z+x)+(2z+x+y)}{2\sqrt{2}} = \text{RHS}.$$

3. If $a, b, c > 0$ and $a+b+c=1$, then

$$\frac{a}{a^2+1}+\frac{b}{b^2+1}+\frac{c}{c^2+1} \leq \frac{9}{10}.$$

Solution. By the HM–AM inequality,

$$\frac{a}{a^2+1} = \frac{a}{a^2+\frac{1}{9}+\frac{8}{9}} \leq \frac{a}{\frac{2a}{3}+\frac{8}{9}} = \frac{a}{\frac{a}{3}+\frac{a}{3}+\underbrace{\frac{1}{9}+\ldots+\frac{1}{9}}_{8}} \leq$$

$$\leq \frac{a}{10^2}\left(\frac{3}{a}+\frac{3}{a}+\underbrace{9+\ldots+9}_{8}\right) = \frac{3+36a}{50}.$$

It follows that

$$\frac{a}{a^2+1}+\frac{b}{b^2+1}+\frac{c}{c^2+1} \leq \frac{(3+36a)+(3+36b)+(3+36c)}{50} = \frac{45}{50} = \frac{9}{10}.$$

4. If $a, b, c > 0$ and $a^2+b^2+c^2=1$, then

$$a+b+c+\frac{1}{abc} \geq 4\sqrt{3}.$$

Solution. By the AM–GM inequality,

$$a+b+c+\frac{1}{abc}=a+b+c+\frac{1}{9abc}+\frac{8}{9abc}\geq 4\sqrt[4]{abc\cdot\frac{1}{9abc}}+\frac{8}{9abc}=$$

$$=\frac{4}{\sqrt{3}}+\frac{8}{9\sqrt{a^2b^2c^2}}\geq\frac{4}{\sqrt{3}}+\frac{8}{9\sqrt{\left(\frac{a^2+b^2+c^2}{3}\right)^3}}=4\sqrt{3}.$$

5. Determine (with proof) all integers $n \geq 3$ such that there exists a positive integer M_n satisfying the condition: for all n positive numbers $a_1, a_2, \ldots, a_n$ we have

$$\frac{a_1+a_2+\ldots+a_n}{\sqrt[n]{a_1a_2\ldots a_n}}\leq M_n\left(\frac{a_2}{a_1}+\frac{a_3}{a_2}+\ldots+\frac{a_n}{a_{n-1}}+\frac{a_1}{a_n}\right).(Taiwan, 2005)$$

Solution. For $n = 3$, let

$$x=\frac{a_2}{a_1}+\frac{a_3}{a_2}+\frac{a_1}{a_3}.$$

Then

$$a_3\geq a_1,a_2\Longrightarrow x>\frac{a_2}{a_1},\frac{a_3}{a_2},\frac{a_1}{a_3}\Longrightarrow a_2>\frac{a_3}{x},a_1>\frac{a_2}{x}>\frac{a_3}{x^2}\Longrightarrow$$

$$\Longrightarrow\frac{a_1+a_2+a_3}{\sqrt[3]{a_1a_2a_3}}\leq\frac{3a_3}{\sqrt[3]{\frac{a_3}{x^2}\cdot\frac{a_3}{x}\cdot a_3}}=3x,$$

so we can take $M_3 = 3$. For $n > 3$, assume there is such M_n. Let $a_i = c^i$, $1 \leq i \leq n$. Then

$$M_n\geq\frac{c+c^2+\ldots+c^n}{\sqrt[n]{c^{\frac{n(n+1)}{2}}}}\left[(n-1)c+\frac{1}{c^{n-1}}\right]^{-1}\geq$$

$$\geq\frac{c^n}{c^{\frac{n+1}{2}}}\cdot\frac{1}{c(n-1+c^{-n})}=\frac{c^{\frac{n-3}{2}}}{n-1+c^{-n}}\to\infty,$$

as $c \to \infty$, so M_n cannot be finite, contradiction.

Minkowski's inequality. Let $p \geq 1$, $x_1, \ldots, x_n, y_1, \ldots, y_n$ be real numbers. Then

$$\left(\sum_{i=1}^{n}|x_i+y_i|^p\right)^{\frac{1}{p}}\leq\left(\sum_{i=1}^{n}|x_i|^p\right)^{\frac{1}{p}}+\left(\sum_{i=1}^{n}|y_i|\right)^{\frac{1}{p}}.$$

For $p < 1$, the reverse inequality holds. Equality is attained only when the sequences are proportional.

Example. For a, b, $c > 0$,

$$\sqrt{a^2 - ab + b^2} + \sqrt{a^2 - ac + c^2} \geq \sqrt{b^2 + bc + c^2}.$$

Solution. By the Minkowski's inequality,

$$\sqrt{\left(\frac{b}{2} - a\right)^2 + \left(\frac{b}{2}\right)^2 + \left(\frac{b}{2}\right)^2 + \left(\frac{b}{2}\right)^2} + \sqrt{\left(\frac{c}{2} - a\right)^2 + \left(\frac{c}{2}\right)^2 + \left(\frac{c}{2}\right)^2 + \left(\frac{c}{2}\right)^2} \geq$$

$$\geq \sqrt{\left(\frac{b}{2} - \frac{c}{2}\right)^2 + \left(\frac{b}{2} + \frac{c}{2}\right)^2 + \left(\frac{b}{2} + \frac{c}{2}\right)^2 + \left(\frac{b}{2} + \frac{c}{2}\right)^2} = \sqrt{b^2 + bc + c^2}.$$

Muirhead's inequality. If the sequence $a_1, a_2, \ldots, a_n$ majorizes the sequence $b_1, b_2, \ldots, b_n$, then for any positive real numbers $x_1, x_2, \ldots, x_n$,

$$\sum_{sym} x_1^{a_1} x_2^{a_2} \ldots x_n^{a_n} \geq \sum_{sym} x_1^{b_1} x_2^{b_2} \ldots x_n^{b_n}.$$

Examples

1. Prove that for any positive real numbers a, b, c

$$40\left(a^4 + b^4 + c^4\right)(a + b + c) - 39\left(a^5 + b^5 + c^5\right) \geq (a + b + c)^5.$$

Solution. We have

$$40\sum a^5 + 40\sum a^4 b + 40\sum ab^4 - 39\sum a^5 \geq$$

$$\geq \sum a^5 + 5\sum a^4 b + 5\sum ab^4 + 10\sum a^3 b^2 + 10\sum a^2 b^3 + 20\sum a^3 bc +$$

$$+30\sum a^2 b^2 c \Leftrightarrow 7\sum a^4 b + 7\sum ab^4 \geq$$

$$\geq 2\sum a^3 b^2 + 2\sum a^2 b^3 + 4\sum a^3 bc + 6\sum a^2 b^2 c.$$

2. Prove that for any positive real numbers a, b, c, d,

$$3\left(a^3 + b^3 + c^3 + d^3\right) \geq$$

$$\geq 2\left(ab\sqrt{cd} + ac\sqrt{bd} + ad\sqrt{bc} + bc\sqrt{ad} + bd\sqrt{ac} + cd\sqrt{ab}\right).$$

Solution. Write the inequality as

$$\sum_{sym} a^3 \geq \sum_{sym} ab\sqrt{cd}.$$

We see that $(3,0,0,0)$ majorizes $\left(1,1,\frac{1}{2},\frac{1}{2}\right)$, so Muirhead's inequality applies.

Nesbitt's inequality. If a, b, c are positive real numbers, then

$$\frac{a}{b+c}+\frac{b}{c+a}+\frac{c}{a+b}\geq\frac{3}{2}.$$

Examples

1. Prove that the inequality for positive real numbers a, b, c,

$$\frac{a+b}{a+b+2c}+\frac{b+c}{2a+b+c}+\frac{c+a}{a+2b+c}\geq\frac{3}{2}$$

is stronger as the Nesbitt's inequality.

Solution. The inequality is true, since it is the Nesbitt's inequality for $a+b$, $b+c$, $c+a$. We have

$$\frac{a}{b+c}+\frac{b}{c+a}+\frac{c}{a+b}-\frac{3}{2}\geq\frac{a+b}{a+b+2c}+\frac{b+c}{2a+b+c}+\frac{c+a}{a+2b+c}-\frac{3}{2}\Leftrightarrow$$

$$\Leftrightarrow\sum_{cyc}\frac{(a-b)^2}{(b+c)(c+a)}\geq\sum_{cyc}\frac{(a-b)^2}{(2a+b+c)(a+2b+c)}.$$

The last inequality is true, since $(2a+b+c)(a+2b+c)\geq(b+c)(c+a)$.

2. (Generalized Nesbitt's inequality). Let n be a positive integer >1, $a_1, a_2, \ldots, a_n$ be positive real numbers and $s=a_1+a_2+\ldots+a_n$. Then

$$\sum_{i=1}^{n}\frac{a_i}{s-a_i}\geq\frac{n}{n-1}.$$

Solution. We have

$$\sum_{i=1}^{n}\frac{a_i}{s-a_i}=\sum_{i=1}^{n}\frac{a_i-s+s}{s-a_i}=-n+s\sum_{i=1}^{n}\frac{1}{s-a_i}\geq\frac{n}{n-1}\Leftrightarrow$$

$$\Leftrightarrow(n-1)s\sum_{i=1}^{n}\frac{1}{s-a_i}\geq n^2.$$

The last inequality follows from AM–HM inequality.

Newton's inequalities. For nonnegative real numbers $a_1, a_2, \ldots, a_n$ the elementary symmetric means S_k satisfy $S_{k-1}S_{k+1}\leq S_k^2, 2\leq k\leq n-1$, with equality if and only if all x_i are equal.

Radon's inequality. If n is a positive integer and a_p, b_p are positive real numbers, then for any $t \geq 0$,

$$\sum_{p=1}^{n} \frac{a_p^{t+1}}{b_p^t} \geq \frac{\left(\sum\limits_{p=1}^{n} a_p\right)^{t+1}}{\left(\sum\limits_{p=1}^{n} b_p\right)^{t}}.$$

M. Bencze has included the intermediate term

$$\frac{1}{\binom{n-1}{k-1}} \sum_{1 \leq i_1 < \cdots < i_k \leq n} \frac{\left(a_{i_1} + \cdots + a_{i_k}\right)^{t+1}}{\left(b_{i_1} + \cdots + b_{i_k}\right)^{t}}$$

Example. If $a, b, c \in (0, \infty)$ and $m \in [1, \infty)$, then

$$a\left(\frac{1}{b^m} + \frac{1}{c^m}\right) + b\left(\frac{1}{c^m} + \frac{1}{a^m}\right) + c\left(\frac{1}{a^m} + \frac{1}{b^m}\right) \geq \frac{2 \cdot 3^m}{(a+b+c)^{m-1}}.$$

Solution. By the Radon's inequality,

$$\text{LHS} = \sum_{\text{cyc}} \frac{(b+c)^{m+1}}{(a(b+c))^m} \geq \frac{2^{m+1}(a+b+c)^{m+1}}{2^m(ab+bc+ca)^m} \geq \text{RHS}.$$

The last inequality is true, since $(a + b + c)^2 \geq 3(ab + bc + ca)$.

Rearrangement inequality. Let $a_1 \leq a_2 \leq \ldots \leq a_n$ and $b_1 \leq b_2 \leq \ldots \leq b_n$ be two nondecreasing sequences of real numbers. Then for any permutation σ of $\{1, 2, \ldots, n\}$, we have

$$a_1b_1 + a_2b_2 + \ldots + a_nb_n \geq a_1b_{\sigma(1)} + a_2b_{\sigma(2)} + \ldots + a_nb_{\sigma(n)} \geq$$
$$\geq a_1b_n + a_2b_{n-1} + \ldots + a_nb_1,$$

with equality on the left and right holding if and only if the sequence $\sigma(1)$, $\sigma(2)$, …, $\sigma(n)$ is decreasing and increasing, respectively.

Example. Let c_1, c_2, …, c_n be distinct positive integers. Prove that

$$c_1 + \frac{c_2}{4} + \ldots + \frac{c_n}{n^2} \geq 1 + \frac{1}{2} + \ldots + \frac{1}{n}. (\textit{IMO}, 1978)$$

Solution. Let a_1, a_2, …, a_n be the numbers c_i in increasing order. Then $a_i \geq i$, since the a_i's are distinct positive integers. By the rearrangement inequality,

$$c_1 + \frac{c_2}{4} + \ldots + \frac{c_n}{n^2} \geq a_1 + \frac{a_2}{4} + \ldots + \frac{a_n}{n^2} \geq 1 + \frac{1}{2} + \ldots + \frac{1}{n}.$$

Schur's inequality. Let a, b, c be nonnegative real numbers and $r > 0$. Then

$$a^r(a-b)(a-c) + b^r(b-c)(b-a) + c^r(c-a)(c-b),$$

with equality only when $a = b = c$ or some two of a, b, c are equal and the other is 0.

Examples

1. Prove that for any positive real numbers x, y, z

$$(x+y)\sqrt{(z+x)(z+y)} + (y+z)\sqrt{(x+y)(x+z)} + (z+x)\sqrt{(y+z)(y+x)} \geq$$
$$\geq 4(xy+yz+zx).$$

Solution. Denote $\sqrt{y+z}=a, \sqrt{z+x}=b, \sqrt{x+y}=c,$ then

$$x = \frac{-a^2+b^2+c^2}{2}, y = \frac{a^2-b^2+c^2}{2}, z = \frac{a^2+b^2-c^2}{2},$$

whence

$$4\sum_{\text{cyc}} xy = \sum_{\text{cyc}} \left[c^4 - (a^2-b^2)^2\right] = -\sum_{\text{cyc}} a^4 + 2\sum_{\text{cyc}} a^2b^2.$$

Now, the inequality becomes

$$\sum_{\text{cyc}} a^4 + abc\sum_{\text{cyc}} a \geq 2\sum_{\text{cyc}} a^2b^2.$$

By the Schur's inequality for $r = 2$, we have

$$\sum_{\text{cyc}} a^2(a-b)(a-c) \geq 0 \Leftrightarrow \sum_{\text{cyc}} a^4 + abc\sum_{\text{cyc}} a \geq \sum_{\text{cyc}} (a^3b + ab^3)$$

and it remains to prove that

$$\sum_{\text{cyc}} (a^3b + ab^3) \geq 2\sum_{\text{cyc}} a^2b^2,$$

which is a consequence of the AM–GM inequality.

2. Let a, b, c be positive real numbers such that $a + b + c = 3$. Prove that

$$abc + \frac{12}{ab + bc + ca} \geq 5.$$

Solution. Denote $p = a + b + c$, $q = ab + bc + ca$, $s = abc$. Then

$$abc + \frac{12}{ab + bc + ca} - 5 = s + \frac{12}{q} - 5 = \frac{4p^5 + 81qs - 15p^3q}{3p^3q} =$$

$$= \frac{4p(p^2 - 3q)^2 + 9q(p^3 + 9s - 4pq)}{3p^3q} \geq 0,$$

since $p^3 + 9s - 4pq \geq 0$, as a consequence of the Schur's inequality for $r = 1$ [3–5], [7–15], [17–20].

1.2 Functional Equations

Examples

1. Show that if a polynomial $P(x)$ satisfies

$$P(2x^2 - 1) = \frac{P(x)^2}{2} - 1,$$

then it must be constant.

Solution. Let $P(x) = a_0x^n + a_1x^{n-1} + \ldots + a_n$, $n \geq 1$ be such a polynomial, then

$$a_0(2x^2 - 1)^n + a_1(2x^2 - 1)^{n-1} + \ldots + a_n = \frac{(a_0x^n + a_1x^{n-1} + \ldots + a_n)^2}{2} - 1.$$

The coefficients of x^{2n} are the same: $2a_0 \cdot 2^n = a_0^2$, whence $a_0 = 2^{n+1}$. If a_0, a_1, …, a_k are rational numbers, then comparing the coefficients of x^{2n-k-1} we have on the left a rational number depending on a_0, …, a_k, and on the right a_0a_{k+1} plus a rational number depending on a_0, …, a_k, hence a_{k+1} is rational. We conclude that all coefficients are rational, but $P(1) = 1 \pm \sqrt{3}$, contradiction.

2. Determine all functions $f, g : (0, \infty) \to (0, \infty)$ such that

$$f(g(x)) = \frac{x}{xf(x) - 2}, g(f(x)) = \frac{x}{xg(x) - 2}, \forall x \in (0, \infty).$$

Solution. Let $F(x) = xf(x)$, $G(x) = xg(x)$. Then $f(g(x)) > 0$, $g(f(x)) > 0$ imply $F(x) > 2$, $G(x) > 2$, for all $x > 0$. Define $a_1 = 2$. We have

$$\frac{G(x)}{F(x)-2}=g(x)f(g(x))=F(g(x))>a_1\Rightarrow$$

$$\Rightarrow G(x)>a_1F(x)-2a_1,(1)$$

and similarly, $F(x) > a_1G(x) - 2a_1$. (2) Multiplying (1) by a_1, adding to (2) and simplifying, we get

$$F(x)<\frac{2a_1}{a_1-1}=4.$$

Define $b_1 = 4$. Similarly we get $G(x) < b_1$. Repeating the above steps, but reversing all inequalities, we obtain

$$F(x)>\frac{2b_1}{b_1-1}=a_2, G(x)>a_2; F(x)<\frac{2a_2}{a_2-1}=b_2, G(x)<b_2.$$

This suggests the definition

$$a_{n+1}=\frac{2b_n}{b_n-1}, b_{n+1}=\frac{2a_n}{a_n-1}, n\geq 1.$$

Repeating the steps above, we can prove by induction that

$$a_n<F(x), G(x)<b_n, n\geq 1.$$

It is easy to see that the sequences (a_n), (b_n) are convergent and

$$\lim_{n\to\infty} a_n=\lim_{n\to\infty} b_n=3,$$

hence $F(x) = G(x) = 3$ for all $x > 0$, that is,

$$f(x)=g(x)=\frac{3}{x}, \forall x>0.$$

3. Find all functions $f\colon [0, \infty) \to [0, \infty)$ such that for every $x \geq 0$ we have $4f(x) \geq 3x$ and $f(4f(x) - 3x) = x$ for all $x \geq 0$.
   Solution. Answer: $f(x) = x$ for all $x \geq 0$.
   Assume that f is a solution with $f(c) \neq c$ for some $c \geq 0$. Define

$$x_0=f(c), x_1=c, x_{n+2}=-3x_{n+1}+4x_n, n\geq 0.$$

We can prove by induction that $x_n = f(x_{n+1}) \geq 0$ for all n. Since

$$z^2 + 3z - 4 = (z - 1)(z + 4),$$

it follows that $x_n = \alpha + (-4)^n\beta$ for some real numbers α, β. Now,

$$n = 0 \Longrightarrow f(c) = \alpha + \beta; n = 1 \Longrightarrow c = \alpha - 4\beta \Longrightarrow \beta = \frac{f(c) - c}{5} \neq 0;$$

$$\beta > 0 \Longrightarrow \lim_{k \to \infty} x_{2k+1} = \lim_{k \to \infty} \left[\alpha + (-4)^{2k+1}\beta\right] = -\infty,$$

$$\beta < 0 \Longrightarrow \lim_{k \to \infty} x_{2k} = \lim_{k \to \infty} \left[\alpha + (-4)^{2k}\beta\right] = -\infty,$$

contradiction [4], [7–10], [12–14], [16], [17].

1.3 Functional Inequalities

Examples

1. Let $f : \mathbb{R} \to \mathbb{R}$ be a function such that

   (i) $f(x) + f(y) + 1 \geq f(x + y) \geq f(x) + f(y)$ for all $x, y \in \mathbb{R}$,
   (ii) $f(0) \geq f(x)$, for all $x \in [0, 1)$.
   (iii) $-f(-1) = f(1) = 1$.

   Find all such functions. (*APMO*, 1994)

   Solution. By (*iii*), $f(-1) = -1, f(1) = 1$, so

$$f(0) = f(-1 + 1) \geq f(1) + f(-1) = 0.$$

By (*i*), $f(1) = f(1 + 0) \geq f(1) + f(0)$, so $f(0) \leq 0$. Then $f(0) = 0$.

Now we will prove that $f(x) = 0, \forall x \in (0, 1)$. Since $f(0) = 0$, by (*ii*) $f(x) \leq 0, \forall x \in (0, 1)$. By (*i*) and (*ii*), $f(x) + f(1 - x) + 1 \geq f(1) = 1$, so

$$f(x) \geq -f(1 - x).$$

If $x \in (0, 1)$, then $1 - x \in (0, 1)$, so $f(1 - x) \leq 0$ and $f(x) \geq -f(1 - x) \geq 0$, hence $f(x) = 0$.

Next by (*i*) and (*iii*), we have $f(x + 1) \geq f(x) + f(1) = f(x) + 1$ and

$$f(x) \geq f(x + 1) + f(-1) = f(x + 1) - 1.$$

These give $f(x + 1) = f(x) + 1$. So $f(x) = 0$ for all $x \in [0, 1)$, $f(x + 1) = f(x) + 1$ for all x, hence $f(x) = \lfloor x \rfloor$. We can check that this function satisfies the required properties.

2. Prove that there is no function $f\colon (0,\infty)\to(0,\infty)$ such that

$$(f(x))^2 \geq f(x+y)(f(x)+y), \forall x, y \in (0, \infty).(Bulgaria, 1998)$$

Solution. Let f be a function satisfying the required condition. Then

$$f(x)-f(x+y)\geq \frac{f(x)y}{f(x)+y},$$

whence it follows that f is strictly decreasing. Now, we will prove that

$$f(x)-f(x+1)\geq \frac{1}{2}, \forall x>0.$$

Fix x and choose a positive integer n such that $n\geq \frac{1}{f(x+1)}$. Then

$$f\left(x+\frac{k}{n}\right)-f\left(x+\frac{k+1}{n}\right)\geq \frac{f\left(x+\frac{k}{n}\right)\cdot\frac{1}{n}}{f\left(x+\frac{k}{n}\right)+\frac{1}{n}}\geq \frac{1}{2n}, 0\leq k\leq n-1.$$

Adding these n inequalities, we get the desired inequality. Let m be a positive integer such that $m \geq 2f(x)$. Then

$$f(x)-f(x+m)=\sum_{i=1}^{m}(f(x+i-1)-f(x+i))\geq \frac{m}{2}\geq f(x),$$

hence $f(x + m) \leq 0$, contradiction.

3. Does there exist a bounded function $f\colon \mathbb{R}\to\mathbb{R}$ such that $f(1) > 0$ and

$$(f(x+y))^2\geq (f(x))^2+2f(xy)+(f(y))^2, \forall x, y\in\mathbb{R}?(Russia, 2005)$$

Solution. Assume that such f exists. Let $a = 2f(1) > 0$. For $x_1 \neq 0$, let $y_1 = \frac{1}{x_1}$. Then $(f(x_1 + y_1))^2 \geq (f(x_1))^2 + 2f(1) + (f(y_1))^2 \geq (f(x_1))^2 + a$. For $n > 1$, let $x_n = x_{n-1} + y_{n-1}, y_n = \frac{1}{x_n}$. We have

$$(f(x_n+y_n))^2\geq (f(x_n))^2+a=(f(x_{n-1}+y_{n-1}))^2+a\geq (f(x_{n-1}))^2+2a\geq$$
$$\geq\ldots\geq (f(x_1))^2+na.$$

As $n\to\infty$, f becomes unbounded, contradiction.

4. Let f be a function from the set of real numbers to itself that satisfies

$$f(x+y)\leq yf(x)+f(f(x))$$

for all real numbers x, y. Prove that $f(x) = 0$ for all $x \leq 0$. (*IMO*, 2011.)

Solution. For $y = t - x$, we have

$$f(t) \leq tf(x) - xf(x) + f(f(x)). \quad (1)$$

For $(t, x) = (f(a), b)$ and $(t, x) = (f(b), a)$, with a, b arbitrary real numbers, we get

$$f(f(a)) - f(f(b)) \leq f(a)f(b) - bf(b),$$
$$f(f(b)) - f(f(a)) \leq f(a)f(b) - af(a).$$

Adding these inequalities, we obtain

$$2f(a)f(b) \geq af(a) + bf(b).$$

For $b = 2f(a)$, we have $2f(a)f(b) \geq af(a) + 2f(a)f(b)$, hence
$f(a) \geq 0$ if $a < 0$. (2)
If $f(x) > 0$ for some x, then (1) implies $f(t) < 0$ for

$$t < \frac{xf(x) - f(f(x))}{f(x)},$$

which contradicts (2), therefore

$$f(x) \leq 0, \forall x \in \mathbb{R}. \quad (3)$$

Now, (2) implies $f(x) = 0$ for all $x < 0$. It remains to find $f(0)$. For $t = x < 0$ in (1), we get $0 \leq 0 - 0 + f(0)$, hence $f(0) \geq 0$. By (3), $f(0) = 0$.

5. Does there exist a function $f\colon \mathbb{R} \to \mathbb{R}$ such that for arbitrary real numbers x, y, we have $f(x - f(y)) \leq x - yf(x)$? (*Ukraine*, 2016)
Solution. Answer: no.
Assume such function exists. If $y = 0$, then $f(x - f(0)) \leq x$. Replacing x by $x + f(0)$, we get $f(x) \leq x + f(0)$. Setting $x = f(y)$, we obtain

$$f(0) \leq f(y) - yf(f(y)) \leq y + f(0) - yf(f(y)),$$

which implies $yf(f(y)) \leq y$. If $y < 0$, then

$$1 \leq f(f(y)) \leq f(y) + f(0) \leq y + 2f(0).$$

The last inequality is satisfied for all $y < 0$, contradiction.

6. Determine all functions $f\colon \mathbb{R} \to \mathbb{R}$ satisfying

$$f(a^2) - f(b^2) \leq (f(a) + b)(a - f(b))$$

for arbitrary $a, b \in \mathbb{R}$. (*Romania*, 2016)

Solution. For $a = b = 0$, we get $(f(0))^2 \leq 0$, hence $f(0) = 0$. If $a = 0$, then $f(b^2) \geq bf(b)$; if $b = 0$, then $f(a^2) \leq af(a)$, so $f(x^2) = xf(x)$ for all x (1). Using this in the enounce, we get $f(a)f(b) \leq ab$ (2). Next,

$$-xf(-x) = f\left(x^2\right) = xf(x),$$

so f is an odd function, which implies $f(a)f(b) = -f(a)f(-b) \geq -(-ab) = ab$. Using (2), we get $f(a)f(b) = ab$. Then $(f(1))^2 = 1$, so $f(1) = \pm 1$. For all x, $f(x)$ $f(1) = x$, hence $f(x) = x$ for all x or $f(x) = -x$ for all x. Both functions are solutions [4], [9], [10], [14], [17].

1.4 Mathematical Induction

Let a be an integer and for any integer $n \geq a$, let $P(n)$ be a statement involving n. If

(i) $P(a)$ is true,
(ii) $P(k) \Rightarrow P(k + 1)$ for any integer $k \geq a$,

then the statement $P(n)$ is true for all $n \geq a$.

Examples

1. Let a, b be nonnegative integers and $q = \frac{a^2+b^2}{ab+1}$. Then

$$q = (\gcd(a, b))^2.$$

Proof. Denote by $S(ab)$ the given statement and apply induction an ab.

Base case. If $ab = 0$, then $a = 0$ or $b = 0$, so $\gcd(0, b) = b$ or $\gcd(a, 0) = 0$ and $q = b^2$ or $q = a^2$.

Inductive step. Assume that $S(ac)$ holds for any c such that $0 \leq c < b$. We can suppose that $a \leq b$. Then

$$q = \frac{a^2 + b^2}{ab + 1} = \frac{a^2 + (aq - b)^2}{a(aq - b) + 1}$$

and $0 \leq c := aq - b < b$. Indeed,

$$q = \frac{a^2 + c^2}{ac + 1} > 0 \Rightarrow ac + 1 > 0 \Rightarrow c \geq 0,$$

$$q = \frac{a^2 + b^2}{ab + 1} < \frac{a^2 + b^2}{ab} \Rightarrow aq < \frac{a^2}{b} + b \leq 2b \Rightarrow c = aq - b < b.$$

Observe that $\gcd(a, c) = \gcd(a, aq - b) = \gcd(a, b)$, so by the induction hypothesis, $q = (\gcd(a, c))^2 = (\gcd(a, b))^2$, which proves $S(ab)$.

2. Let a, b be positive integers and let d be an odd factor of $a^2 + ab + b^2$. Then d divides $a^{2^n} + b^{2^n} + (a+b)^{2^n}$.

   Proof. Set $c = a + b$ and $s_n = a^{2n} + b^{2n} + c^{2n}$.

   Base case. $s_1 = a^2 + b^2 + c^2 = 2(a^2 + ab + b^2)$, implies that d divides s_1.

   Inductive step. We have $a^2b^2 + b^2c^2 + c^2a^2 = (a^2 + ab + b^2)^2$, so $a^2b^2 + b^2c^2 + c^2a^2$ is divisible by d^2. Also,

$$s_n = \left(a^2 + b^2 + c^2\right)s_{n-1} - \left(a^2b^2 + b^2c^2 + c^2a^2\right)s_{n-2} + a^2b^2c^2s_{n-3}, n \geq 3.$$

   For $n = 2$, $s_{3n-5} = s_1$ is divisible by d, hence s_{3n-2} is divisible by d for all $n \geq 1$, by the induction hypothesis. For $n = 2$, $s_{3n-4} = s_2 = a^4 + b^4 + c^4 = 2(a^2 + ab + b^2)^2$ is divisible by d^2. Also, s_{3n-2}, $a^2 + b^2 + c^2$ are divisible by d and $a^2b^2 + b^2c^2 + c^2a^2$ is divisible by d^2, hence by the induction hypothesis, s_{3n-1} is divisible by d^2 for all $n \geq 1$.

   Observe that $2^n \equiv 2, 4 \pmod 6$ for all $n \geq 1$, which ends the proof.

3. Let p be an odd positive integer and let a, b be the solutions of the equation $x^2 + px - 1 = 0$. Set $s_n = a^n + b^n$ for any nonnegative integer n. Then s_n is an integer for all $n \geq 0$ and $\gcd(s_n, s_{n+1}) = 1$, $n \geq 0$.

   Proof. Let $S(n)$ be the given statement.

   Base case. We have $s_0 = 2$, $s_1 = -p$ and $\gcd(s_0, s_1) = 1$, since p is odd, so $S(0)$ is true.

   Inductive step. Assume that $S(k)$ is true. By the Viète's relations,

$$s_{k+2} = -ps_{k+1} + s_k.$$

   so s_{k+2} is an integer, if s_k, s_{k+1} are integers. If d is a common divisor of s_{k+2} and s_{k+1}, then by the above recurrence relation, d divides s_k, hence $d = \pm 1$, by the induction hypothesis. It follows that $\gcd(s_{k+1}, s_{k+2}) = 1$, which proves $S(k+1)$.

4. For every $n > 1$, the following inequality holds:

$$1 + \frac{1}{2^3} + \frac{1}{3^3} + \ldots + \frac{1}{n^3} < \frac{1}{2}\left(3 - \frac{1}{n^2}\right).$$

   Proof. Let $S(n)$ be the given statement.

   Base case. $S(2) : 1 + \frac{1}{8} < \frac{3}{2} - \frac{1}{8}$ is true.

   Inductive step. Assume that $S(k)$ is true for some $k \geq 2$. Then

$$1 + \frac{1}{2^3} + \ldots + \frac{1}{k^3} + \frac{1}{(k+1)^3} < \frac{1}{2}\left(3 - \frac{1}{k^2}\right) + \frac{1}{(k+1)^3}.$$

   We want to have on the right-hand side the expression for $S(k+1)$. The equivalences

$$\frac{1}{2}\left(3-\frac{1}{k^2}\right)+\frac{1}{(k+1)^3}<\frac{1}{2}\left(3-\frac{1}{(k+1)^2}\right)\Leftrightarrow\frac{2}{(k+1)^3}<\frac{1}{k^2}-\frac{1}{(k+1)^2}\Leftrightarrow$$

$$\Leftrightarrow 2k^2<(k+1)(2k+1)$$

show that $S(k + 1)$ is true.

5. Let $(a_n)_{n \geq 1}$ be a sequence of positive integers satisfying: (1) $a_{2n} = a_n + n$ for any n; (2) if a_n is a prime number, then so is n. Prove that $a_n = n$, $\forall n \geq 1$.

   Proof. We will prove by induction that $a_n = a_1 + n - 1$ for all n.

   Base case. We have $a_2 = a_1 + 1$, $a_4 = a_2 + 2 = a_1 + 3$, so $a_3 = a_1 + 2$, since the sequence is increasing.

   Inductive step. For $n = 2^k$, use induction on k. For $2^k < n < 2^{k+1}$,

$$a_1+2^k-1=a_{2^k}<a_{2^k+1}<\ldots<a_n<\ldots<a_{2^{k+1}}=a_1+2^{k+1}-1,$$

   which is possible only if $a_n = a_1 + n - 1$.

   Now, we will prove that $a_1 = 1$: assume $a_1 > 1$, let p, q be consecutive primes and $a_1 < p < q$. Then $a_q - a_1 + 1 = a_1 + q - 1 - a_1 + 1 = q$, so $q - a_1 + 1$ is a prime number. It follows that $q - a_1 + 1 \leq p$, hence $q - p \leq \leq a_1 - 1$. The numbers $(a_1 + 1)! + 2, \ldots, (a_1 + 1)! + a_1 + 1$ are all composite, therefore

$$p<(a_1+1)!+2<\ldots<(a_1+1)!+a_1+1<q,$$

   whence $q - p > a_1 - 1$, contradiction.

6. Let $(a_n)_{n \geq 1}$ be the sequence defined by $a_1 = 0, a_n = 1 + a_{\lfloor \frac{n}{2} \rfloor}, n \geq 2$. Then

$$a_n=\lfloor \log_2 n\rfloor, \forall n\geq 1.$$

   Proof. Let $A(n)$ be the assertion that $a_n = \lfloor \log_2 n \rfloor$. We will prove first $A(n)$ when n is a power of 2.

   Base case. $a_1 = \lfloor \log_2 1 \rfloor = 0$.

   Inductive step. Assume $A(2^k)$ and prove $A(2^{k+1})$ for some $k \geq 0$:

$$a_{2^{k+1}}=1+a_{2^k}=1+k=\lfloor \log_2 2^{k+1}\rfloor.$$

   Let $B(k)$ be the assertion that for $0 \leq m \leq 2^k - 1$, we have $a_{2^k} = a_{2^k+m}$.

   Base case. When $k = 0$, we have $m = 0$.

   Inductive step. If $B(r)$ holds for some $r \geq 0$, then

$$a_{2^r}=a_{2^r+1}=\ldots=a_{2^{r+1}-1}.$$

   For $0 \leq m \leq 2^{r+1} - 1$, we have to prove that

$$a_{2^{r+1}} = a_{2^{r+1}+m}.$$

We have

$$a_{2^{r+1}+m} = 1 + a_{2^r + \lfloor \frac{m}{2} \rfloor} = 1 + a_{2^r} = a_{2^{r+1}}.$$

Now, we will prove $A(n)$ for all n. Let $n = 2^k + i$, $0 \leq i < 2^k$, then

$$a_n = \lfloor \log_2 (2^k + i) \rfloor = \lfloor \log_2 2^k \rfloor = k = a_{2^k}.$$

7. Let P be a polynomial of degree $n \geq 1$ with real coefficients such that $|P(x)| \leq 1$ for all $x \in [0, 1]$.
   Show that

$$\left| P\left(-\frac{1}{n}\right) \right| \leq 2^{n+1} - 1.$$

Proof. Denote $f(x) = P\left(\frac{x}{n}\right)$. Then we have to prove that if $|f(x)| \leq 1$ for $x \in [0, n]$, then $|f(-1)| \leq 2^{n+1} - 1$. Apply induction on n.

Base case. For $n = 1$, we have $f(x) = ax + b$, $a, b \in \mathbb{R}$, and

$$|f(0)| = |b| \leq 1, |f(1)| = |a + b| \leq 1,$$

so $|f(-1)| = |-a + b| = |-a - b + 2b| \leq |a + b| + 2|b| \leq 3$.

Inductive step. If $\deg f = n$ and $g(x) = f(x) - f(x + 1)$, then $\frac{1}{2}g$ has degree at most $n - 1$ and satisfies the given condition, therefore

$$|f(-1)| = |f(0) + g(-1)| \leq 1 + 2(2^n - 1) = 2^{n+1} - 1.$$

8. Show that for each $n \geq 1$, the greatest distance among a set of n points is realized by at most n different pairs of points.

   Proof. Let $A(n)$ be the assertion that among n points in the plane, the greatest distance is realized by at most n different pairs of points.

   Base case. For $1 \leq n \leq 3$, the number of pairs is at most n, hence $A(n)$ is true.

   Inductive step. Assume that $A(m - 1)$ is true for some $m \geq 4$ and let $\{P_1, \ldots, P_m\}$ a set of m points with maximum distance between two points at most 1.

   Case 1. If each point is at distance 1 from at most two other points, then there are at most m pairs with distance 1, so $A(m)$ is true.

   Case 2. Some point, say P_1, is at distance 1 from P_2, P_3, P_4. Then at least one of P_2, P_3, P_4 has only P_1 at distance 1: all angles $\angle P_iP_1P_j$ are acute and suppose that P_3 is interior for the angle $\angle P_2P_1P_4$. If P_3 is at a distance 1 from some P_k, then $[P_3P_k]$ is not disjoint from either $[P_1P_2]$ or $[P_1P_4]$. But if two segments of length 1 are disjoint, then two endpoints are at distance >1, so $[P_3P_k]$ intersects both $[P_1P_2]$ and $[P_1P_4]$, hence $k = 1$.

Delete P_3 to obtain a set of $m - 1$ points which satisfies the induction hypothesis. This set together P_3 form a set satisfying $A(m)$.

9. Given $3n$ points in a plane such that no three lie on a line. Then these points are the vertices of n disjoint triangles.

 Proof.

 Base case. For $n = 1$, the assertion is obvious.

 Inductive step. Consider $3k + 3$ points in the plane, $A_1, A_2, \ldots, A_{3k+3}$, no three on the same line. Let $A_1A_2\ldots A_s$, $3 \leq s \leq 3k + 3$, be the convex hull of the set containing these points.

 If $\Delta A_1A_2A_3$ does not contain any point from $A_4, \ldots, A_{3k+3}$, then these points are the vertices of k disjoint triangles, by the induction hypothesis, so we have together $\Delta A_1A_2A_3$, $k + 1$ disjoint triangles.

 If $\Delta A_1A_2A_3$ contains some of the points $A_4, \ldots, A_{3k+3}$, then there exists among them a unique point B for which $\angle BA_1A_2$ is minimal. The half-plane opposite to the half-plane A_1BA_2 contains $3k$ points, which are the vertices of k disjoint triangles, by the induction hypothesis. Together with ΔA_1A_2B, there are $k + 1$ disjoint triangles.

10. An infinite chessboard has squares labelled (m, n), with m, n positive integers. Then starting at $(1, 1)$, a knight can reach any square of the chessboard in a finite number of moves.

 Proof. Let $S(m + n)$ be the statement that the square (m, n) can be reached in a finite number of moves starting at $(1, 1)$.

 Base case. $S(2)$ is true (no moves).

 Inductive step. The sequences

$$(m,n) \rightarrow (m+2,n+1) \rightarrow (m+1,n+3) \rightarrow (m,n+1),$$
$$(m,n) \rightarrow (m+1,n+2) \rightarrow (m+3,n+1) \rightarrow (m+1,n),$$

 show that $S(m + n)$ implies $S(m + n + 1)$ [4], [6], [9], [10], [14], [17].

1.5 Polynomials

A polynomial P is an expression of the form

$$a_nx^n + a_{n-1}x^{n-1} + \ldots + a_1x + a_0, a_n \neq 0,$$

where n is a nonnegative integer (the degree of P), $a_0, a_1, \ldots, a_n$ are numbers from the sets $\mathbb{Z}$, $\mathbb{Q}$, $\mathbb{R}$ or $\mathbb{C}$, called coefficients, and x is a symbol which does not have any value (is an indeterminate). When P is viewed as a function, then x is a variable. A polynomial is irreducible if it cannot be expressed as a product of polynomials of lower degrees.

Viète's formulas. A root of a polynomial P is a number r such that $P(r) = 0$. A polynomial of degree $n \geq 1$ with complex coefficients has exactly n roots.

Let $x_1, x_2, \ldots, x_n$ be the roots of the polynomial

$$P = a_n x^n + a_{n-1} x^{n-1} + \ldots + a_1 x + a_0.$$

Then

$$x_1 + x_2 + \ldots + x_n = -\frac{a_{n-1}}{a_n},$$

$$x_1x_2 + x_1x_3 + \ldots + x_{n-1}x_n = \frac{a_{n-2}}{a_n},$$

$$\vdots$$

$$x_1x_2 \ldots x_n = (-1)^n \frac{a_0}{a_n}$$

are the Viète's formulas. The left-hand sides are elementary symmetric polynomials of the roots.

Example. Let a, b be complex numbers and x_1, x_2, x_3 be the roots of the polynomial $f = x^3 + ax + b$. Calculate $x_1^2x_2 + x_2^2x_3 + x_3^2x_1$.

Solution. The Viète's relations are

$$x_1 + x_2 + x_3 = 0, x_1x_2 + x_2x_3 + x_3x_1 = a, x_1x_2x_3 = -b.$$

Denote $u = x_1^2x_2 + x_2^2x_3 + x_3^2x_1, v = x_1x_2^2 + x_2x_3^2 + x_3x_1^2$. Then

$$u + v = x_1x_2(x_1 + x_2) + x_2x_3(x_2 + x_3) + x_3x_1(x_3 + x_1) =$$
$$= x_1x_2(-x_3) + x_2x_3(-x_1) + x_3x_1(-x_2) = -3x_1x_2x_3 = 3b,$$
$$uv = x_1^3x_2^3 + x_2^3x_3^3 + x_3^3x_1^3 + x_1x_2x_3\left(x_1^3 + x_2^3 + x_3^3\right) + 3x_1^2x_2^2x_3^2 =$$
$$= (-ax_1 - b)(-ax_2 - b) + (-ax_2 - b)(-ax_3 - b) + (-ax_3 - b)(-ax_1 - b) +$$
$$+ (-b)[-a(x_1 + x_2 + x_3) - 3b] + 3(-b)^2 =$$
$$= a^2(x_1x_2 + x_2x_3 + x_3x_1) + 2ab(x_1 + x_2 + x_3) + 3b^2 + 3b^2 + 3b^2 = a^3 + 9b^2,$$

so u, v are the roots of the quadratic polynomial $g(t) = t^2 - 3bt + a^3 + 9b^2$.

Divisibility of Polynomials

Examples

1. Find positive integers n for which the polynomial $f = x^n + x + 1$ is divisible by the polynomial $g = x^3 - x^2 + 1$ in $\mathbb{Z}[x]$.
   Solution. Answer: $n = 5$.
   We have $x^5 + x + 1 = (x^2 + x + 1)g$. From $g(-1) = -1$, $g(0) = 1$, it follows that there exists $a \in (-1, 0)$ such that $g(a) = 0$. If g divides f for some n, then $f(a) = a^n + a + 1 = a^5 + a + 1 = 0$, whence $n = 5$.
2. Find integers a, b such that the polynomial $f = x^{15} + ax + b(a + 2)$ is divisible by $g = x^2 + x + b$ in $\mathbb{Z}[x]$.
   Solution. Answer: $(a, b) \in \{(-1, 0), (89, 2)\}$.
   Case 1. $b = 0$. Then the roots of g are -1 and 0. From $f(-1) = 0$, it follows that $a = -1$.
   Case 2. $b \neq 0$. If $z \in \mathbb{C}$ is a root of g, then $z^2 + z + b = 0$, $z^3 + z^2 + bz = z^3 + (b - 1)z - b = 0$, …, $z^n + a_n z + b_n = 0$, with $a_n, b_n \in \mathbb{Z}$. From

$$z^{n+1} + a_n z^2 + b_n z = z^{n+1} + (-a_n + b_n)z + (-ba_n) = z^{n+1} + a_{n+1}z + b_{n+1} = 0,$$

it follows that $a_{n+1} = -a_n + b_n$, $b_{n+1} = -ba_n$, therefore

$$a_2 = 1, a_3 = b - 1, a_{n+1} = -a_n - ba_{n-1}, n \geq 3.$$

The polynomial $h = x^{15} + a_{15}x + b_{15}$ has the root $z \in \mathbb{C}$ (arbitrarily chosen), so g divides h in $\mathbb{C}[x]$. If g divides f in $\mathbb{Z}[x]$, then g divides f in $\mathbb{C}[x]$, so that g divides $f - h$ in $\mathbb{C}[x]$. We deduce

$$a = a_{15}, b(a + 2) = b_{15}, 2 = a + 2 - a = \frac{b_{15}}{b} - a_{15} = -a_{14} - a_{15} = ba_{13}.$$

As a_{13} is an integer, this implies $b \in \{-2, -1, 1, 2\}$. The first terms of the sequence (a_n) are

$$a_4 = -2b + 1, a_5 = -b^2 + 3b - 1, a_6 = 3b^2 - 4b + 1, \ldots,$$
$$a_{12} = -6b^5 + 35b^4 - 56b^3 + 36b^2 - 10b + 1,$$
$$a_{13} = -b^6 + 21b^5 - 70b^4 + 84b^3 - 45b^2 + 11b - 1.$$

The equalities

$$-ba_{13} + 2 = (b - 2)\left(b^6 - 19b^5 + 32b^4 - 20b^3 + 5b^2 - b - 1\right) = 0$$

show that $b = 2$ is the only solution, since for $b = \pm 1$ the second parenthesis is nonzero. For $b = 2$, we have $a_{12} = 45$, $a_{13} = 1$, $a_{14} = -91$, $a_{15} = 89$.

3. Prove that there are infinitely many positive integers n such that the polynomial $x^{5n} + (x - 1)^n$ has at least three irreducible factors in $\mathbb{Z}[x]$.
 Application. Find a 5 digit divisor of the number $2^{25 \cdot 2025} + 31^{2025}$.
 Solution. If $n > 1$ is odd, then the given polynomial is divisible by

$$x^5 + x - 1 = (x^2 - x + 1)(x^3 + x^2 - 1),$$

 and the quotient is a nonconstant monic polynomial with integer coefficients. The number $2^{25 \cdot 2025} + 31^{2025}$ is divisible by

$$2^{25} + 2^5 - 1 = (2^{10} - 2^5 + 1)(2^{15} + 2^{10} - 1),$$

 so it has the prime divisor $2^{15} + 2^{10} - 1 = 33791$.

Roots of Polynomials

Examples

1. Let $P(x)$ be a nonconstant polynomial with integer coefficients ≥ -1. Show that if $P(2) = 0$, then $P(1) \neq 0$.
 Solution. From $P(2) = 0$, it follows that $P(x) = (x - 2)(a_n x^n + \ldots + a_0)$, with $n \geq 0$, $a_n \neq 0$, $a_n, \ldots, a_0 \in \mathbb{Z}$. If $n = 0$, then $P(1) = -a_0 \neq 0$. If $n > 0$, then

$$-2a_0 \geq -1, a_{i-1} - 2a_i \geq -1, 1 \leq i \leq n, a_n \geq -1.$$

 It follows that $a_0 \leq \frac{1}{2}$ is negative or 0, and inductively,

$$a_{i-1} \leq 0 \Rightarrow a_i \leq \frac{a_{i-1} + 1}{2} \leq \frac{1}{2} \Rightarrow a_i \leq 0.$$

 Then $a_n = -1$ and $P(1) = -(a_n + \ldots + a_0) \geq -a_n = 1$.
2. If the equation $ax^2 + (c - b)x + e - d = 0$ has real roots >1, show that the equation $ax^4 + bx^3 + cx^2 + dx + e = 0$ has at least one real root. (*Greece*, 1995)
 Solution. Suppose that $p(x) = ax^4 + bx^3 + cx^2 + dx + e$ has no real roots and let $y > 1$ be a real root of the equation $ax^2 + (c - b)x + e - d = 0$ and let $z = \sqrt{y}$. Then

$$p(x) = ax^4 + (c - b)x^2 + e - d + (x - 1)(bx^2 + d) \Rightarrow$$
$$\Rightarrow p(z) = (z - 1)(bz^2 + d), p(-z) = (-z - 1)(bz^2 + d).$$

From $z > 1$, it follows that $p(z)p(-z) < 0$, hence p has a root in the interval $(-z, z)$, contradiction.

3. Let a, b be real numbers such that $a \geq 1, b > a^2 - a + 1$. Show that the polynomial $p = x^5 - ax^3 + a^2x - b$ has a unique real root x_0, and

$$2b - a^3 < x_0^6 < b^2 + a - a^3.$$

Solution. The derivative $p'(x) = 5x^4 - 3ax^2 + a^2$ has the discriminant, as a quadratic function in x^2, $9a^2 - 20a^2 = -11a^2$, so p is increasing and has at most one real root. Since the degree of p is odd, it follows that p has exactly one real root x_0.

From $x_0^5 - ax_0^3 + a^2x_0 = b > a^2 - a + 1$, with equality for $x_0 = 1$, we deduce that $x_0 > 1$, because LHS is strictly increasing as function of x_0. The above equation has the equivalent form

$$b\left(x_0^2 + a\right) = x_0\left(x_0^6 + a^3\right).$$

If $x_0^6 \geq b^2 + a - a^3$, then

$$b\left(x_0^2 + a\right) \geq x_0\left(b^2 + a\right), \quad (bx_0 - a)(x_0 - b) \geq 0.$$

Since $b - a > (a - 1)^2 \geq 0$, and $x_0 > 1$, we have $bx_0 - a > 0$, so $x_0 > b$. The polynomial function p is strictly increasing, hence

$$0 = p(x_0) > p(b) = \left(b^2 - a\right)b^3 + \left(a^2 - 1\right)b,$$

contradiction, because $b^2 - a > b - a > 0$ and $a^2 - 1 \geq 0$.

If $2b - a^3 \geq x_0^6$, then

$$b\left(x_0^2 + a\right) = x_0\left(x_0^6 + a^3\right) \leq 2bx_0 < b\left(x_0^2 + 1\right) \leq b\left(x_0^2 + a\right),$$

contradiction.

4. For a given positive integer n, consider the polynomials

$$f_0 = x(x+3)(x+6)\ldots(x+3n),$$
$$f_1 = (x+1)(x+4)\ldots(x+3n-2),$$
$$f_2 = (x+2)(x+5)\ldots(x+3n-1).$$

Find n for which the roots of the polynomial $p = f_0 + f_1 + f_2$ are all real and irrational.

Solution. Answer: $n \neq 2$.

For $n = 1$, one has $p = x^2 + 5x + 3$, with irrational roots

$$x_{1,2}=\frac{-5\pm\sqrt{13}}{2}.$$

For $n = 2$, the polynomial

$$p=x(x+3)(x+6)+(x+1)(x+4)+(x+2)(x+5)$$

has the root $x = -7$.

We will prove that for $n \geq 3$, each interval $(-3k-1, -3k)$ contains at least one root of p. As $\deg p = n + 1$, it follows that each interval contains exactly one root. If p has a rational root, then this root is an integer, since p is monic, contradiction. We deduce that all the roots of p are irrational.

Let $a_k = 3 \cdot 6 \cdot \ldots \cdot 3k$, $b_k = 1 \cdot 4 \cdot \ldots \cdot (3k-2)$, $c_k = 2 \cdot 5 \cdot \ldots \cdot (3k-1)$, $k \in \mathbb{N}^*$. We have

$$0<b_k<c_k<a_k, \forall k\in\mathbb{N}^*, p(0)=b_n+c_n>0, p(-1)=-b_n+c_n<0,$$

$$p(-3k)=(-1)^k(b_{n-k}c_k+b_kc_{n-k}), p(-3k-1)=(-1)^{k+1}(b_{k+1}c_{n-k}-b_{n-k}c_k),$$

$$1\leq k\leq n-1, p(-3n)=(-1)^n(c_n+b_n),$$

$$p(-3n-1)=(-1)^{n+1}(b_{n+1}-a_n-c_n).$$

Now, we show that $b_{k+1}c_{n-k} - b_{n-k}c_k > 0$, $1 \leq k \leq n-1$. Indeed, this inequality is

$$\frac{(3k+1)b_k}{c_k}>\frac{b_{n-k}}{c_{n-k}},$$

with equivalent forms,

$$k<n-k: \quad 3k+1>1>\frac{3k+1}{3k+2}\cdot\ldots\cdot\frac{3(n-k)-2}{3(n-k)-1},$$

$$k=n-k: \quad 3k+1>1,$$

$$k>n-k: \quad \frac{3k+1}{3k-1}\cdot\frac{3k-2}{3k-4}\cdot\ldots\cdot\frac{3(n-k)+4}{3(n-k)+2}\cdot(3(n-k)+1)>1,$$

which are true. Now, we show that $b_{n+1} - a_n - c_n > 0$, $\forall n \geq 3$. Indeed, from

$$b_4-a_3-c_3=1\cdot4\cdot7\cdot10-3\cdot6\cdot9-2\cdot5\cdot8>0,$$

$$b_{n+1}-a_n-c_n=3n(b_n-a_{n-1}-c_{n-1})+b_n+c_{n-1},$$

we obtain by induction the above inequality.

The values of the polynomial at $-3k-1$ and $-3k$ have opposite sign, so each interval $(-3k-1, -3k)$, $0 \le k \le n$ contains one root of p, which ends the proof.

5. Let n be a positive integer, a, b be nonzero real numbers and let $z_0 \in \mathbb{C} - \mathbb{R}$ be a root of the polynomial $p = x^{n+1} + ax + nb$. Show that $|z_0| \ge \sqrt[n+1]{|b|}$.

   Solution. Let $z_0 = r(\cos\alpha + i\sin\alpha)$, $r > 0$, $\sin\alpha \ne 0$. Then

$$r^{n+1}\cos(n+1)\alpha + ar\cos\alpha + nb = 0, \quad r^{n+1}\sin(n+1)\alpha + ar\sin\alpha = 0,$$

whence

$$r^{n+1}[\sin(n+1)\alpha\cos\alpha - \cos(n+1)\alpha\sin\alpha] - nb\sin\alpha = 0,$$

that is,

$$r^{n+1}\sin n\alpha = nb\sin\alpha.$$

But $\sin\alpha \ne 0$ implies $\sin n\alpha \ne 0$, hence

$$r^{n+1} = \frac{nb\sin\alpha}{\sin n\alpha}.$$

By induction we can prove the inequality $|\sin n\alpha| \le n|\sin\alpha|$, therefore

$$r^{n+1} = |b| \cdot \left|\frac{n\sin\alpha}{\sin n\alpha}\right| \ge |b|.$$

We deduce $r = |z_0| \ge \sqrt[n+1]{|b|}$.

Lagrange interpolation. Let y_i, z_i, $1 \le i \le n$ be arbitrary complex numbers such that z_i are distinct. Define

$$l_k(z) := \frac{\prod\limits_{i=1, i\ne k}^{n} (z - z_i)}{\prod\limits_{i=1, i\ne k}^{n} (z_k - z_i)}, 1 \le k \le n.$$

Then there exists a unique polynomial function p such that

$$p(z_i) = y_i, 1 \le i \le n.$$

This polynomial function is

$$p(z) = \sum_{k=1}^{n} y_k l_k(z),$$

called *Lagrange interpolation*. If

$$\omega(z) := \prod_{i=1}^{n} (z - z_i),$$

then

$$l_k(z) = \frac{\omega(z)}{(z - z_k)\omega'(z_k)}$$

and

$$p(z) = \sum_{k=1}^{n} \frac{y_k \omega(z)}{(z - z_k)\omega'(z_k)}.$$

Examples

1. If $f(x)$ is a polynomial of degree at most n, and

$$f(k) = \frac{n-k+1}{k+1}, 0 \leq k \leq n,$$

then find $f(n+1)$.
Solution. Define

$$P_k(x) = \frac{x(x-1)\ldots(x-n)}{x-k},$$

then

$$P_k(n+1) = \frac{(n+1)!}{n+1-k}, P_k(k) = (-1)^{n-k} k!(n-k)!,$$

$$f(n+1) = (-1)^{n+1} \sum_{k=0}^{n} (-1)^{k+1} \frac{(n+1)!}{(k+1)!(n-k)!} = (-1)^n.$$

2. Determine all polynomials $P(x)$ with real coefficients such that for every rational number r, the equation $P(x) = r$ has a rational solution.
Solution. If $P(x) = ax + b$, $a, b \in \mathbb{Q}$, $a \neq 0$ and $P(x) = r \in \mathbb{Q}$, then

$$x = \frac{r - b}{a} \in \mathbb{Q}.$$

Conversely, if $P(x)$ satisfies the required condition, let $n =$ deg P. For each $r = 0, 1, \ldots, n$, let $P(x_r) = r$ for some $x_r \in \mathbb{Q}$. By the Lagrange interpolation formula,

$$P(x) = \sum_{r=0}^{n} \left(r \prod_{0 \leq r \leq n, r \neq i} \frac{x - x_i}{x_r - x_i} \right).$$

It follows that $P(x) \in \mathbb{Q}[x]$. Then $Q(x) = MP(x) \in \mathbb{Z}[x]$, where M is the product of the denominators. Let k be the leading coefficient of $Q(x)$ and c be the constant term of $P(x)$. Let (p_i) be the sequence of prime numbers and let

$$P(t_i) = c + \frac{p_i}{M}, t_i \in \mathbb{Q}.$$

Then $Q(x) - (cM + p_i)$ has k as the leading coefficient and $-p_i$ as constant term. Now,

$$Q(t_i) = 0 \Longrightarrow t_i \in \left\{ \frac{1}{d_i}, \frac{p_i}{d_i} \right\}, d_i | k.$$

Since the values $P(t_i)$ are distinct, it follows that t_i are distinct, so the number of indices i for which $t_i = \frac{1}{d_i}$ is less than or equal to the number of divisors of k. Then there exists a divisor d of k for which $t_i = \frac{p_i}{d}$ infinitely many times, hence the polynomial

$$P(x) - \left(c + \frac{dx}{M} \right)$$

has infinitely many roots, so it is the zero polynomial.

3. Let P be a point on the plane of ΔABC. Prove that

$$\frac{PA}{BC} + \frac{PB}{CA} + \frac{PC}{AB} \geq \sqrt{3}.$$

Solution. Take $P(z)$, $A(z_1)$, $B(z_2)$, $C(z_3)$ in the complex plane. The only polynomial $P(x)$ of degree at most 2 that equals 1 at z_1, z_2, z_3 is the constant polynomial, so

$$\frac{(x-z_1)(x-z_2)}{(z_3-z_1)(z_3-z_2)}+\frac{(x-z_2)(x-z_3)}{(z_1-z_2)(z_1-z_3)}+\frac{(x-z_3)(x-z_1)}{(z_2-z_3)(z_2-z_1)}=1.$$

Set $x = z$ and apply the triangle inequality:

$$\frac{PA}{BC}\cdot\frac{PB}{CA}+\frac{PB}{CA}\cdot\frac{PC}{AB}+\frac{PC}{AB}\cdot\frac{PA}{BC}\geq 1.$$

Now the required inequality follows from the well-known inequality for positive real numbers: $(a + b + c)^2 \geq 3(ab + bc + ca)$.

4. Let a_i, b_j, $1 \leq i, j \leq 4$ be real numbers such that $a_i - b_j \neq 0$, $1 \leq i, j \leq 4$. Suppose that the system of equations

$$\begin{cases}\dfrac{x_1}{b_1-a_1}+\dfrac{x_2}{b_1-a_2}+\dfrac{x_3}{b_1-a_3}+\dfrac{x_4}{b_1-a_4}=1\\ \dfrac{x_1}{b_2-a_1}+\dfrac{x_2}{b_2-a_2}+\dfrac{x_3}{b_2-a_3}+\dfrac{x_4}{b_2-a_4}=1\\ \dfrac{x_1}{b_3-a_1}+\dfrac{x_2}{b_3-a_2}+\dfrac{x_3}{b_3-a_3}+\dfrac{x_4}{b_3-a_4}=1\\ \dfrac{x_1}{b_4-a_1}+\dfrac{x_2}{b_4-a_2}+\dfrac{x_3}{b_4-a_3}+\dfrac{x_4}{b_4-a_4}=1\end{cases}$$

has a unique solution (x_1, x_2, x_3, x_4). Find $x_1 + x_2 + x_3 + x_4$.

Solution. The coefficient of x^3 in the polynomial

$$P(x)=\prod_{i=1}^{4}(x-a_i)-\prod_{i=1}^{4}(x-b_i)$$

is

$$-\sum_{i=1}^{4}a_i+\sum_{i=1}^{4}b_i.$$

Define f_1, f_2, f_3, f_4 using a_1, a_2, a_3, a_4 to get the Lagrange interpolation formula

$$P(x)=\sum_{i=1}^{4}P(a_i)\cdot\frac{f_i(x)}{f_i(a_i)}.$$

The coefficient of x^3 in $f_i(x)$ is 1, hence the coefficient of x^3 in $P(x)$ is also

$$\sum_{i=1}^{4}\frac{P(a_i)}{f_i(a_i)}.$$

Observe that

$$\frac{P(b_j)}{f_i(b_j)} = b_j - a_i.$$

Setting $x = b_j$ in the interpolation formula for $1 \le j \le 4$ and dividing by $P(b_j)$ we get

$$1 = \sum_{i=1}^{4} \frac{P(a_i)}{P(b_j)} \cdot \frac{f_i(b_j)}{f_i(a_i)} = \sum_{i=1}^{4} \frac{P(a_i)}{f_i(a_i)} \cdot \frac{1}{b_j - a_i}.$$

By the uniqueness of the solution, we deduce

$$x_i = \frac{P(a_i)}{f_i(a_i)}, 1 \le i \le 4,$$

therefore

$$\sum_{i=1}^{4} x_i = \sum_{i=1}^{4} \frac{P(a_i)}{f_i(a_i)} = \sum_{i=1}^{4} (b_i - a_i).$$

5. Prove that there exists a positive constant c such that for any positive integer n and all real numbers a_1, a_2, ..., a_n, if

$$P(x) = (x - a_1)(x - a_2)\ldots(x - a_n),$$

then

$$\max_{x \in [0,2]} |P(x)| \le c^n \max_{x \in [0,1]} |P(x)|.$$

Solution. Let M be the maximum of $|P(x)|$ for $x \in [0,1]$. Define

$$b_i = \frac{i}{n}, f_i(x) = \frac{(x - b_0)\ldots(x - b_n)}{x - b_i}, 0 \le i \le n.$$

By the Lagrange interpolation formula,

$$P(x) = \sum_{i=0}^{n} P(b_i) \cdot \frac{f_i(x)}{f_i(b_i)}, \forall x \in \mathbb{R}.$$

If $y \in [0,2]$, then $|y - b_k| \le |2 - b_k|$, $0 \le k \le n$, so

$$|f_i(y)| \le |f_i(2)| = \prod_{i=0}^{n} \left(2 - \frac{i}{n}\right) = \frac{(2n)!}{n!n^n},$$

$$P(b_i) \le M, |f_i(b_i)| = \frac{i!(n-i)!}{n^n}.$$

By the triangle inequality,

$$|P(y)| \le \sum_{i=0}^{n} |P(b_i)| \cdot \left|\frac{f_i(y)}{f_i(b_i)}\right| \le M \sum_{i=0}^{n} \binom{2n}{i}\binom{2n-i}{n}.$$

Finally,

$$\sum_{i=0}^{n} \binom{2n}{i}\binom{2n-i}{n} \le \sum_{i=0}^{n} \binom{2n}{i}\binom{2n}{n} = 2^{2n}\binom{2n}{n} \le 2^{4n} \Rightarrow$$

$$\Rightarrow \max_{y \in [0,2]} |P(y)| \le 2^{4n} M = 16^n \max_{x \in [0,1]} |P(x)|.$$

Irreducible Polynomials

Examples

1. If $f(x)$ is a polynomial of degree $2m + 1$ with integral coefficients for which there are $2m + 1$ integers $k_1, k_2, \ldots, k_{2m+1}$ such that $f(k_i) = 1$, $1 \le i \le 2m + 1$, prove that $f(x)$ is not the product of two nonconstant polynomials with integral coefficients
Solution. If P, Q are nonconstant polynomials with integer coefficients and $f = PQ$, then $f(k_i) = P(k_i)Q(k_i) = 1$, hence $P(k_i) = Q(k_i) = 1$ for at least $m + 1$ values or $P(k_i) = Q(k_i)$ for at least $m + 1$ values. Since $\deg f = 2m + 1$, one of P, Q has degree at most m, so this factor is constant, contradiction.
2. For $n \ge 4$, prove that $x^n - x^{n-1} - x^{n-2} - \ldots - x - 1$ cannot be factored into a product of two polynomials with rational coefficients, both with degree >1.
 Solution. Let $P_n(x) = x^n - x^{n-1} - x^{n-2} - \ldots - x - 1$, $Q_n(x) = (x - 1)P_n(x) = x^{n+1} - 2x^n + 1$. The cases $n = 2, 3$ follow directly from the rational root theorem. For $n \ge 4$, the Descartes' rule of signs shows that there is a positive root r. It is easy to see that $P_n(\sqrt{3}))$, so $r > \sqrt{3}$. If $P_n(s) = 0$ with $|s| > 1$, then $Q_n(s) = 0$, which implies $|s|^n|s - 2| = 1$. We have

$$2 \le |s - 2| + |s| = |s|^{-n} + |s|,$$

so $Q_n(|s|) \ge 0$. Since $Q_n(x) < 0$ for $1 < x < r$, we must have $|s| \ge r$. If $P_n(t) = 0$

and $|t| < 1$, then $1 = |t-2||t|^n \leq 3|t|^n$. It follows that the absolute value of the product of all roots t of $P_n(x)$ with $|t| < 1$ is at least $\frac{1}{3}$, so r is the only root of $P_n(x)$ with absolute value >1. If $f(x)$, $g(x)$ are monic polynomials with integer coefficients and positive degrees, $P_n(x) = f(x)g(x)$ and $f(r) = 0$, then the roots of g have absolute value <1, hence $|g(0)| < 1$, contradiction, since $|g(0)| = |P_n(0)| = 1$.

Eisenstein's criterion. This is a sufficient condition asserting that a polynomial with integer coefficients is irreducible in $\mathbb{Q}[x]$.

Let $P = a_n x^n + a_{n-1}x^{n-1} + \ldots + a_1 x + a_0$ be a polynomial with integer coefficients and let p be a prime number. If p divides $a_0, \ldots, a_{n-1}$, but not divide a_n, and moreover, p^2 does not divide a_0, then P is irreducible in $\mathbb{Q}[x]$.

Example. For any positive integer n, the polynomial

$$f(x) = 1 + x + \frac{x^2}{2!} + \ldots + \frac{x^n}{n!}$$

is irreducible in $\mathbb{Q}[x]$.

Solution. We have to show that the polynomial

$$n!f(x) = x^n + nx^{n-1} + n(n-1)x^{n-2} + \ldots + n!$$

is irreducible in $\mathbb{Z}[x]$. This is clear for $n = 1, 2$. For $n \geq 3$, it suffices to find a prime number p such that $p|n!, p^2 \nmid n!$, that is, $p \leq n < 2p$. By the Bertrand's postulate, there is a prime number p such that $\frac{n}{2} < p \leq n$.

If $n = 2m$, then $m < p \leq 2m$, so $p \leq n < 2p$.

If $n = 2m + 1$, then $p \leq n - 1$ and $n - 1 < 2p$. But $n - 1$ is even, so $n < 2p$. Also, $p \leq n - 1 < n$ [1], [2], [4], [9], [10], [14], [17].

Chapter 2
Problems

2.1 1968

1. (*Yugoslavia*). Let $n > 1$ be an integer and x a real number.

(i) Evaluate $S(x, n) = \sum (x + p)(x + q)$, where the summation is over all pairs (p, q) of different numbers from $\{1, 2, \dots, n\}$.
(ii) Do there exist integers x, n for which $S(x, n) = 0$?

2.2 1969

2. (*Yugoslavia*). Given real numbers a_i, b_i $(i = 1, 2, \dots, n)$ such that

$$a_1 \geq a_2 \geq \dots \geq a_n > 0, b_1 \geq a_1, b_1 b_2 \geq a_1 a_2, \dots, b_1 b_2 \dots b_n \geq a_1 a_2 \dots a_n.$$

Prove that $b_1 + b_2 + \dots + b_n \geq a_1 + a_2 + \dots + a_n$.

2.3 1972

3. (*Yugoslavia*). Assume that the numbers from the table

$$\begin{pmatrix} a_{11} & a_{12} & \dots & a_{1n} \\ a_{21} & a_{22} & \dots & a_{2n} \\ \vdots & \vdots & \vdots & \vdots \\ a_{n1} & a_{n2} & \dots & a_{nn} \end{pmatrix}$$

satisfy the inequality

C. Mănescu-Avram, *Selection Tests in Algebra for Mathematical Olympiads*, Problem Books in Mathematics, https://doi.org/10.1007/978-3-032-01032-2_2

$$\sum_{j=1}^{n} \left| a_{j1}x_1 + a_{j2}x_2 + \ldots + a_{jn}x_n \right| \le M$$

for each choice $x_j = \pm 1$. Prove that

$$|a_{11} + a_{22} + \ldots + a_{nn}| \le M.$$

2.4 1975

4. (*Romania*). Given the sets of real numbers $A = \{a_n | n \in \mathbb{N}^*\}$, $B = \{b_n | n \in \mathbb{N}^*\}$ such that

$$a_1 = b_1, b_{n+1} = b_n(1 - a_{n+1}) + (1 - b_n)a_{n+1}.$$

Show that $\frac{1}{2} \in A$ if and only if $\frac{1}{2} \in B$.

5. (*Romania*). Find all functions $f : \mathbb{Q} \to \mathbb{Q}$ which satisfy

$$f(x+y) + f(x-y) = 2[f(x) + f(y) + 1]$$

for all $x, y \in \mathbb{Q}$.

6. (*Romania*). Solve the system of equations

$$\sin x_1 \cos x_2 = \sin x_2 \cos x_3 = \ldots = \sin x_{n-1} \cos x_n = \sin x_n \cos x_1,$$

where $x_1 + x_2 + \ldots + x_n = \pi, x_i \in \left[0, \frac{\pi}{2}\right], i = 1, 2, \ldots, n.$

2.5 1976

7. (*Romania*). Solve the system of equations:

$$(x_1 + x_2 + x_3)x_4 = (x_1 + x_2 + x_4)x_3 = (x_1 + x_3 + x_4)x_2 = (x_2 + x_3 + x_4)x_1 = a.$$

8. (*Yugoslavia*). Given that $x + z = y + t = 1$ and $x, y, z, t \ge 0$. Find the minimum and maximum values of the function

$$f(x, y, z, t) = \frac{ax^2 + by^2}{ax + by} + \frac{az^2 + bt^2}{az + bt}, a > 0, b > 0.$$

2.6 1977

9. (*Germany*). We consider two sequences of real numbers
$x_1 \geq x_2 \geq \ldots \geq x_n$ and $y_1 \geq y_2 \geq \ldots \geq y_n$.
Let $z_1, z_2, \ldots, z_n$ be a permutation of the numbers $y_1, y_2, \ldots, y_n$. Prove that

$$\sum_{i=1}^{n} (x_i - y_i)^2 \leq \sum_{i=1}^{n} (x_i - z_i)^2.$$

10. (*Germany*). Determine the polynomials P of two variables so that:

1. for any real numbers t, x, y we have $P(tx, ty) = t^n P(x, y)$, where n is a positive integer, the same for all t, x, y;
2. for any real numbers a, b, c we have $P(a+b, c) + P(b+c, a) + P(c+a, b) = 0$;
3. $P(1, 0) = 1$.

11. (*Romania*). Given $x_i \in [1, 2]$, $i = 1, 2, \ldots, n$. Show that

$$\left(\sum_{i=1}^{n} x_i\right)\left(\sum_{i=1}^{n} \frac{1}{x_i}\right)^2 \leq n^3.$$

The condition $x_i \leq 2$ for every $i = 1, 2, \ldots, n$ is necessary?

12. (*Yugolavia*). Determine the set of all real numbers α with the following property: for each positive real number c, there exists a rational number $\frac{m}{n}, m \in \mathbb{Z}, n \in \mathbb{N}^*$, different from α such that

$$\left|\alpha - \frac{m}{n}\right| < \frac{c}{n}.$$

2.7 1978

13. (*Austrian – Polish Competition*). Determine all functions $f : (0, \infty) \to \mathbb{R}$ that satisfy

$$f(x+y) = f(x^2+y^2), \forall x, y \in (0, \infty).$$

14. (*Romania*). A sequence $(x_n)_{n \geq 0}$ of real numbers satisfies $x_0 > 1 = x_{n+1}(x_n - \lfloor x_n \rfloor)$ for each $n \geq 1$. Prove that if $(x_n)_{n \geq 0}$ is periodic, then x_0 is a root of a quadratic equation with integer coefficients. Study the converse.

15. (*Romania*). Let P, Q, R be polynomials of third degree with real coefficients such that $P(x) \le Q(x) \le R(x)$ for any real number x and, moreover, there exists a real number a with $P(a) = R(a)$. Prove that there exists a constant $\lambda \in [0, 1]$ for which

$$Q = \lambda P + (1 - \lambda)R.$$

The assertion remains valid for polynomials of fourth degree?

16. (*Romania*). Consider the function $f : \mathbb{N} \rightarrow \mathbb{N}, f(n) = n^2$. Prove that there exists a function $F : \mathbb{N} \rightarrow \mathbb{N}$ such that $F \circ F = f$.

2.8 1979

17. (*Austrian – Polish Competition*). Find all polynomials of the form

$$P_n(x) = n!x^n + a_{n-1}x^{n-1} + \ldots + a_1 x + (-1)^n(n+1)$$

with integer coefficients, having n real roots $x_1, \ldots, x_n$ satisfying $k \le x_k \le k + 1$ for $k = 1, \ldots, n$.

18. (*Austrian – Polish competition*). Find all positive integers n such that the inequality

$$\left(\sum_{i=1}^{n} a_i^2\right)\left(\sum_{i=1}^{n} a_i\right) - \sum_{i=1}^{n} a_i^3 \ge 6\prod_{i=1}^{n} a_i$$

holds for any n positive numbers $a_1, \ldots, a_n$.

19. (*Romania*). Find a polynomial $P \in \mathbb{R}[x]$ of second degree such that

$$\left|P(x) + \frac{1}{x-4}\right| \le 0,01, \forall x \in [-1, 1].$$

There are polynomials of first degree having this property?

20. (*Romania*). Let n be a positive integer. Find the polynomial $P \in \mathbb{R}[X]$ such that

$$P\left(x + \frac{1}{n}\right) + P\left(x - \frac{1}{n}\right) = 2P(x), \forall x \in \mathbb{Q}.$$

21. (*Yugoslavia*). Let $n \ge 1$ be an integer and $a_1, a_2, \ldots, a_n$ be distinct positive integers. Show that

$$\sum_{i=1}^{n} a_i^3 \geq \left(\sum_{i=1}^{n} a_i\right)^2.$$

2.9 1980

22. (*Austrian – Polish Competition*). Prove that

$$\sum \frac{1}{i_1 i_2 \ldots i_k} = n,$$

where the sum is taken over all nonempty subsets $\{i_1, i_2, \ldots, i_k\}$ of $\{1, 2, \ldots, n\}$.

23. (*Austrian – Polish Competition*). Let a_1, a_2, a_3, ... be a sequence of real numbers satisfying the inequality

$|a_{k+m} - a_k - a_m| \leq 1$ for all k, m.

Show that the following inequality holds for all positive integers k and m:

$$\left|\frac{a_k}{k} - \frac{a_m}{m}\right| < \frac{1}{k} + \frac{1}{m}.$$

24. (*Romania*). Define

$$\left(1 + x + x^2\right)^n = a_0 + a_1 x + a_2 x^2 + \ldots + a_{2n} x^{2n}.$$

Prove that

- $a_0 \leq a_1 \leq \ldots \leq a_n$, $a_n \geq a_{n+1} \geq \ldots \geq a_{2n}$.
- Three of the sums

$$a_0 + a_4 + a_8 + \ldots, a_1 + a_5 + a_9 + \ldots, a_2 + a_6 + a_{10} + \ldots, a_3 + a_7 + a_{11} + \ldots$$

are equal and the fourth differs from these by 1.

25. (*Romania*). Find positive integers n for which it is possible to choose signs such that

$$1 \pm 2 \pm 3 \pm \ldots \pm n = 0.$$

26. (*Yugoslavia*). Let a, b, c, m integers, where $m > 1$. Prove that if

$$a^n + bn + c \equiv 0 (\text{mod } m)$$

for each positive integer n, then $b^2 \equiv 0(\text{mod } m)$. Must $b \equiv 0(\text{mod } m)$ also hold?

2.10 1981

27. (*Austrian – Polish competition*). Let $P(x) = x^4 + a_1x^3 + a_2x^2 + a_3x + a_4$ be a polynomial with rational coefficients. Show that if $P(x)$ has exactly one real root ξ, then ξ is a rational number.

28. (*Romania*). For positive real numbers $x_1, x_2, \ldots, x_n$, show that there exist $a_1, a_2, \ldots, a_n \in \{-1, 1\}$ such that

$$a_1x_1^2 + a_2x_2^2 + \ldots + a_nx_n^2 \geq (a_1x_1 + a_2x_2 + \ldots + a_nx_n)^2.$$

2.11 1982

29. (*Austrian – Polish Competition*). Define

$$S_n = \sum_{j,k=1}^{n} \frac{1}{\sqrt{j^2 + k^2}}.$$

Find a positive constant C such that $n \leq S_n \leq Cn$ holds for all $n \geq 3$.

30. (*Romania*). Let $f: \mathbb{R} \to \mathbb{R}$ be a function such that

$$f(f(x)) = \frac{x^9}{(x^2+1)(x^6+x^4+2x^2+1)}, \forall x \in \mathbb{R}.$$

Show that there exists a unique real number a such that $f(a) = a$.

2.12 1983

31. (*Austrian – Polish Competition*). Nonnegative real numbers a, b, x, y satisfy $a^5 + + b^5 \leq 1$ and $x^5 + y^5 \leq 1$. Prove that $a^2x^3 + b^2y^3 \leq 1$.

32. (*Romania*). Let $P \in \mathbb{R}[X]$ be a polynomial such that $P(\sin t) = P(\cos t)$, $\forall\, t \in \mathbb{R}$. Prove that there exists a polynomial $Q \in \mathbb{R}[X]$ such that $P(X) = Q(X^4 - X^2)$.

2.13 1984

33. (*Austrian – Polish Competition*). The functions $f_0, f_1 : (1, \infty) \to (1, \infty)$ are given by

$$f_0(x) = 2x, f_1(x) = \frac{x}{x-1}.$$

Show that for any real numbers a, b with $1 \leq a < b$ there exist a positive integer k and indices $i_1, i_2, \ldots, i_k \in \{0, 1\}$ such that

$$a < f_{i_k}(f_{i_{k-1}}(\ldots(f_{i_1}(2))\ldots)) < b.$$

2.14 1985

34. (*Austrian – Polish Competition*). Solve the system of equations

$$\begin{cases} x^4 + y^2 - xy^3 - \frac{9}{8}x = 0 \\ y^4 + x^2 - yx^3 - \frac{9}{8}y = 0 \end{cases}.$$

35. (*Vietnam*). Find all real values of a for which the equation

$$\left(a - 3x^2 + \cos\frac{9\pi x}{2}\right)\sqrt{3 - ax} = 0$$

has an odd number of solutions in the interval $[-1, 5]$.

36. (*Yugoslavia*). Prove the inequality for positive a, b, c, d:

$$\frac{a}{b+c} + \frac{b}{c+d} + \frac{c}{d+a} + \frac{d}{a+b} \geq 2.$$

2.15 1986

37. (*Austrian – Polish Competition*). The monic polynomial $P(x) = x^n + a_{n-1}x^{n-1} + \ldots + a_0$ of degree $n > 1$ has n distinct negative roots. Prove that $a_1P(1) > 2n^2a_0$.

38. (*China*). Let $a_1, a_2, \ldots, a_n$ and $b_1, b_2, \ldots, b_n$ be $2n$ real numbers. Prove that the following statements are equivalent:

- For any n real numbers x_1, x_2, …, x_n satisfying $x_1 \leq x_2 \leq \ldots \leq x_n$, we have

$$\sum_{k=1}^{n} a_k x_k \leq \sum_{k=1}^{n} b_k x_k.$$

- We have

$$\sum_{k=1}^{s} a_k \leq \sum_{k=1}^{s} b_k$$

for every $s \in \{1, 2, \ldots, n-1\}$ and

$$\sum_{k=1}^{n} a_k = \sum_{k=1}^{n} b_k.$$

39. (*China*). Given a positive integer A written in decimal expansion, $A = \overline{a_n a_{n-1} \ldots a_0}$. Define

$$f(A) = \sum_{k=0}^{n} 2^{n-k} a_k, A_1 = f(A), A_2 = f(A_1), \ldots$$

Prove that

- There exists positive integer k for which $A_{k+1} = A_k$.
- Find such A_k for $A = 19^{86}$.

2.16 1987

40. (*Austrian – Polish Competition*). Let n be the square of an integer whose each prime divisor has an even number of decimal digits. Consider $P(x) = x^n - 1987x$. Show that if x, y are rational numbers with $P(x) = P(y)$, then $x = y$.

41. (*Romania*). Let $P(X) = a_n X^n + a_{n-1} X^{n-1} + \ldots + a_1 X + a_0$ be a real polynomial of degree n. Suppose n is an even number and

- $a_0 > 0$, $a_n > 0$;
- $a_1^2 + a_2^2 + \ldots + a_{n-1}^2 \leq \frac{4 \min\{a_0^2, a_n^2\}}{n-1}$.

Prove that $P(x) \geq 0$ for all real values x.

42. (*Romania*). Prove that for all real numbers $\alpha_1, \ldots, \alpha_n$:

$$\sum_{i=1}^{n}\sum_{j=1}^{n} ij\cos\left(\alpha_i - \alpha_j\right) \geq 0.$$

43. (*Yugoslavia*). Let

$$f(x) = \frac{\sqrt{2+\sqrt{2}}x + \sqrt{2-\sqrt{2}}}{-\sqrt{2-\sqrt{2}}x + \sqrt{2+\sqrt{2}}}.$$

Find $\underbrace{f(f(\ldots(f}_{1987\text{ times}}(x))\ldots))$.

2.17 1988

44. (*China*). Given real numbers A, B, C such that for all real numbers x, y, z the following inequality holds:

$$A(x-y)(x-z) + B(y-z)(y-x) + C(z-x)(z-y) \geq 0.$$

Find the necessary and sufficient condition that A, B, C must satisfy.

45. (*Romania*). Find all tuples $(x_1, x_2, \ldots, x_n)$ of real numbers with the property that each number is the sum of reciprocals of remaining numbers.

46. (*Romania*). Let $p > 2$ be a prime number. Find the least positive number a which can be represented as

$$a = (X-1)f(X) + \left(X^{p-1} + X^{p-2} + \ldots + X + 1\right)g(X),$$

where $f(X)$ and $g(X)$ are integer polynomials.

47. (*Romania*). Let T denote the set of all plane triangles. Consider the function $f: T \to (0, \infty)$ defined by

$$f(\Delta ABC) = \min\left\{\frac{b}{a}, \frac{c}{b}\right\},$$

for any triangle ΔABC with $a \leq b \leq c$. Find the set of values of f.

2.18 1989

48. (*Austrian – Polish Competition*). Functions $f_0, f_1, f_2, \ldots$ are recursively defined by $f_0(x) = x, f_{2k+1}(x) = 3^{f_{2k}(x)}$ and $f_{2k+2}(x) = 2^{f_{2k+1}(x)}, k = 0, 1, 2, \ldots$ for all real x. Find the greatest one of the numbers $f_{10}(1)$ and $f_9(2)$.

49. (*China*). Find the greatest n such that $(z + 1)^n = z^n + 1$ has all its nonzero roots on the unit circle.

50. (*Romania*). Find all monic polynomials $P(X)$, $Q(X)$ with integer coefficients such that

$$Q(0) = 0, P(Q(X)) = (X - 1)(X - 2)\ldots(X - 15).$$

51. (*Romania*). Prove that for any positive integer n,

$$\sqrt{1 + \sqrt{2 + \ldots + \sqrt{n}}} < 2.$$

2.19 1990

52. (*China*). In a set S there is an operation "$\circ$" such that $\forall a, b \in S$, a unique $a \circ b \in S$ exists and

- $(a \circ b) \circ c = a \circ (b \circ c), \forall a, b, c \in S.$
- $a \circ b \neq b \circ a$ when $a \neq b$.

Prove that

1. $(a \circ b) \circ c = a \circ c, \forall\, a, b, c \in S.$
2. If $S = \{1, 2, \ldots, 1990\}$, try to define an operation "$\circ$" in S with the above properties.

53. (*China*). Find the least integer n such that among every n distinct numbers a_1, a_2, …, a_n chosen from $[1, 1000]$, there always exist a_i, a_j such that

$$0 < a_i - a_j < 1 + 3\sqrt[3]{a_i a_j}.$$

54. (*China*). Find all functions $f, g, h : \mathbb{R} \to \mathbb{R}$ such that

$$f(x) - g(y) = (x - y)h(x + y)$$

for all reals x, y.

55. (*Romania*). Let $f : N \to N$ be a function such that the set $A = \{k | f(k) \leq k\}$ is finite. Prove that for all functions $g : N \to N$, the set $B = \{k | g(f(k)) \geq g(k)\}$ is infinite.

56. (*Romania*). Find all real polynomials $P(X)$ such that $2P(2X^2 - 1) = (P(X))^2 - 2$.

57. (*Turkey*). For which integers $n \geq 2$ the following statement is always true:

$$x_1 + x_2 + \ldots + x_n = 0 \Longrightarrow x_1x_2 + x_2x_3 + \ldots + x_{n-1}x_n + x_nx_1 \leq 0.$$

2.20 1991

58. (*Austrian – Polish Competition*). For a given positive integer n determine the maximum value of the function

$$f(x) = \frac{x + x^2 + \ldots + x^{2n-1}}{(1 + x^n)^2}$$

over all $x \geq 0$ and find all positive x for which the maximum is attained.

59. (*China*). Let $n \geq 2$ be an integer and $f(x) = x^n + a_1x^{n-1} + \ldots + a_n$ a polynomial with real coefficients and real roots $b_1, b_2, \ldots, b_n$. Prove that if $x >$ max $\{b_1, b_2, \ldots, b_n\}$, then

$$f(x+1) \geq \frac{2n^2}{\frac{1}{x-b_1} + \frac{1}{x-b_2} + \ldots + \frac{1}{x-b_n}}.$$

60. (*Romania*). Prove the identity

$$\sum_{j+h=n, j\geq h} (-1)^h \frac{1}{h} \binom{j}{h} 2^{j-h} = \frac{2}{n}, n \in N^*.$$

61. (*Romania*). Let n be a positive integer. Prove that if $x_1, x_2, \ldots, x_{2n}$ are positive real numbers and $x_1 + x_2 + \ldots + x_{2n} = 1$, then

$$x_1^2x_2^2\ldots x_n^2 + x_2^2x_3^2\ldots x_{n+1}^2 + \ldots + x_{2n}^2x_1^2\ldots x_{n-1}^2 < \frac{1}{n^{2n}}.$$

2.21 1992

62. (*Romania*). Let $x_1, x_2, \ldots, x_n$ be real numbers such that

$$1 \geq x_1 \geq x_2 \geq \ldots \geq x_n \geq 0, \sum_{i=1}^{n} x_i^2 = 1, \left\lceil \sum_{i=1}^{n} x_i \right\rceil = m.$$

Prove that

$$\sum_{i=1}^{m} x_i \geq 1.$$

63. (*Romania*). Let $a_1, a_2, \ldots, a_k$ be distinct positive integers such that all 2^k sums

$$\sum_{i=1}^{k} \varepsilon_i a_i, \varepsilon_i \in \{0, 1\}$$

are distinct.

1. Prove that

$$\frac{1}{a_1} + \frac{1}{a_2} + \ldots + \frac{1}{a_k} \leq 2\left(1 - 2^{-k}\right).$$

2. Find all sequences of distinct positive integers for which the above inequality is an equality.

64. (*Romania*). Let x, y be real numbers such that $1 \leq x^2 - xy + y^2 \leq 2$. Show that

1. $\frac{2}{9} \leq x^4 + y^4 \leq 8$;
2. $x^{2n} + y^{2n} \geq \frac{2}{3^n}$, for all $n \geq 3$.

65. (*Turkey*). Let $x_1, x_2, \ldots, x_{n+1}$ be positive real numbers satisfying

$$\frac{1}{1 + x_1} + \frac{1}{1 + x_2} + \ldots + \frac{1}{1 + x_{n+1}} = 1.$$

Prove that $x_1 x_2 \ldots x_{n+1} \geq n^{n+1}$.

66. (*Yugoslavia*). Periodic sequences (a_n), (b_n), (c_n) and (d_n) satisfy the following conditions:

$$a_{n+1} = a_n + b_n, b_{n+1} = b_n + c_n, c_{n+1} = c_n + d_n, d_{n+1} = d_n + a_n$$

for $n = 1, 2, \ldots$. Prove that $a_2 = b_2 = c_2 = d_2 = 0$.

2.22 1993

67. (*Austrian – Polish Competition*). If $a,\ b \geq 0$ are real numbers, prove the inequalities:

$$\left(\frac{\sqrt{a}+\sqrt{b}}{2}\right)^2 \leq \frac{a+\sqrt[3]{a^2b}+\sqrt[3]{ab^2}+b}{4} \leq \frac{a+\sqrt{ab}+b}{3} \leq \sqrt{\left(\frac{\sqrt[3]{a^2}+\sqrt[3]{b^2}}{2}\right)^3}.$$

For each of the inequalities, find the cases of equality.

68. (*Italy*). Let $n \geq 2$ be an integer and $x_1, x_2, \ldots, x_n$ positive real numbers with sum 1. Prove that

$$\sum_{i=1}^{n} \frac{1}{\sqrt{1-x_i}} \geq n\sqrt{\frac{n}{n-1}}.$$

69. (*Moldova*). Solve in positive integers the equation

$$\left\lfloor \sqrt{1} \right\rfloor + \left\lfloor \sqrt{2} \right\rfloor + \left\lfloor \sqrt{3} \right\rfloor + \ldots + \left\lfloor \sqrt{x^2-1} \right\rfloor = 125.$$

70. (*Moldova*). Prove that if $x_i > 0,\ 1 \leq i \leq n$ and $x_1 + x_2 + \ldots + x_n = \frac{1}{2}$, then

$$\frac{(1-x_1)(1-x_2)\ldots(1-x_n)}{(1+x_1)(1+x_2)\ldots(1+x_n)} \geq \frac{1}{3}.$$

71. (*Romania*). Prove that the sequence of the imaginary parts of the complex numbers

$$z_n = (1+i)(2+i)\ldots(n+i)$$

contains infinitely many positive and infinitely many negative numbers.

72. (*Romania*). Show that for any function $f : \mathscr{P}(\{1,2,\ldots,n\}) \to \{1,2,\ldots,n\}$ there exist two subsets A and B of the set $\{1,2,\ldots,n\}$ such that $A \neq B$ and

$$f(A) = f(B) = \max\{t | t \in A\Delta B\}.$$

73. (*Romania*). Given integers $m > n > 1$ such that $\gcd(m,n) = 1$. Find

$$\gcd\left(X^{m+n} - X^{m+1} - X + 1, X^{m+n} + X^{n+1} - X + 1\right).$$

2.23 1994

74. (*Austrian – Polish Competition*). Let $n > 1$ be an odd positive integer. Assume that $x_1, x_2, \ldots, x_n$ are nonnegative integers and satisfy the system of equations

$$\begin{cases} (x_2 - x_1)^2 + 2(x_2 + x_1) + 1 = n^2 \\ (x_3 - x_2)^2 + 2(x_3 + x_2) + 1 = n^2 \\ \cdots \\ (x_1 - x_n)^2 + 2(x_1 + x_n) + 1 = n^2 \end{cases}.$$

Show either $x_1 = x_n$,or there exists j with $1 \leq j \leq n - 1$ such that $x_j = x_{j+1}$.

75. (*China*). Given $5n$ real numbers $r_i, s_i, t_i, u_i, v_i \geq 1$, $1 \leq i \leq n$, let

$$R = \frac{1}{n}\sum_{i=1}^{n} r_i, S = \frac{1}{n}\sum_{i=1}^{n} s_i, T = \frac{1}{n}\sum_{i=1}^{n} t_i, U = \frac{1}{n}\sum_{i=1}^{n} u_i, V = \frac{1}{n}\sum_{i=1}^{n} v_i.$$

Prove that

$$\prod_{i=1}^{n} \frac{r_i s_i t_i u_i v_i + 1}{r_i s_i t_i u_i v_i - 1} \geq \left(\frac{\mathrm{RSTUV} + 1}{\mathrm{RSTUV} - 1}\right)^n.$$

76. (*China*). Given distinct prime numbers p and q and a positive integer $n \geq 3$, find all $a \in \mathbb{Z}$ such that the polynomial $f(x) = x^n + ax^{n-1} + pq$ can be factored into two polynomials with integer coefficients of degree at least 1.

77. (*Hong Kong*). Let x, y, z be positive real numbers such that $xy + yz + zx = 1$. Prove that

$$x(1 - y^2)(1 - z^2) + y(1 - z^2)(1 - x^2) + z(1 - x^2)(1 - y^2) \leq \frac{4\sqrt{3}}{9}.$$

2.24 1995

78. (*Belarus*). Prove that the number of odd coefficients in the polynomial $(1 + x)^n$ is a power of 2 for every positive integer n.

79. (*Belarus*). If $0 < a, b < 1$ and $p, q \geq 0$, $p + q = 1$ are real numbers, then prove that:

$$a^p b^q + (1-a)^p (1-b)^q \leq 1.$$

80. (*Italy*). A function $f : \mathbb{R} \to \mathbb{R}$ satisfies the conditions:

$$f(x+24) \leq f(x) + 24, f(x+77) \geq f(x) + 77, \forall x \in \mathbb{R}.$$

Prove that $f(x + 1) = f(x) + 1$ for all real x.

81. (*Romania*). Let $a_1, a_2, \ldots, a_n$ be distinct positive integers. Prove that

$$\left(a_1^5 + \ldots + a_n^5\right) + \left(a_1^7 + \ldots + a_n^7\right) \geq 2\left(a_1^3 + \ldots + a_n^3\right)^2$$

and find the cases of equality.

82. (*Turkey*). Given real numbers $b \geq a > 0$, find all real solutions of the system

$$\begin{cases} x_1^2 + 2ax_1 + b^2 = x_2 \\ x_2^2 + 2ax_2 + b^2 = x_3 \\ \ldots \\ x_n^2 + 2ax_n + b^2 = x_1 \end{cases}.$$

83. (*Vietnam*). Find all integers k such that for infinitely many integers $n \geq 3$ the polynomial

$$P(X) = X^{n+1} + kX^n - 870X^2 + 1945X + 1995$$

is reducible in $\mathbb{Z}[X]$.

2.25 1996

84. (*Bosnia and Herzegovina*).

1. Let a, b, c be positive real numbers. Prove that for all positive integers m, $(a + b)^m + (b + c)^m + (c + a)^m \leq 2^m(a^m + b^m + c^m)$.
2. Does the inequality hold for 1) arbitrary real numbers a, b, c; 2) any integer m?

85. (*Estonia*). Find all functions $f : \mathbb{R} \to \mathbb{R}$ which satisfy for all x:

1. $f(x) = f(-x)$; *ii)* $f(x + 1) = f(x) + 1$;
2. $f\left(\frac{1}{x}\right) = \frac{1}{x^2} f(x), x \neq 0.$

86. (*Moldova*). Let $f : (0, \infty) \to (0, \infty)$ be a function satisfying the following condition:

$$f(xy) \le f(x)f(y), \forall x, y \in (0, \infty).$$

Prove that

$$f(x^n) \le f(x)f(x^2)^{\frac{1}{2}} \dots f(x^n)^{\frac{1}{n}}$$

for any positive real number x and positive integer n.

87. (*Moldova*). Find all polynomials $P(X)$ of fourth degree with real coefficients that verify the properties:

$P(-x) = P(x), \forall x \in \mathbb{R}$;
$P(x) \ge 0, \forall x \in \mathbb{R}$;
$P(0) = 1$;
$P(X)$ has exactly two minimum points such that $|x_1 - x_2| = 2$.

88. (*Romania*). Let $a \in \mathbb{R}$ and $f_1, f_2, \dots, f_n : \mathbb{R} \to \mathbb{R}$ be additive functions such that

$$f_1(x)f_2(x)\dots f_n(x) = ax^n, \forall x \in \mathbb{R}.$$

Prove that there exist $b \in \mathbb{R}$ and $i \in \{1, 2, \dots, n\}$ such that $f_i(x) = bx$ for all $x \in \mathbb{R}$.

89. (*Romania*). Let $n \ge 3$ and $x_1, x_2, \dots, x_{n-1}$ be positive integers such that

(a) $x_1 + x_2 + \dots + x_{n-1} = n$;
(b) $x_1 + 2x_2 + \dots + (n-1)x_{n-1} = 2n - 2$.

Find the minimum of the function

$$F(x_1, x_2, \dots, x_{n-1}) = \sum_{k=1}^{n-1} k(2n-k)x_k.$$

90. (*Romania*). Let x, y, z be real numbers. Prove that the following conditions are equivalent:

(a) $x, y, z > 0, \frac{1}{x} + \frac{1}{y} + \frac{1}{z} \le 1$.
(b) For every quadrilateral with sidelengths a, b, c, d, $a^2x + b^2y + c^2z > d^2$.

91. (*Vietnam*). Find the minimum value of the expression

$$(a+b)^4 + (b+c)^4 + (c+a)^4 - \frac{4}{7}\left(a^4 + b^4 + c^4\right).$$

2.26 1997

92. (*Austrian – Polish Competition*). Prove that there does not exist a function $f:\mathbb{Z}\to\mathbb{Z}$ such that

$$f(x+f(y))=f(x)-y$$

for all integers x and y.

93. (*Austrian – Polish Competition*).

(a) Prove that for all real numbers p and q the inequality $p^2+q^2+1\geq p(q+1)$ holds.
(b) Determine the greatest real number b such that for all real numbers p and q the inequality $p^2+q^2+1\geq bp(q+1)$ holds.
(c) Determine the greatest real number c such that for all integers p and q the inequality $p^2+q^2+1\geq cp(q+1)$ holds.

94. (*China*). Find all polynomials $f(x)$ with real coefficients which satisfy the following conditions:

(i) $f(x)=a_0x^{2n}+a_2x^{2n-2}+\ldots+a_{2n-2}x^2+a_{2n},\ a_0>0$;
(ii) $\sum\limits_{j=0}^{n}a_{2j}a_{2n-2j}\leq\binom{2n}{n}a_0a_{2n}$;
(iii) All the roots of $f(x)$ are imaginary numbers with no real part.

95. (*Croatia*). Let a, b, c, d be real numbers such that at least one is different from zero. Prove that all roots of the polynomial $P(x)=x^6+ax^3+bx^2+cx+d$ cannot be real.

96. (*Italy*). Let x, y, z, t be real numbers with x, y, z nonzero and not all equal such that

$$x+\frac{1}{y}=y+\frac{1}{z}=z+\frac{1}{x}=t.$$

Find all possible values of t such that $xyz+t=0$.

97. (*Latvia*). Does there exist a function $f(x)$ which is defined for all reals and for which the identities

$$f(f(x))=x, f(f(x)+1)=1-x$$

hold?

98. (*Latvia*). For positive a and b and natural n prove

$$\frac{1}{a+b}+\frac{1}{a+2b}+\ldots+\frac{1}{a+nb}<\frac{n}{\sqrt{a(a+nb)}}.$$

99. (*Romania*). Let p, $q \in \mathbb{Q}[X]$ be monic irreducible polynomials. Suppose that p, q have the roots α, β,respectively, and that $\alpha + \beta$ is a rational number. Prove that the polynomial $p^2 - q^2$ has a rational root.

100. (*Swizerland*). Let u, v be two randomly chosen roots of the equation $z^{1997} - 1 = 0$ (all roots are equiprobable). Find the probability that $\sqrt{2+\sqrt{3}} \le |u+v|$.

101. (*Turkey*). The sequences $(a_n)_{n \ge 1}$, $(b_n)_{n \ge 1}$ are defined by

$$a_1 = \alpha, b_1 = \beta, a_{n+1} = \alpha a_n - \beta b_n, b_{n+1} = \beta a_n + \alpha b_n, n \ge 1.$$

How many pairs (α, β) of real numbers are such that $(a_{1997}, b_{1997}) = (b_1, a_1)$?

102. (*Turkey*). Given an integer $n \ge 2$, find the minimal value of

$$\frac{x_1^5}{x_2 + x_3 + \ldots + x_n} + \frac{x_2^5}{x_1 + x_3 + \ldots + x_n} + \ldots + \frac{x_n^5}{x_1 + x_2 + \ldots + x_{n-1}}$$

subject to $x_1^2 + x_2^2 + \ldots + x_n^2 = 1$, where $x_1, x_2, \ldots, x_n$ are positive real numbers.

103. (*Vietnam*). Find all pairs (a, b) of positive real numbers such that for every $n \in \mathbb{N}^*$ and every real root x_n of the equation $4n^2x = \log_2(2n^2x + 1)$, we always have

$$a^{x_n} + b^{x_n} \ge 2 + 3x_n.$$

2.27 1998

104. (*Brazil*).

(i) Show that for each positive integer n the number of monic polynomials of degree n with integer coefficients having all roots on the unit circle is finite.
(ii) Let $P(x)$ be a monic polynomial with integer coefficients having all its roots on the unit circle. Show that there exists a positive integer m such that $y^m = 1$ for each root y of $P(x)$.

105. (*Greece*). If $x, y, z > 0$ and $k > 2$, then show that

$$\frac{x}{x + ky + kz} + \frac{y}{kx + y + kz} + \frac{z}{kx + ky + z} \ge \frac{3}{2k+1}.$$

106. (*Romania*). The volume of a parallelepiped is 216 cm^3 and its total area is 216 cm^2. Prove that the parallelepiped is a cube.

107. (*Romania*). Show that for any positive integer n the polynomial $f = (X^2 + X)^{2^n} + 1$ is irreducible in $\mathbb{Z}[X]$.

108. (*Singapore*). Let $a_1 \geq \ldots \geq a_n \geq a_{n+1} = 0$ be a sequence of real numbers. Prove that

$$\sqrt{\sum_{k=1}^{n} a_k} \leq \sum_{k=1}^{n} \sqrt{k}\left(\sqrt{a_k} - \sqrt{a_{k+1}}\right).$$

109. (*Switzerland*). Given positive numbers a, b, c, find the minimum of the function

$$f(x) = \sqrt{a^2 + x^2} + \sqrt{(b - x)^2 + c^2}.$$

2.28 1999

110. (*Armenia*). Solve the equation

$$\frac{1}{x^2} + \frac{1}{\left(4 - \sqrt{3}x\right)^2} = 1.$$

111. (*Bosnia and Herzegovina*). Let a, b and c be the sidelengths of triangle ABC. Prove that at least one of the equations

$$x^2 - 2bx + 2ac = 0, x^2 - 2cx + 2ab = 0, x^2 - 2ax + 2bc = 0$$

does not have real solutions.

112. (*Bosnia and Herzegovina*). Given the polynomial $P(x) = x^4 + 3x^3 + 3x + p$, $p \in \mathbb{R}$.

(a) Find p for which P has an imaginary root x_1 such that $|x_1| = 1$ and $Re(x_1) = \frac{\sqrt{17} - 3}{4}$.
(b) Find all the roots of the polynomial P.
(c) Prove that does not exist positive integer n such that $x_1^n = 1$.

113. (*China*). For nonnegative real numbers x_1, x_2, …, x_n which satisfy $x_1 + x_2 + \ldots + x_n = 1$ find the largest possible value of

$$\sum_{j=1}^{n} \left(x_j^4 - x_j^5\right).$$

114. (*Italy*).

(i) Determine all strictly monotone functions $f: \mathbb{R} \to \mathbb{R}$ such that

$$f(x+f(y))=f(x)+y, \forall x, y \in \mathbb{R}.$$

(ii) Prove that for every integer $n > 1$ there do not exist strictly monotone function $f: \mathbb{R} \to \mathbb{R}$ such that

$$f(x+f(y))=f(x)+y^n, \forall x, y \in \mathbb{R}.$$

115. (*Moldova*). Let a, b, c be positive real numbers. Prove that a triangle with sidelengths a, b, c exists if and only if the system of equations

$$\begin{cases} \frac{y}{z}+\frac{z}{y}=\frac{a}{x} \\ \frac{z}{x}+\frac{x}{z}=\frac{b}{y} \\ \frac{x}{y}+\frac{y}{x}=\frac{c}{z} \end{cases}$$

has a real solution.

116. (*Romania*). Show that for all positive real numbers x_1, x_2, ..., x_n which satisfy $x_1 x_2 \ldots x_n = 1$, the following inequality holds:

$$\frac{1}{x_1+n-1}+\frac{1}{x_2+n-1}+\ldots+\frac{1}{x_n+n-1} \leq 1.$$

117. (*Romania*). Let x_1, x_2, ..., x_n be distinct positive integers. Prove that

$$x_1^2+x_2^2+\ldots+x_n^2 \geq \frac{2n+1}{3}(x_1+x_2+\ldots+x_n).$$

118. (*Romania*). Prove that for any integer $n \geq 3$ there exist n positive integers a_1, a_2, ..., a_n in arithmetic progression and n positive b_1, b_2, ..., b_n in geometric progression such that

$$b_1 < a_1 < b_2 < a_2 < \ldots < b_n < a_n.$$

Give an example of such progressions having at least five terms.

119. (*Romania*). Let a be a positive real number and $(x_n)_{n \geq 1}$ a sequence of real numbers such that

$$x_1 = a, x_{n+1} \geq (n+2)x_n - \sum_{k=1}^{n} kx_k, n \geq 1.$$

Show that there exists a positive integer n such that $x_n > 1999\,!$.

120. (*Romania*). Let $n \geq 2$, a be integers and $p > |a| + 1$ a prime number. Prove that the polynomial $f = X^n + aX + p$ is irreducible in $\mathbb{Z}[X]$.

121. (*Serbia*). Let n be a positive integer and $P(x)$ a polynomial of degree $2n$ such that $P(0) = 1$ and $P(k) = 2^{k-1}$ for $1 \leq k \leq 2n$. Prove that

$$2P(2n+1) - P(2n+2) = 1.$$

122. (*Switzerland*). Find all n for which there are real numbers $0 < a_1 \leq a_2 \leq \ldots \leq a_n$ with

$$\sum_{k=1}^{n} a_k = 96, \sum_{k=1}^{n} a_k^2 = 144, \sum_{k=1}^{n} a_k^3 = 216.$$

123. (*Turkey*). Determine all functions $f: \mathbb{R} \to \mathbb{R}$ such that the set

$$\left\{ \frac{f(x)}{x} | x \in \mathbb{R}^* \right\}$$

is finite and $f(x - 1 - f(x)) = f(x) - x - 1$ for all $x \in \mathbb{R}$.

2.29 2000

124. (*Austrian – Polish Competition*). Find all polynomials $P(x)$ with real coefficients having the following property: There exists a positive integer n such that the equality

$$\sum_{k=1}^{2n+1} (-1)^k \left\lfloor \frac{k}{2} \right\rfloor P(x+k) = 0$$

holds for infinitely many real numbers x.

125. (*Austrian – Polish Competition*). For each integer $n \geq 3$, solve the system of equations:

$$x_1^3 = x_2 + x_3 + 1, \ldots, x_{n-1}^3 = x_n + x_1 + 1, x_n^3 = x_1 + x_2 + 1.$$

126. (*Belarus*). Real numbers a, b, c satisfy the equation

$$2a^3 - b^3 + 2c^3 - 6a^2b + 3ab^2 - 3ac^2 - 3bc^2 + 6abc = 0.$$

If $a < b$, find which of the numbers b, c is larger.

127. (*Belarus*).

(a) Prove that

$$\left\{n\sqrt{3}\right\} > \frac{1}{n\sqrt{3}}$$

for any positive integer n.

(b) Is there a constant $c > 1$ such that

$$\left\{n\sqrt{3}\right\} > \frac{c}{n\sqrt{3}}$$

for any positive integer n?

128. (*Bosnia and Herzegovina*). Find the real roots x_1, x_2 of the equation $x^5 - 55x + \ + 21 = 0$, if we know that $x_1x_2 = 1$.

129. (*Bosnia and Herzegovina*). Prove that for all positive real numbers a, b and c holds:

$$\frac{bc}{a^2 + 2bc} + \frac{ca}{b^2 + 2ca} + \frac{ab}{c^2 + 2ab} \leq 1 \leq \frac{a^2}{a^2 + 2bc} + \frac{b^2}{b^2 + 2ca} + \frac{c^2}{c^2 + 2ab}.$$

130. (*Brazil*). Show that if the sidelengths a, b, c of a triangle satisfy the equation

$$2\left(ab^2 + bc^2 + ca^2\right) = a^2b + b^2c + c^2a + 3abc$$

then the triangle is equilateral. Show also that the equation is satisfied by positive real numbers that are not the sidelengths of a triangle.

131. (*China*). Given positive integers k, m, n such that $1 \leq k \leq m \leq n$. Evaluate

$$\sum_{i=0}^{n} \frac{(-1)^i}{n+k+i} \cdot \frac{(m+n+i)!}{i!(n-i)!(m+i)!}.$$

132. (*France*). A function $f : N^* \to N^*$ satisfies

(a) $f(ab) = f(a)f(b)$ for any two coprime positive integers a, b.
(b) $f(p + q) = f(p) + f(q)$ for any two primes p, q.

Prove that $f(2) = 2, f(3) = 3, f(1999) = 1999$.

133. (*France*). Let a, b, c, d be positive real numbers with sum 1. Show that

$$\frac{a^2}{a+b}+\frac{b^2}{b+c}+\frac{c^2}{c+d}+\frac{d^2}{d+a}\geq\frac{1}{2},$$

with equality if and only if $a = b = c = d$.

134. (*Italy*). Given positive numbers a_1, b_1, consider the sequences defined by

$$a_{n+1}=a_n+\frac{1}{b_n}, b_{n+1}=b_n+\frac{1}{a_n}, n\geq 1.$$

Prove that $a_{25}+b_{25}\geq 10\sqrt{2}$.

135. (*Romania*). Let $n \geq 1$ be a positive integer and x_1, x_2, …, x_n be real numbers such that $|x_{k+1} - x_k| \leq 1$ for any $k = 1, 2, \ldots, n - 1$. Show that

$$\sum_{k=1}^{n}|x_k|-\left|\sum_{k=1}^{n}x_k\right|\leq\frac{n^2-1}{4}.$$

136. (*Romania*). Prove that if $P, Q \in \mathbb{C}[X]$ are monic polynomials and $P(P(X)) = Q(-Q(X))$, then $P = Q$.

137. (*Switzerland*). Real numbers a_1, a_2, …, a_{16} satisfy

$$\sum_{i=1}^{16}a_i=100, \sum_{i=1}^{16}a_i^2=1000.$$

What is the greatest possible value of a_{16}?

138. (*Turkey*). Suppose that $f : \mathbb{R} \to \mathbb{R}$ is a function such that $|f(x + y) - f(x) - f(y)| \leq 1$ for all $x, y \in \mathbb{R}$. Prove that there is a function $g : \mathbb{R} \to \mathbb{R}$ such that $|f(x) - g(x)| \leq 1$ and $g(x + y) = g(x) + g(y)$ for all $x, y \in \mathbb{R}$.

2.30 2001

139. (*Austrian – Polish Competition*). Let n be a positive integer >2. Solve in nonnegative real numbers the following system of equations:

$$x_k+x_{k+1}=x_{k+2}^2, \quad k=1,2,\ldots,n,$$

where $x_{n+1} = x_1$ and $x_{n+2} = x_2$.

140. (*China*). For a given positive integer $k > 1$, find all functions $f: \mathbb{R} \to \mathbb{R}$ such that

$$f(x^k + f(y)) = y + (f(x))^k, \forall x, y \in \mathbb{R}.$$

141. (*India*). Let $x, y, z > 0$. Prove that if $xyz \geq xy + yz + zx$, then $xyz \geq 3(x + y + z)$.

142. (*India*). Find all functions $f: \mathbb{R}_+ \to \mathbb{R}_+$ satisfying $f(f(x) - x) = 2x$, $\forall\, x > 0$.

143. (*India*). Let $P(x)$ be a polynomial of degree n with real coefficients and let $a \geq 3$. Prove that

$$\max_{0 \leq i \leq n+1} \left| a^i - P(i) \right| \geq 1.$$

144. (*Italy*). Let $0 \leq a \leq b \leq c$ be real numbers. Prove that $(a + 3b)(b + 4c)(c + 2a) \geq 60abc$.

145. (*Romania*). Show that if a, b, c are complex numbers such that

$$\begin{cases} (a+b)(a+c) = b \\ (b+c)(b+a) = c, \\ (c+a)(c+b) = a \end{cases}$$

then a, b, c are real numbers.

146. (*Romania*). Find the integer n such that the number

$$\sqrt{\frac{4n-2}{n+5}}$$

is rational.

147. (*Singapore*). Determine all integers $n > 1$ such that

$$\sum_{i=1}^{n} x_i^2 \geq x_n \sum_{i=1}^{n-1} x_i.$$

for all real numbers $x_1, x_2, \ldots, x_n$.

148. (*Vietnam*). Consider real numbers a, b, c satisfying $21ab + 2bc + 8ca \leq 12$. Find the minimal value of

$$\frac{1}{a} + \frac{1}{b} + \frac{1}{c}.$$

2.31 2002

149. (*Austrian – Polish Competition*). Determine the number of real solutions of the system

$$\cos x_1 = x_2, \ldots, \cos x_{n-1} = x_n, \cos x_n = x_1.$$

150. (*Bosnia and Herzegovina*). Let x, y, z be real numbers that satisfy

$$x + y + z = 3, xy + yz + zx = a,$$

where a is a real parameter. Find the value of a for which the difference between the maximum and the minimum possible values of x equals 8.

151. (*Bosnia and Herzegovina*). Let a, b, c be real numbers such that $a^2 + b^2 + c^2 = 1$. Prove that

$$\frac{a^2}{1+2bc} + \frac{b^2}{1+2ca} + \frac{c^2}{1+2ab} \geq \frac{3}{5}.$$

152. (*China*). Let

$$f(x_1, x_2, x_3) = -2\left(x_1^3 + x_2^3 + x_3^3\right) + 3\left(x_1^2(x_2 + x_3) + x_2^2(x_1 + x_3) + x_3^2(x_1 + x_2)\right) - 12x_1x_2x_3.$$

For any real numbers r, s, t denote

$$g(r, s, t) = \max_{t \leq x_3 \leq t+2} |f(r, r+2, x_3) + s|.$$

Find the minimum value of $g(r, s, t)$.

153. (*Estonia*). Let $0 < a < \frac{\pi}{2}$ and $x_1, x_2, \ldots, x_n$ be real numbers such that

$$\sin x_1 + \sin x_2 + \ldots + \sin x_n \geq n \sin a.$$

Prove that

$$\sin(x_1 - a) + \sin(x_2 - a) + \ldots + \sin(x_n - a) \geq 0.$$

154. (*France*). Let n be a positive integer and $(a_1, a_2, \ldots, a_{2n})$ a permutation of 1, 2, …, $2n$ such that the numbers $|a_{i+1} - a_i|$ are distinct for $i = 1, \ldots, 2n - 1$. Prove that

$$\{a_2, a_4, \ldots, a_{2n}\} = \{1, 2, \ldots, n\}$$

if and only if $a_1 - a_{2n} = n$.

155. (*Greece*). Let a, x, y be real numbers such that

$$x+y=x^3+y^3=x^5+y^5=a.$$

Determine all possible values of a.

156. (*Greece*). Prove that the following inequality holds for every triple (a,b,c) of nonnegative real numbers with $a^2+b^2+c^2=1$:

$$\frac{a}{b^2+1}+\frac{b}{c^2+1}+\frac{c}{a^2+1}\geq\frac{3}{4}\left(a\sqrt{a}+b\sqrt{b}+c\sqrt{c}\right)^2.$$

157. (*India*). Let a, b, c be positive real numbers such that $a^2+b^2+c^2=3abc$. Prove that

$$\frac{a}{b^2c^2}+\frac{b}{c^2a^2}+\frac{c}{a^2b^2}\geq\frac{9}{a+b+c}.$$

158. (*India*). Let x_1, x_2, …, x_n be arbitrary real numbers. Prove the inequality

$$\frac{x_1}{1+x_1^2}+\frac{x_2}{1+x_1^2+x_2^2}+\ldots+\frac{x_n}{1+x_1^2+\ldots+x_n^2}<\sqrt{n}.$$

159. (*India*). Let a, b, c be positive real numbers. Prove that

$$\frac{a}{b}+\frac{b}{c}+\frac{c}{a}\geq\frac{a+b}{a+c}+\frac{b+c}{b+a}+\frac{c+a}{c+b}.$$

160. (*Moldova*). Positive numbers $a, b, x_1, \ldots, x_n$ satisfy $x_1+x_2+\ldots+x_n=1$. Prove that

$$\sum_{i=1}^{n}\frac{x_i}{ax_i+bx_{i+1}}\geq\frac{1}{n(a+b)},$$

where $x_{n+1}=x_1$.

161. (*Romania*). Find all pairs (A,B) of sets which satisfy the conditions:

(a) $A\cup B=\mathbb{Z}$;
(b) if $x\in A$, then $x-1\in B$;
(c) if $x, y\in B$, then $x+y\in A$.

162. (*Romania*). Let $n\geq 4$ be an integer and $a_1, a_2, \ldots, a_n$ be positive real numbers such that $a_1^2+a_2^2+\ldots+a_n^2=1$. Show that the following inequality holds:

$$\frac{a_1}{a_2^2+1}+\frac{a_2}{a_3^2+1}+\ldots+\frac{a_n}{a_1^2+1}\geq\frac{4}{5}\left(a_1\sqrt{a_1}+a_2\sqrt{a_2}+\ldots+a_n\sqrt{a_n}\right)^2.$$

163. (*Serbia*). What is the maximal value of the expression $a + b + c + abc$, if a, b, c are nonnegative real numbers such that $a^2 + b^2 + c^2 + abc \leq 4$?

164. (*Turkey*). A positive integer n and real numbers $a_1, \ldots, a_n$ are given. Show that there exist integers $k, m \in \{1, \ldots, n\}$ such that

$$\left|\sum_{i=1}^{m} a_i - \sum_{i=m+1}^{n} a_i\right| \leq |a_k|.$$

2.32 2003

165. (*Austrian – Polish Competition*). Find all real polynomials $p(x)$ such that

$$p(x-1)p(x+1) = p\left(x^2 - 1\right).$$

166. (*Bosnia and Herzegovina*). Let a, b and c be real numbers such that $|a| > 2$ and $a^2 + b^2 + c^2 = abc + 4$. Prove that numbers x and y exist such that $a = x + \frac{1}{x}, b = y + \frac{1}{y}$ and $c = xy + \frac{1}{xy}$.

167. (*Bulgaria*). For any positive integer n define

$$A_n = \{j | 1 \leq j \leq n, \gcd(j, n) = 1\}.$$

Find all n such that the polynomial

$$P_n(x) = \sum_{j \in A_n} x^{j-1}$$

is irreducible over $\mathbb{Z}[x]$.

168. (*Bulgaria*). Find all $f : \mathbb{R} \to \mathbb{R}$ such that $f(x^2 + y + f(y)) = 2y + (f(x))^2$.

169. (*Estonia*). Let a, b, c be positive real numbers satisfying the condition $a + b + c = abc$. Prove that

$$\frac{a}{\sqrt{1+a^2}} + \frac{b}{\sqrt{1+b^2}} + \frac{c}{\sqrt{1+c^2}} \leq \frac{3\sqrt{3}}{2}.$$

When does the equality hold?

170. (*Germany*). Find all functions $f: \mathbb{R} \to \mathbb{R}$ such that

$$f(f(x)+y) = 2x + f(f(y) - x), \forall x, y \in \mathbb{R}.$$

171. (*India*). Let p be a polynomial with integer coefficients such that $p(n) > n$ for every positive integer n. Let (x_k) be the sequence defined by $x_1 = 1$, $x_{i+1} = p(x_i)$, $i \geq 1$. Prove that if for every positive integer N one of x_i is divisible by N, then

$$p(x) = x + 1.$$

172. (*Moldova*). Find all polynomials with real coefficients

$$P(X) = X^{2n} + (2n-10)X^{2n-1} + a_2X^{2n-2} + \ldots + a_{2n-2}X^2 + (2n-10)X + 1,$$

which have only positive real roots.

173. (*Romania*). Let $(a_n)_{n \geq 1}$ be the sequence defined by $a_1 = \frac{1}{2}$ and $a_{n+1} = \frac{a_n^2}{a_n^2 - a_n + 1}$ for $n \geq 1$ Prove that for any integer $n \geq 1$ we have

$$\sum_{k=1}^{n} a_k < 1.$$

174. (*Romania*). Let $f \in \mathbb{Z}[X]$ be a monic irreducible polynomial such that $|f(0)|$ is not the square of an integer. Prove that the polynomial $g(X) = f(X^2)$ is also irreducible.

175. (*Romania*). Find all integers a, b, m, n with $m > n > 1$, for which the polynomial $f(X) = X^n + aX + b$ divides the polynomial $g(X) = X^m + aX + b$.

176. (*Serbia*). If $p(x)$ is a polynomial, denote by $p^n(x)$ the polynomial $p(p(\ldots p(x)\ldots))$, where p is iterated n times. Prove that the polynomial $p^{2003}(x) - 2p^{2002}(x) + p^{2001}(x)$ is divisible by $p(x) - x$.

177. (*Switzerland*). Find the greatest real number C_1 and the smallest real number C_2 such that

$$C_1 < \frac{a}{a+b} + \frac{b}{b+c} + \frac{c}{c+d} + \frac{d}{d+e} + \frac{e}{e+a} < C_2$$

for all positive real numbers a, b, c, d, e.

2.33 2004

178. (*Bosnia and Herzegovina*). For $0 \leq x < \frac{\pi}{2}$ prove the inequality: $a^2 \tan x \sqrt[3]{\cos x} + b^2 \sin x \geq 2abx$, where a and b are real numbers.

179. (*Bulgaria*). Find all $k > 0$ such that there exists a function $f\colon [0, 1] \times [0, 1] \to [0, 1]$ satisfying the following conditions:

$$f(f(x, y), z) = f(x, f(y, z)), f(x, y) = f(y, x), f(x, 1) = x, f(zx, zy) = z^k f(x, y),$$

for any x, y, $z \in [0, 1]$.

180. (*Bulgaria*). Prove that if a, b, $c \geq 1$ and $a + b + c = 9$, then

$$\sqrt{ab + bc + ca} \leq \sqrt{a} + \sqrt{b} + \sqrt{c}.$$

181. (*China*). Given integer n larger than 5, solve the system of equations (assuming $x_i \geq 0$ for all i):

$$\begin{cases} x_1 + x_2 + x_3 + \ldots + x_n = n + 2 \\ x_1 + 2x_2 + 3x_3 + \ldots + nx_n = 2n + 2 \\ x_1 + 2^2x_2 + 3^2x_3 + \ldots + n^2x_n = n^2 + n + 4 \\ x_1 + 2^3x_2 + 3^3x_3 + \ldots + n^3x_n = n^3 + n = 8 \end{cases}.$$

182. (*Croatia*). Prove that if a, b, c are positive real numbers with $abc = 1$, then

$$\frac{a}{b} + \frac{b}{c} + \frac{c}{a} \geq a + b + c.$$

183. (*Estonia*). Denote

$$f(m) = \sum_{k=1}^{m} (-1)^k \cos \frac{k\pi}{2m + 1}.$$

For which positive integers m is $f(m)$ rational?

184. (*France*). Let n be a positive integer and $a_1, \ldots, a_n, b_1, \ldots, b_n$ be $2n$ positive real numbers such that $a_1 + \ldots + a_n = b_1 + \ldots + b_n = 1$. Find the minimal value of

$$\frac{a_1^2}{a_1 + b_1} + \frac{a_2^2}{a_2 + b_2} + \ldots + \frac{a_n^2}{a_n + b_n}.$$

185. (*India*). For a, b, c positive real numbers, find the minimum value of

$$\frac{a^2+b^2}{c^2+ab}+\frac{b^2+c^2}{a^2+bc}+\frac{c^2+a^2}{b^2+ca}.$$

186. (*India*). Determine all functions $f\colon \mathbb{R} \to \mathbb{R}$ such that

$$f(x+y) = f(x)f(y) - c\sin x \sin y, \forall x, y \in \mathbb{R},$$

where $c > 1$ is a given constant.

187. (*India*). Let $x_1, x_2, \ldots, x_n$ be n real numbers such that $0 < x_i < \frac{1}{2}, i = 1, 2, \ldots, n$. Prove that

$$\frac{\prod_{i=1}^{n} x_i}{\left(\sum_{i=1}^{n} x_i\right)^n} \leq \frac{\prod_{i=1}^{n} (1-x_i)}{\left(\sum_{i=1}^{n} (1-x_i)\right)^n}.$$

188. (*Moldova*). Let a, b, c be positive real numbers. Prove that

$$4\left|\frac{a^3-b^3}{a+b}+\frac{b^3-c^3}{b+c}+\frac{c^3-a^3}{c+a}\right| \leq (a-b)^2+(b-c)^2+(c-a)^2.$$

189. (*New Zealand*). A function f is defined on the interval $[0, 1]$ so that

$$f(0) = f(1) = 0, f\left(\frac{a+b}{2}\right) \leq f(a)+f(b), a, b \in [0, 1].$$

(a) Show that the equation $f(x) = 0$ has infinitely many solutions in $[0, 1]$.
(b) Are there functions f on $[0, 1]$ which satisfy the above conditions and are not identically zero?

190. (*Romania*). Consider an integer $n \geq 2$ and a disc $\mathcal{D}$ in the complex plane. Prove that for any $z_1, z_2, \ldots, z_n \in \mathcal{D}$, there exists $z \in \mathcal{D}$ such that $z^n = z_1 z_2 \ldots z_n$.

191. (*Romania*). Let $n \geq 2$ be an integer and a_1, a_2, …, a_n real numbers. Prove that for any nonempty subset $\subset \{1, 2, \ldots, n\}$, the following inequality holds:

$$\left(\sum_{k\in S} a_k\right)^2 \leq \sum_{1\leq i\leq j\leq n} (a_i + \ldots + a_j)^2.$$

192. (*Serbia*). Let $P(x)$ be a polynomial of degree n whose roots are $i-1$, $i-2$, …, $i - - n$ (where $i^2 = - 1$), and let $R(x)$ and $S(x)$ be the polynomials with real coefficients such that $P(x) = R(x) + iS(x)$. Show that the polynomial R has n real roots.

193. (*Singapore*). Let $0 < a, b, c < 1$ with $ab + bc + ca = 1$. Prove that

$$\frac{a}{1-a^2}+\frac{b}{1-b^2}+\frac{c}{1-c^2}\geq\frac{3\sqrt{3}}{2}.$$

Determine when equality holds.

194. (*Switzerland*). Let a, b, c be positive real numbers such that $abc = 1$. Prove that

$$\frac{ab}{a^5+ab+b^5}+\frac{bc}{b^5+bc+c^5}+\frac{ca}{c^5+ca+a^5}\leq 1.$$

195. (*Switzerland*). Real numbers a, b, c, d satisfy

$$a=\sqrt{45-\sqrt{21-a}}, b=\sqrt{45+\sqrt{21-b}}, c=\sqrt{45-\sqrt{21+c}}, d$$
$$=\sqrt{45+\sqrt{21+d}}.$$

Show that $abcd = 2004$.

2.34 2005

196. (*Argentina*). Find all functions $f\colon \mathbb{R}\to\mathbb{R}$ such that

$$f(xf(x)+f(y))=(f(x))^2+y, \forall x, y\in\mathbb{R}.$$

197. (*Austrian – Polish Competition*). Determine all polynomials P with integer coefficients satisfying

$$P(P(P(P(P(x)))))=x^{28}\cdot P(P(x)), \forall x\in\mathbb{R}.$$

198. (*Austrian – Polish Competition*). Determine all monotone functions $f\colon\mathbb{Z}\to\mathbb{Z}$ so that for all x, $y \in \mathbb{Z}$,

$$f\left(x^{2005}+y^{2005}\right)=(f(x))^{2005}+(f(y))^{2005}.$$

199. (*Austrian – Polish Competition*). Determine all pairs (k, n) of nonnegative integers such that the following equality holds $\forall x$, $y > 0$:

$$1+\frac{y^n}{x^k} \geq \frac{(1+y)^n}{(1+x)^k}.$$

200. (*Bosnia and Herzegovina*). If a, b and c are nonnegative real numbers for which $a + b + c = 1$, then prove the inequality

$$a\sqrt{b}+b\sqrt{c}+c\sqrt{a} \leq \frac{1}{\sqrt{3}}.$$

201. (*Bulgaria*). Let a_1, a_2, ..., a_m be arbitrary positive integers. Prove that there exist distinct positive integers b_1, b_2, ..., b_n, $n \leq m$, such that the following conditions are satisfied:

1. All subsets of $\{b_1, b_2, \ldots, b_n\}$ have distinct sums of elements;
2. Every number $a_1, a_2, \ldots, a_m$ is the sum of the elements of some subset of $\{b_1, b_2, \ldots, b_n\}$.

202. (*Bulgaria*). Find all functions $f\colon \mathbb{R}^* \to \mathbb{R}^*$ such that

$$f(x^2+y) = (f(x))^2 + \frac{f(xy)}{f(x)}, x, y \in \mathbb{R}^*, x^2+y \neq 0.$$

203. (*Colombia*). Prove the following inequality for positive real numbers a_1, a_2, ..., a_n and b_1, b_2, ..., b_n:

$$\left(\sum_{i=1}^{n} a_i\right)\left(\sum_{i=1}^{n} b_i\right) \geq \left(\sum_{i=1}^{n} (a_i+b_i)\right)\left(\sum_{i=1}^{n} \frac{a_i b_i}{a_i+b_i}\right).$$

204. (*Estonia*). Find all pairs (a, b) of real numbers such that all roots of the polynomials

$$6x^2 - 24x - 4a, \quad x^3 + ax^2 + bx - 8$$

are nonnegative real numbers.

205. (*France*). Let P be a polynomial of degree $n \geq 5$ with integer coefficients,

$$P(x) = a_n x^n + a_{n-1}x^{n-1} + \ldots + a_0, a_n \neq 0.$$

Suppose that P has n integer roots 0, α_2, ..., α_n. Find all integers k such that $P(P(k)) = 0$.

206. (*Georgia*). Let x, y, z be positive real numbers satisfying $x^2 + y^2 + z^2 = 25$. Find the minimal value of the expression

$$\frac{xy}{z}+\frac{yz}{x}+\frac{zx}{y}.$$

207. (*Germany*). Given the positive real numbers a, b and the positive integer n, find the greatest among $n + 1$ monomials in the binomial expansion of $(a + b)^n$.

208. (*Germany*). Let $a_0, a_1, a_2, \ldots$ be an infinite sequence of real numbers satisfying $a_n = |a_{n+1} - a_{n+2}|$ for all $n \geq 0$, where $a_0 \neq a_1$. Can the sequence be bounded?

209. (*Germany*). If a, b, c are positive real numbers such that $a + b + c = 1$, then prove that

$$\frac{1+a}{1-a}+\frac{1+b}{1-b}+\frac{1+c}{1-c}\leq 2\left(\frac{a}{c}+\frac{b}{a}+\frac{c}{b}\right).$$

210. (*Germany*). If a, b, c are positive real numbers such that $ab + bc + ca = 1$, then prove that

$$\sqrt[3]{6a+\frac{1}{c}}+\sqrt[3]{6b+\frac{1}{a}}+\sqrt[3]{6c+\frac{1}{b}}\leq\frac{1}{abc}.$$

211. (*Germany*). Find all monotonically increasing or monotonically decreasing functions $f: \mathbb{R}_+ \to \mathbb{R}_+$ which satisfy the equation

$$f(xy)f\left(\frac{f(y)}{x}\right)=1, \forall x, y\in\mathbb{R}_+.$$

212. (*Germany*). For any positive integer n, prove that there exists a polynomial P of degree n such that all coefficients of P are integers and the numbers $P(0)$, $P(1)$, $P(2)$, …, $P(n)$ are distinct powers of 2.

213. (*Germany*). Let $n \geq 3$ be a positive integer and $a_1, a_2, \ldots, a_n, b_1, b_2, \ldots, b_n$ be positive real numbers satisfying $a_1 + a_2 + \ldots + a_n = 1, b_1^2 + b_2^2 + \ldots + b_n^2 = 1$. Prove that $a_1(a_2 + b_1) + a_2(a_3 + b_2) + \ldots + a_{n-1}(a_n + b_{n-1}) + a_n(a_1 + b_n) < 1$.

214. (*Greece*). The side lengths of a triangle are the roots of a cubic polynomial with rational coefficients. Prove that the altitudes of this triangle are roots of a polynomial of sixth degree with rational coefficients.

215. (*India*). For real numbers a, b, c, d not all equal to zero, define the function

$$f(x)=a+b\cos 2x+c\sin 5x+d\cos 8x.$$

Suppose that $f(t) = 4a$ for some real number t. Prove that there exists a real number s such that $f(s) < 0$.

216. (*Iran*). Let $a_1, a_2, \ldots, a_n$ be positive real numbers such that $a_1 \leq a_2 \leq \ldots \leq a_n$. Let

$$\frac{a_1 + a_2 + \ldots + a_n}{n} = m, \quad \frac{a_1^2 + a_2^2 + \ldots + a_n^2}{n} = 1.$$

Suppose that for some i we have $a_i \leq m$. Prove that $n - i \geq n(m - a_i)^2$.

217. (*Lithuania*). If a, b, c are nonzero real numbers, what values can be taken by the expression

$$\frac{a^2 - b^2}{a^2 + b^2} + \frac{b^2 - c^2}{b^2 + c^2} + \frac{c^2 - a^2}{c^2 + a^2}?$$

218. (*Switzerland*). Find the maximum value of the expression

$$\frac{xyz}{(1 + x)(x + y)(y + z)(z + 16)},$$

where x, y, z are positive real numbers.

219. (*Switzerland*). Let $n \geq 2$ be a positive integer. Prove that the following polynomial

$$\left(x^2 - 1^2\right)\left(x^2 - 2^2\right)\left(x^2 - 3^2\right)\ldots\left(x^2 - n^2\right) + 1$$

is irreducible in $\mathbb{Z}[x]$.

2.35 2006

220. (*Austrian – Polish Competition*). Find all polynomials $P(x)$ with real coefficients satisfying the equation

$$(x + 1)^3 P(x - 1) - (x - 1)^3 P(x + 1) = 4\left(x^2 - 1\right)P(x)$$

for all real numbers x.

221. (*Bosnia nd Herzegovina*). Prove that for every positive integer n holds the inequality $\left\{n\sqrt{7}\right\} > \frac{3\sqrt{7}}{14n}$, where $\{x\}$ is the fractional part of x.

222. (*Bosnia and Herzegovina*). Let $a_1, a_2, \ldots, a_n$ be constant real numbers and x be variable real number. Let

$$f(x) = \cos(x + a_1) + \frac{\cos(x + a_2)}{2} + \frac{\cos(x + a_3)}{2^2} + \ldots + \frac{\cos(x + a_n)}{2^{n-1}}.$$

If $f(x_1) = f(x_2) = 0$, prove that $x_1 - x_2 = m\pi$, where m is an integer.

223. (*Bulgaria*). Prove that if x, y and a are real numbers from the interval $(0, 1)$, then

$$\frac{|x - y|}{1 - xy} \leq \frac{|x^a - y^a|}{1 - x^a y^a}.$$

224. (*Bulgaria*). Find all pairs (P, Q) of polynomials with real coefficients such that

$$\frac{P(x)}{Q(x)} - \frac{P(x+1)}{Q(x+1)} = \frac{1}{x(x+2)}$$

for infinitely many $x \in \mathbb{R}$.

225. (*China*). Let $x_1, x_2, \ldots, x_n$ be positive numbers such that $x_1 + x_2 + \ldots + x_n = 1$. Prove that

$$\left(\sum_{i=1}^{n} \sqrt{x_i}\right)\left(\sum_{i=1}^{n} \frac{1}{\sqrt{1 + x_i}}\right) \leq \frac{n^2}{\sqrt{n+1}}.$$

226. (*Croatia*). Assume that a, b and c are positive real numbers for which

$$(a + b)(b + c)(c + a) = 1.$$

Prove that

$$ab + bc + ca \leq \frac{3}{4}.$$

227. (*France*). Let a, b, c be positive real numbers such that $abc = 1$. Show that

$$\frac{a}{(a+1)(b+1)} + \frac{b}{(b+1)(c+1)} + \frac{c}{(c+1)(a+1)} \geq \frac{3}{4}.$$

When there is equality?

228. (*Germany*). Find all functions $f : \mathbb{R}^+ \to \mathbb{R}^+$ which have the property

$$f(x)f(y) = 2f(x + yf(x)), \forall x, y \in \mathbb{R}^+.$$

229. (*Germany*). Four real numbers p, q, r, s satisfy

$$p+q+r+s=9, p^2+q^2+r^2+s^2=21.$$

Prove that there exists a permutation (a,b,c,d) of (p,q,r,s) such that $ab-cd\geq 2$.

230. (*Germany*). Find all functions $f:\mathbb{R}\to\mathbb{R}$ such that

$$f(x+y)+f(x)f(y)=f(xy)+2xy+1, \forall x,y\in\mathbb{R}.$$

231. (*Germany*). Let n be a positive integer and let b_1, b_2, ..., b_n be positive real numbers. Set

$$a_1=\frac{b_1}{b_1+b_2+\ldots+b_n}, \quad a_k=\frac{b_1+b_2+\ldots+b_k}{b_1+b_2+..+b_{k-1}}, k>1.$$

Prove the inequality:

$$a_1+a_2+\ldots+a_n\leq\frac{1}{a_1}+\frac{1}{a_2}+\ldots+\frac{1}{a_n}.$$

232. (*Hong Kong*). Let x, y, z be positive real numbers such that $x+y+z=1$. For a positive integer n, let $S_n=x^n+y^n+z^n$. Let $P=S_2S_{2005}$, $Q=S_3S_{2004}$.

(a) Find the smallest possible value of Q.
(b) If x, y, z are distinct, determine which of P or Q is the larger.

233. (*Italy*). Let $P(x)$ be a polynomial with complex coefficients such that $P(0)\neq 0$. Prove that there exists a multiple of $P(x)$ with real positive coefficients if and only if $P(x)$ has no real positive root.

234. (*Lithuania*). Let a_1, a_2, ..., a_n be positive real numbers whose sum is 1. Prove that

$$\frac{a_1^2}{a_1+a_2}+\frac{a_2^2}{a_2+a_3}+\ldots+\frac{a_{n-1}^2}{a_{n-1}+a_n}+\frac{a_n^2}{a_n+a_1}\geq\frac{1}{2}.$$

235. (*Romania*). Let $p\geq 5$ be a prime number. Find the number of irreducible polynomials in $\mathbb{Z}[X]$ of the form $x^p+px^k+px^l+1$, $k>l$, $k,l\in\{1,2,\ldots,p-1\}$.

236. (*Romania*). Let a_1, a_2, ..., a_n be real numbers such that $|a_i|\leq 1$ for all $i\in\{1,2,\ldots,n\}$, and $a_1+a_2+\ldots+a_n=0$.

(a) Prove that there exists $k\in\{1,2,\ldots,n\}$ such that

$$|a_1+2a_2+\ldots+ka_k|\leq\frac{2k+1}{4}.$$

(b) Prove that for $n>2$ the bound above is the best possible.

237. (*Romania*). Let $x_1, x_2, \ldots, x_n$ be real numbers. Prove that

$$\sum_{1 \le i < j \le n} |x_i + x_j| \ge \frac{n-2}{2} \sum_{i=1}^{n} |x_i|.$$

238. (*Romania*). Let $q \ge p \ge 0$ be integers. Let $n \ge 2$ be an integer and $a_0 = 0$, $a_1 \ge 0$, $a_2, \ldots, a_{n-1}$, $a_n = 1$ be real numbers such that

$$a_k \le \frac{a_{k-1} + a_{k+1}}{2}, k = 1, 2, \ldots, n-1.$$

Prove that

$$(p+1) \sum_{k=1}^{n-1} a_k^p \ge (q+1) \sum_{k=1}^{n-1} a_k^q.$$

239. (*Romania*). Let $(x_n)_{n \ge 1}$ be the sequence of real numbers defined by

$$x_1 = 1, x_{n+1} = x_n + \frac{1}{2x_n}, n \ge 1.$$

Prove that $\lfloor 25x_{625} \rfloor = 625$.

240. (*Singapore*). Let $n > 1$ be an integer and let $x_1, x_2, \ldots, x_n$ be real numbers such that

$$|x_1| + |x_2| + \ldots + |x_n| = 1, x_1 + x_2 + \ldots + x_n = 0.$$

Prove that

$$\left| \frac{x_1}{1} + \frac{x_2}{2} + \ldots + \frac{x_n}{n} \right| \le \frac{1}{2}\left(1 - \frac{1}{n}\right).$$

241. (*Switzerland*). The roots of the polynomial $P(x) = x^3 - 2x^2 - x + 1$ are $a > b > c$. Find the value of $a^2b + b^2c + c^2a$.

242. (*Vietnam*). Prove that for all real numbers x, y, $z \in [1, 2]$ the following inequality holds:

$$(x + y + z)\left(\frac{1}{x} + \frac{1}{y} + \frac{1}{z}\right) \ge 6\left(\frac{x}{y+z} + \frac{y}{z+x} + \frac{z}{x+y}\right).$$

2.36 2007

243. (*Argentina*). Find all real values $x > 1$ which satisfy

$$\frac{x^2}{x-1} + \sqrt{x-1} + \frac{\sqrt{x-1}}{x^2} = \frac{x-1}{x^2} + \frac{1}{\sqrt{x-1}} + \frac{x^2}{\sqrt{x-1}}.$$

244. (*Bosnia and Herzegovina*). Find all $x \in \mathbb{Z}$ and $a \in \mathbb{R}$ satisfying

$$\sqrt{x^2-4} + \sqrt{x+2} = \sqrt{x-a} + a.$$

245. (*Bosnia and Herzegovina*). Let $P(x)$ be a polynomial such that $P(x) = x^3 - 2x^2 + ax + b$. Roots of $P(x)$ belong to the interval $(0, 1)$. Prove that $8a + 9b \leq 8$. When does the equality hold?

246. (*Bulgaria*). Find all $a \in \mathbb{R}$ for which there exists a nonconstant function $f: (0, 1] \to \mathbb{R}$ such that

$$a + f(x+y-xy) + f(x)f(y) \leq f(x) + f(y), \forall x, y \in (0, 1].$$

247. (*Bulgaria*). Let $n \geq 2$ be an integer. Find the best constant $C(n)$ such that

$$\sum_{i=1}^{n} x_i \geq C(n) \sum_{1 \leq j < i \leq n} \left(2x_ix_j + \sqrt{x_ix_j}\right)$$

is true for all real numbers $x_i \in (0, 1)$, $i = 1, 2, \ldots, n$ for which $4(1 - x_i)(1 - x_j) \geq 1$, $1 \leq j < i \leq n$.

248. (*China*). Does there exist any function $f: \mathbb{R} \to \mathbb{R}$ satisfying $f(0) > 0$ and

$$f(x+y) \geq f(x) + yf(f(x)), \forall x, y \in \mathbb{R}?$$

249. (*China*). Find all functions $f: \mathbb{Q}^+ \to \mathbb{Q}^+$ such that

$$f(x) + f(y) + 2xyf(xy) = \frac{f(xy)}{f(x+y)}.$$

250. (*Croatia*). Let $a, b, c > 0$ such that $a + b + c = 1$. Prove that

$$\frac{a^2}{b} + \frac{b^2}{c} + \frac{c^2}{a} \geq 3\left(a^2 + b^2 + c^2\right).$$

251. (*France*). Let a, b, c, d be positive reals such that $a + b + c + d = 1$. Prove that

$$6(a^3 + b^3 + c^3 + d^3) \geq a^2 + b^2 + c^2 + d^2 + \frac{1}{8}.$$

252. (*India*). Let a, b, c be nonnegative real numbers such that

$$a + b \leq c + 1, b + c \leq a + 1, c + a \leq b + 1.$$

Show that

$$a^2 + b^2 + c^2 \leq 2abc + 1.$$

253. (*Iran*). Find all monic polynomials $f \in \mathbb{Z}[x]$ such that $f(\mathbb{Z})$ is closed under multiplication.

254. (*Moldova*). Let a_1, a_2, ..., $a_n \in [0, 1]$. If $S = a_1^3 + a_2^3 + \ldots + a_n^3$, then prove that

$$\frac{a_1}{2n + 1 + S - a_1^3} + \frac{a_2}{2n + 1 + S - a_2^3} + \ldots + \frac{a_n}{2n + 1 + S - a_n^3} \leq \frac{1}{3}.$$

255. (*Peru*). Let a, b, c be positive real numbers such that $abc(a + b + c) \geq ab + bc + ca$. Prove that

$$a + b + c \geq \frac{3}{a + b + c} + \frac{2}{abc}.$$

256. (*Romania*). Show that any function $f : \mathbb{Q} \rightarrow \mathbb{R}$ with the property $|f(x) - f(y)| \leq (x - y)^2$ for all $x, y \in \mathbb{Q}$, is constant.

257. (*Romania*). Show that for n, $p \geq 4$ positive integers, the proposition $\mathcal{P}(n, p)$ below is false:

$$\sum_{i=1}^{n} \frac{1}{x_i^p} \geq \sum_{i=1}^{n} x_i^p, x_i > 0, 1 \leq i \leq n, \sum_{i=1}^{n} x_i = n.$$

258. (*Romania*). For $n \in \mathbb{N}$, $n \geq 2$, a_i, $b_i \in \mathbb{R}$, $1 \leq i \leq n$, such that

$$\sum_{i=1}^{n} a_i^2 = \sum_{i=1}^{n} b_i^2 = 1, \sum_{i=1}^{n} a_i b_i = 0,$$

prove that

$$\left(\sum_{i=1}^{n} a_i\right)^2 + \left(\sum_{i=1}^{n} b_i\right)^2 \leq n.$$

259. (*Singapore*). Prove the inequality

$$\sum_{i<j} \frac{a_i a_j}{a_i + a_j} \leq \frac{n}{2(a_1 + a_2 + \ldots + a_n)} \sum_{i<j} a_i a_j$$

for all positive real numbers $a_1, a_2, \ldots, a_n$.

260. (*Switzerland*). Call a pair (r, s) of positive integers *good* if there is a polynomial P with integer coefficients and there are distinct integers $a_1, \ldots, a_r, b_1, \ldots, b_s$ such that

$$P(a_1) = \ldots = P(a_r) = 2, P(b_1) = \ldots = P(b_s) = 5.$$

Find all good pairs.

2.37 2008

261. (*Argentina*). Find all functions $f\colon \mathbb{R}^+ \longrightarrow \mathbb{R}^+$ that satisfy

$$x^2(f(x) + f(y)) = (x + y)f(yf(x)).$$

262. (*Bosnia and Herzegovina*). Find all functions $f\colon \mathbb{R} \to \mathbb{R}$ satisfying

$$f(f(x) + y) = f\left(x^2 - y\right) + 4f(x)y, \forall x, y \in \mathbb{R}.$$

263. (*Bulgaria*). Find all real numbers a for which there exists a function $f\colon \mathbb{R}^+ \to \mathbb{R}^+$ such that

$$3(f(x))^2 = 2f(f(x)) + ax^4, \forall x \in \mathbb{R}^+.$$

264. (*China*). Let z_1, z_2, z_3 be complex numbers of moduli ≤ 1 and w_1, w_2 the roots of the equation $(z - z_1)(z - z_2) + (z - z_2)(z - z_3) + (z - z_3)(z - z_1) = = 0$. Prove that for $j = 1, 2, 3$, $\min\{|z_j - w_1|, |z_j - w_2|\} \leq 1$ holds.

265. (*Croatia*). Let x, y, z be positive numbers. Find the minimum value of

(i) $\frac{x^2+y^2+z^2}{xy+yz}$;

(ii) $\frac{x^2+y^2+2z^2}{xy+yz}$.

266. (*Germany*). Show that there is a digit unequal to 2 in the decimal representation of $\sqrt[3]{3}$ between the 1000000th and 3141592th position after decimal point.

267. (*Germany*). Consider those functions $f : \mathbb{N} \to \mathbb{N}$ which satisfy the condition

$$f(m+n) \geq f(m) + f(f(n)) - 1$$

for all $m, n \in \mathbb{N}$. Find all possible values of $f(2007)$.

268. (*Germany*). Find all polynomials f with real coefficients such that

$$2yf(x+y) + (x-y)(f(x) + f(y)) \geq 0, \forall x, y \in \mathbb{R}.$$

269. (*Germany*). Let $a_1, a_2, \ldots, a_{100}$ be nonnegative real numbers such that $a_1^2 + a_2^2 + \ldots + a_{100}^2 = 1$. Prove that

$$a_1^2 a_2 + a_2^2 a_3 + \ldots + a_{100}^2 a_1 < \frac{12}{25}.$$

270. (*Greece*). Find all possible values of $a \in \mathbb{R}$ and $n \in \mathbb{N}^*$ such that

$$f(x) = (x-1)^n + (x-2)^{2n+1} + \left(1-x^2\right)^{2n+1} + a$$

is divisible by $\phi(x) = x^2 - x + 1$.

271. (*Hong Kong*). Let $\alpha_1, \alpha_2, \ldots, \alpha_{2008}$ be real numbers. Find the maximum value of

$$\sin \alpha_1 \cos \alpha_2 + \sin \alpha_2 \cos \alpha_3 + \ldots + \sin \alpha_{2007} \cos \alpha_{2008} + \sin \alpha_{2008} \cos \alpha_1.$$

272. (*Hong Kong*). Let $f : \mathbb{Z} \to \mathbb{Z}$ be a function such that

$$f(1) = 1, f(2) = 20, f(-4) = -4,$$
$$f(x+y) = f(x) + f(y) + axy(x+y) + bxy + c(x+y) + 4, \forall x, y \in \mathbb{Z}, \forall x, y \in \mathbb{Z},$$

where a, b, c are constants.

(i) Find a formula for f.
(ii) If $f(x) \geq mx^2 + (5m+1)x + 4m$ for all nonnegative integers x, then find the greatest value of m.

273. (*Moldova*). Find all solutions $(x, y) \in \mathbb{R} \times \mathbb{R}$ of the system of equations

$$\begin{cases} x^3 + 3xy^2 = 49 \\ x^2 + 8xy + y^2 = 17x + 8y \end{cases}.$$

274. (*Romania, IMAR*). Show that for any function $f: (0, \infty) \to (0, \infty)$ there exist positive real numbers x, y such that $f(x + y) < yf(f(x))$.

275. (*Singapore*). Find all functions $f: \mathbb{R} \to \mathbb{R}$ such that

$$(x+y)(f(x)-f(y)) = (x-y)f(x+y)$$

for all $x, y \in \mathbb{R}$.

276. (*Switzerland*). Let $P(x) = x^4 - 2x^3 + px + q$ a polynomial with real coefficients, whose roots are all real. Prove that the greatest of these roots belongs to the interval $[1, 2]$

277. (*Turkey*). The equation $x^3 - ax^2 + bx - c = 0$ has three positive real roots. Find the minimal possible value of

$$\frac{1+a+b+c}{3+2a+b} - \frac{c}{b}.$$

278. (*Ukraine*). For positive a, b, c, d,prove that

$$(a+b)(b+c)(c+d)(d+a)\left(1+\sqrt[4]{abcd}\right)^4 \geq 16abcd(1+a)(1+b)(1+c) \times (1+d).$$

2.38 2009

279. (*Albania*). Find all functions $f: \mathbb{R} \to \mathbb{R}$ with the following property:

$$f(x) = f\left(\frac{x}{2}\right) + \frac{x}{2}f'(x), \forall x \in \mathbb{R}.$$

280. (*Argentina*). Let $a_1, a_2, \ldots, a_{300}$ be nonnegative real numbers such that

$$\sum_{i=1}^{300} a_i = 1.$$

Find the maximum possible value of

$$\sum_{i \neq j,\, i|j} a_i a_j.$$

281. (*Bosnia and Herzegovina*). Let a_1, a_2, …, a_{100} be real numbers such that

$$a_1 \geq a_2 \geq \ldots \geq a_{100} \geq 0, a_1^2 + a_2^2 \geq 100, a_3^2 + a_4^2 + \ldots + a_{100}^2 \geq 100.$$

What is the minimum value of the sum $a_1 + a_2 + \ldots + a_{100}$?

282. (*China*). Find all complex polynomials $P(x)$ such that for any integers a, b, c satisfying $a + b + c \neq 0$,

$$\frac{P(a) + P(b) + P(c)}{a + b + c}$$

is an integer.

283. (*Croatia*). For all positive real numbers a, b, c, d, prove the inequality:

$$\frac{a-b}{b+c} + \frac{b-c}{c+d} + \frac{c-d}{d+a} + \frac{d-a}{a+b} \geq 0.$$

284. (*Croatia*). Solve in the set of real numbers:

$$3\left(x^2 + y^2 + z^2\right) = 1, x^2y^2 + y^2z^2 + z^2x^2 = xyz(x + y + z)^3.$$

285. (*Germany*). Let a, b, c, d be positive real numbers such that $abcd = 1$ and

$$a + b + c + d > \frac{a}{b} + \frac{b}{c} + \frac{c}{d} + \frac{d}{a}.$$

Prove that

$$a + b + c + d < \frac{a}{d} + \frac{b}{a} + \frac{c}{b} + \frac{d}{c}.$$

286. (*Germany*). Find all functions $f: \mathbb{R} \to \mathbb{R}$ such that $\forall x, y, z \in \mathbb{R}$, we have: if $x^3 + f(y)x + f(z) = 0$, then $(f(x))^3 + yf(x) + z = 0$.

287. (*Greece*). Find all triples $(x, y, z) \in \mathbb{R}^3$ such that x, y, $z > 3$ and

$$\frac{(x+2)^2}{y+z-2} + \frac{(y+4)^2}{z+x-4} + \frac{(z+6)^2}{x+y-6} = 36.$$

288. (*Iran*). Find all polynomials $P \in \mathbb{R}[x, y]$ such that

$$P(x^2, y^2) = P\left(\frac{(x+y)^2}{2}, \frac{(x-y)^2}{2}\right).$$

289. (*Moldova*). Let f, $g \in \mathbb{Z}[x]$ be nonconstant polynomials such that g divides f and the polynomial $f(x) + 2009$ has 50 integer roots. Prove that the degree of g is at least 5.

290. (*Netherlands*). Let a, b and c be positive real numbers such that $a + b + c \geq abc$. Prove that

$$a^2 + b^2 + c^2 \geq \sqrt{3}abc.$$

291. (*Netherlands*). Find all functions $f : \mathbb{Z} \to \mathbb{Z}$ satisfying

$$f(m+n) + f(mn-1) = f(m)f(n) + 2, \forall m, n \in \mathbb{Z}.$$

292. (*Serbia*). Let x, y, z be positive real numbers such that $xy + yz + zx = x + y + z$. Prove the inequality

$$\frac{1}{x^2+y+1} + \frac{1}{y^2+z+1} + \frac{1}{z^2+x+1} \leq 1.$$

When does the equality hold?

293. (*Switzerland*). Find all positive integers n for which there is a polynomial P with integer coefficients such that if d is a positive divisor of n, then

$$P(d) = \left(\frac{n}{d}\right)^2.$$

294. (*Vietnam*). Find k such that

$$\left(k + \frac{a}{b+c}\right)\left(k + \frac{b}{c+a}\right)\left(k + \frac{c}{a+b}\right) \geq \left(k + \frac{1}{2}\right)^3$$

for every positive real numbers a, b, c.

2.39 2010

295. (*Albania*). Find all continuous functions $f : \mathbb{R} \to \mathbb{R}$ such that

$$(1 + f(x)f(y))f(x+y) = f(x) + f(y), \forall x, y \in \mathbb{R}.$$

296. (*Albania*). A point of the plane is called *rational* if both coordinates are rational and *irrational* if both coordinates are irrational. Check whether the following statements are true or false:

(a) Every point of the plane is in a line that can be defined by two rational points.
(b) Every point of the plane is in a line that can be defined by two irrational points.

297. (*Argentina*). Find all functions $f: \mathbb{R} \to \mathbb{R}$ such that

$$f(x + xy + f(y)) = \left(f(x) + \frac{1}{2}\right)\left(f(y) + \frac{1}{2}\right)$$

holds for all real numbers x, y.

298. (*Bosnia and Herzegovina*). Find all functions $f: \mathbb{Z} \to \mathbb{Z}$ such that the following conditions hold:

(a) $f(n)f(-n) = f(n^2)$ for all $n \in \mathbb{Z}$.
(b) $f(m + n) = f(m) + f(n) + 2mn$ for all $m, n \in \mathbb{Z}$.

299. (*Bulgaria*). The sequence $(a_i)_{i \geq 1}$ is such that $a_1 > 0$ and

$$a_{n+1} = \frac{a_n}{1 + a_n^2}, n \geq 1.$$

(i) Prove that

$$a_n \leq \frac{1}{\sqrt{2n}}, n \geq 2$$

(ii) Prove that there exists n such that

$$a_n > \frac{7}{10\sqrt{n}}.$$

300. (*China*). Find all positive real numbers λ such that for all integers $n \geq 2$ and all positive real numbers $a_1, a_2, \ldots, a_n$ with $a_1 + a_2 + \ldots + a_n = n$, the following inequality holds:

$$\sum_{i=1}^{n} \frac{1}{a_i} - \lambda \prod_{i=1}^{n} \frac{1}{a_i} \leq n - \lambda.$$

301. (*Estonia*). Let $P(x, y)$ be a nonconstant homogenous polynomial with real coefficients such that $P(\sin t, \cos t) = 1$ for every real number t. Prove that there exists a positive integer k such that $P(x, y) = (x^2 + y^2)^k$.

302. (*Greece*). Solve in positive real numbers the system of equations

$$x + y + z + w = 4, \frac{1}{x} + \frac{1}{y} + \frac{1}{z} + \frac{1}{w} + \frac{1}{xyzw} = 5.$$

303. (*India*). Let a, b, c be positive real numbers such that $ab + bc + ca \leq 3abc$. Prove that

$$\sqrt{\frac{a^2+b^2}{a+b}}+\sqrt{\frac{b^2+c^2}{b+c}}+\sqrt{\frac{c^2+a^2}{c+a}}+3\leq\sqrt{2}\left(\sqrt{a+b}+\sqrt{b+c}+\sqrt{c+a}\right).$$

304. (*Indonesia*). Let a_1, a_2, ... be a sequence of real numbers such that

$$a_1=1, a_2=\frac{4}{3}, a_{n+1}=\sqrt{1+a_na_{n-1}}, \forall n\geq 2.$$

Prove that for all $n \geq 2$,

$$a_n^2>a_{n-1}^2+\frac{1}{2}, 1+\frac{1}{a_1}+\frac{1}{a_2}+\ldots+\frac{1}{a_n}>2a_n.$$

305. (*Indonesia*). Define the set $\mathbb{H}$ as follows

1. $\frac{1}{2} \in \mathbb{H}$;
2. if $x \in \mathbb{H}$,then $\frac{1}{1+x} \in \mathbb{H}$ and $\frac{x}{1+x} \in \mathbb{H}$.

Prove that there exists a bijective function $f: \mathbb{Z} \to \mathbb{H}$.

306. (*Iran*). Find all two-variable polynomials $p(x, y)$ such that for each $a, b, c \in \mathbb{R}$:

$$p\left(ab, c^2+1\right)+p\left(bc, a^2+1\right)+p\left(ca, b^2+1\right)=0.$$

307. (*Moldova*). Prove that for any real number x the following inequality is true:

$$\max\{|\sin x|, |\sin(x+2010^\circ)|\}>\frac{1}{\sqrt{17}}.$$

308. (*New Zealand*). Let $a_1, a_2, \ldots, a_8$ be eight distinct integers from $\{1, 2, \ldots, 17\}$. Show that there is an integer $k > 0$ such that there are at least three distinct (not necessarily disjoint) pairs (i, j) such that $a_i - a_j = k$.

Find a set of seven integers from $\{1, 2, \ldots, 17\}$ such that there is no integer $k > 0$ with that property.

309. (*Swtzerland*). Find all positive real solutions (a, b, c, d) of the equation

$$\frac{a^2-bd}{b+2c+d}+\frac{b^2-ca}{c+2d+a}+\frac{c^2-db}{d+2a+b}+\frac{d^2-ac}{a+2b+c}=0.$$

310. (*Turkey*). Show that for all positive real numbers a, b, c, holds

$$\sum_{cyc}\sqrt[4]{\frac{(a^2+b^2)(a^2-ab+b^2)}{2}}\le\frac{2}{3}(a^2+b^2+c^2)\left(\frac{1}{a+b}+\frac{1}{b+c}+\frac{1}{c+a}\right).$$

2.40 2011

311. (*Albania*). The given parabola $y = ax^2 + bx + c$ does not intersect the x-axis and passes through the points $A(-2,1)$ and $B(2,9)$. Find all possible values of the x coordinate of the vertex of this parabola.

312. (*Belarus*). Find all real a such that there exists a function $f\colon \mathbb{R}\to\mathbb{R}$ satisfying the equation

$$f(\sin x)+af(\cos x)=\cos 2x,\quad \forall x\in\mathbb{R}.$$

313. (*Bosnia and Herzegovina*). Let a, b, c be positive real numbers such that $a + b + c = 1$. Prove the inequality

$$a\sqrt[3]{1+b-c}+b\sqrt[3]{1+c-a}+c\sqrt[3]{1+a-b}\le 1.$$

314. (*China*). Let n be a positive integer. Find the largest real number λ such that for all positive real numbers x_1, x_2, ..., x_{2n} satisfying the inequality

$$\frac{1}{2n}\sum_{i=1}^{2n}(x_i+2)^n\ge\prod_{i=1}^{2n}x_i,$$

the following inequality also holds

$$\frac{1}{2n}\sum_{i=1}^{2n}(x_i+1)^n\ge\lambda\prod_{i=1}^{2n}x_i.$$

315. (*Estonia*). Does there exist an operation $*$ on the set of all integers such that the following conditions hold simultaneously:

1. For any integers x, y, z, $(x*y)*z=x*(y*z)$;
2. For any integers x, y, $x*x*y=y*x*x=y$?

316. (*Greece*). Find all functions f, $g:\mathbb{Q}\to\mathbb{Q}$ such that the following two conditions hold

$$f(g(x)-g(y))=f(g(x))-y \quad (1)$$
$$g(f(x)-f(y))=g(f(x))-y \quad (2)$$

for all $x, y \in \mathbb{Q}$.

317. (*Kosovo*). Find all functions $f: \mathbb{R} \to \mathbb{R}$ such that $\forall x \notin \{-1,1\}$ holds:

$$f\left(\frac{x-3}{x+1}\right)+f\left(\frac{3+x}{1-x}\right)=x.$$

318. (*Mongolia*). Let t, k, m be positive integers and $t > \sqrt{km}$. Prove that

$$\binom{2m}{0}+\binom{2m}{1}+\ldots+\binom{2m}{m-t-1}<\frac{2^{2m}}{2k}.$$

319. (*Netherlands*). Find all integers n for which the polynomial $P(x) = 3x^3 - nx - -n - 2$ is reducible in $\mathbb{Z}[x]$.

320. (*New Zealand*). Prove that for any distinct positive real numbers a, b, c:

$$\frac{(a^2-b^2)^3+(b^2-c^2)^3+(c^2-a^2)^3}{(a-b)^3+(b-c)^3+(c-a)^3}>8abc.$$

321. (*Romania*). Prove that the set $S = \{\lfloor n\pi \rfloor | n \in N\}$ contains arithmetic progressions of any finite length, but no infinite arithmetic progressions.

322. (*Saudi Arabia*). Consider a nonzero real number a such that

$$\{a\}+\left\{\frac{1}{a}\right\}=1,$$

where $\{x\}$ denotes the fractional part of x. Prove that for any positive integer n,

$$\{a^n\}+\left\{\frac{1}{a^n}\right\}=1.$$

323. (*Saudi Arabia*). Let n be a positive integer. Prove that all roots of the equation

$$x(x+2)(x+4)\ldots(x+2n)+(x+1)(x+3)\ldots(x+2n-1)=0$$

are real and irrational.

324. (*Vietnam*). Let $n \geq 3$ be an integer. There are n real numbers $x_1, x_2, \ldots, x_n$ that satisfy

$$\begin{cases} x_1 \geq x_2 \geq \ldots \geq x_n, \\ x_1 + x_2 + \ldots + x_n = 0, \\ x_1^2 + x_2^2 + \ldots + x_n^2 = n(n-1) \end{cases}.$$

Find the maximum and the minimum value of the sum $S = x_1 + x_2$.

2.41 2012

325. (*Albania*). Find the greatest value of the expression

$$\frac{1}{x^2 - 4x + 9} + \frac{1}{y^2 - 4y + 9} + \frac{1}{z^2 - 4z + 9},$$

where x, y, z are nonnegative real numbers such that $x + y + z = 1$.

326. (*Albania*). Let $f : \mathbb{R}^+ \to \mathbb{R}^+$ be a function such that $f(x + f(y)) = yf(xy + 1)$, $\forall\, x, y > 0$.

(a) Show that $(y - 1)(f(y) - 1) \leq 0$ for all $y > 0$.
(b) Find all functions that satisfy the given condition.

327. (*Bosnia and Herzegovina*). Positive real numbers a, b, c satisfy $a^2 + b^2 + c^2 = 1$. Prove the inequality

$$\frac{a^3}{b^2 + c} + \frac{b^3}{c^2 + a} + \frac{c^3}{a^2 + b} \geq \frac{\sqrt{3}}{1 + \sqrt{3}}.$$

328. (*China*). Complex numbers x_i, y_i satisfy $|x_i| = |y_i| = 1$ for $i = 1, 2, \ldots, n$. Let

$$x = \frac{1}{n}\sum_{i=1}^{n} x_i, y = \frac{1}{n}\sum_{i=1}^{n} y_i$$

and $z_i = xy_i + yx_i - x_iy_i$. Prove that

$$\sum_{i=1}^{n} |z_i| \leq n.$$

329. (*Estonia*). Prove that for any positive integer k there exist k distinct integers for which the sum of their squares equals the sum of their cubes.

330. (*Estonia*). Let x, y, z be positive real numbers whose sum is 2012. Find the maximum value of

$$\frac{(x^2+y^2+z^2)(x^3+y^3+z^3)}{x^4+y^4+z^4}.$$

331. (*France*). Let $k > 1$ be an integer. A function $f : N^* \to N^*$ is called $k-$*tastrophic* when for every positive integer n we have $f_k(n) = n^k$, where f_k is the kth iteration of f:

$$f_k(n) = \underbrace{f \circ f \circ \ldots \circ f}_{k \text{ times}}(n).$$

For which k does there exist a k-tastrophic function?

332. (*France*). Determine all nonconstant polynomials $X^n + a_{n-1}X^{n-1} + \ldots + a_1X + a_0$ with integer coefficients for which the roots are exactly the numbers $a_0, a_1, \ldots, a_{n-1}$ (with multiplicity).

333. (*Greece*). Let a, b, c be positive real numbers satisfying $a + b + c = 3$. Prove that

$$\sum_{sym} \frac{a^2}{(b+c)^3} \geq \frac{3}{8}.$$

334. (*India*). Let $a \geq b$ and $c \geq d$ be real numbers. Prove that the equation

$$(x+a)(x+d)+(x+b)(x+c)=0$$

has real roots.

335. (*India*). Find all functions $f : (0, \infty) \to \mathbb{R}$ satisfying

$$f(x)+f(y) \leq \frac{f(x+y)}{2}, \quad \frac{f(x)}{x}+\frac{f(y)}{y} \geq \frac{f(x+y)}{x+y}$$

for all $x, y \in (0, \infty)$.

336. (*Moldova*). Prove that the polynomial $X^8 + 98X^4 + 1$ is reducible in $\mathbb{Z}[X]$.

337. (*Netherlands*). Do there exist quadratic polynomials $P(x)$ and $Q(x)$ with real coefficients such that the polynomial $P(Q(x))$ has the roots 2, 3, 5, 7?

338. (*New Zealand*). Find all real numbers x such that

$$x^3 = \left\{(x+1)^3\right\},$$

where $\{y\}$ denotes the fractional part of y.

339. (*Switzerland*). Let f, g be polynomials with integer coefficients and let a, b be fixed points for $f \circ g$. Prove that there exist fixed points c, d for $g \circ f$ such that

$$a + c = b + d.$$

340. (*Turkey*). For all positive real numbers a, b, c satisfying $ab + bc + ca \leq 1$, prove that

$$a + b + c + \sqrt{3} \geq 8abc\left(\frac{1}{a^2 + 1} + \frac{1}{b^2 + 1} + \frac{1}{c^2 + 1}\right).$$

341. (*Ukraine*). We call a triple (a, b, c) of numbers of the interval $[-1, 1]$ *qualitative* if these numbers satisfy the inequality $1 + 2abc \geq a^2 + b^2 + c^2$. Prove that when the triples (a, b, c) and (x, y, z) are qualitative, then (ax, by, cz) is also qualitative.

2.42 2013

342. (*Albania*). Let a, b, c, d be positive real numbers such that $abcd = 1$. Find with proof that $x = 3$ is the minimal value for which the following inequality holds:

$$a^x + b^x + c^x + d^x \geq \frac{1}{a} + \frac{1}{b} + \frac{1}{c} + \frac{1}{d}.$$

343. (*Albania*). Find all functions $f\colon \mathbb{R} \to \mathbb{R}$ such that

$$f(x^3) + f(y^3) = (x + y)\big(f(x^2) + f(y^2) - f(xy)\big), \quad \forall x, y \in \mathbb{R}.$$

344. (*Bosnia and Herzegovina*). Let $x_1, x_2, \ldots, x_n$ be nonnegative real numbers of sum equal to 1. Let

$$F_n = x_1^2 + x_2^2 + \ldots + x_n^2 - 2(x_1x_2 + x_2x_3 + \ldots + x_nx_1).$$

Find: (a) $\min F_3$; (b) $\min F_4$; (c) $\min F_5$.

345. (*China*). For the positive integer n, define $f(n) = \min_{m \in \mathbb{Z}} \left|\sqrt{2} - \frac{m}{n}\right|$. Let $\{n_i\}$ be a strictly increasing sequence of positive integers. Let C be a constant such that $f(n_i) < \frac{C}{n_i^2}$ for all positive integers i. Show that there exists a real number $q > 1$ such that $n_i \geq q^{i-1}$ for all positive integers i.

346. (*China*). Let n and k be two integers which are >1. Let $a_1, a_2, \ldots, a_n, c_1, c_2, \ldots, c_n$ be nonnegative real numbers such that

(a) $a_1 \geq a_2 \geq \ldots \geq a_n$ and $a_1 + a_2 + \ldots + a_n = 1$;
(b) for any integer $m \in \{1, 2, \ldots, n\}$, we have $c_1 + c_2 + \ldots + c_m \leq m^k$.

Find the maximum of $c_1 a_1^k + c_2 a_2^k + \ldots + c_n a_n^k$.

347. (*China*). Find all positive real numbers $r < 1$ such that there exists a set S with the properties:

(a) For any real number t, exactly one of t, $t + r$ and $t + 1$ belongs to S.
(b) For any real number t, exactly one of t, $t - r$ and $t - 1$ belongs to S.

348. (*Croatia*). Determine all functions $f : \mathbb{R} \to \mathbb{R}$ such that for all real numbers x, y we have $f(1) \geq 0$ and $f(x) - f(y) \geq (x - y)f(x - y)$.

349. (*Estonia*). Let $x_1, x_2, \ldots, x_n$ be nonnegative real numbers, not all of which are zeros.

(a) Prove that

$$1 \leq \frac{\left(x_1 + \frac{x_2}{2} + \frac{x_3}{3} + \ldots + \frac{x_n}{n}\right)\left(x_1 + 2x_2 + 3x_3 + \ldots + nx_n\right)}{\left(x_1 + x_2 + x_3 + \ldots + x_n\right)^2} \leq \frac{(n+1)^2}{4n}.$$

(b) Show that for each $n \geq 1$ both inequalities can hold as equalities.

350. (*France*). Let $0 \leq x_1 \leq x_2 \leq \ldots \leq x_n \leq x_{n+1} = 1$ and $0 \leq y_1 \leq y_2 \leq \ldots \leq y_n = 1$ be real numbers. Prove that

$$\sum_{i=1}^{n} (x_i - y_i) + n \sum_{i=1}^{n} (x_{i+1} - x_i) y_i \geq 0.$$

351. (*Greece*). Find the largest possible value of M for which

$$\frac{x}{1 + \frac{yz}{x}} + \frac{y}{1 + \frac{zx}{y}} + \frac{z}{1 + \frac{xy}{z}} \geq M$$

for all $x, y, z > 0$ with $xy + yz + zx = 1$.

352. (*Greece*). Determine whether the polynomial

$$P(x) = \left(x^2 - 2x + 5\right)\left(x^2 - 4x + 20\right) + 1$$

is irreducible over $\mathbb{Z}[x]$.

353. (*India*). We define an operation $\oplus$ on the set $\{0, 1\}$ by

$$0 \oplus 0 = 0, 0 \oplus 1 = 1, 1 \oplus 0 = 1, 1 \oplus 1 = 0.$$

For two positive integers a, b which are written in base 2 as

$$a=(a_1a_2\ldots a_k)_2, \quad b=(b_1b_2\ldots b_k)_2$$

(possibly with leading zeros), we define $a \bigoplus b = c$ where $c = (c_1c_2\ldots c_k)_2$ by

$$c_i = a_i \oplus b_i, 1 \le i \le k.$$

For example, $7 \oplus 3 = 4$, since $7 = (111)_2$, $3 = (011)_2$. Prove that the function

$$f : \mathbb{N} \to \mathbb{N}, f(n) = n \oplus \left\lfloor \frac{n}{2} \right\rfloor$$

is a bijection.

354. (*Moldova*). Find all pairs (x, y) of real numbers satisfying

$$\begin{cases} 2x^2+xy=1 \\ \dfrac{9x^2}{2(1-x)^4} = 1 + \dfrac{3xy}{2(1-x)^2} \end{cases}.$$

355. (*Netherlands*). Show that

$$\sum_{n=0}^{2013} \frac{4026!}{(n!(2013-n)!)^2}$$

is the square of an integer.

356. (*Netherlands*). Determine all 4-tuples (a, b, c, d) of real numbers satisfying the equations

$$\begin{cases} ab + c + d = 3 \\ bc + d + a = 5 \\ cd + a + b = 2 \\ da + b + c = 6 \end{cases}.$$

357. (*Norh Korea*). Let a_1, a_2, …, a_k be numbers such that $a_i \in \{0, 1, 2, 3\}$ for $1 \le i \le k$ and

$$z = \overline{x_k x_{k-1} \ldots x_1}_{(4)}$$

be a base 4 expansion of $z \in \{0, 1, 2, \ldots, 4^k - 1\}$. Define

$$p(z) = \sum_{i=1}^{k} a_i x_i 4^{i-1}, A = \{z | p(z) = z\}.$$

Prove that $|A|$ is a power of 2.

358. (*Peru*). Let $a \geq 3$ be a real number and P be a polynomial of degree n with real coefficients. Prove that at least one of the following numbers is ≥ 1:

$$\left|a^0 - P(0)\right|, \left|a^1 - P(1)\right|, \left|a^2 - P(2)\right|, \ldots, \left|a^{n+1} - P(n+1)\right|.$$

359. (*Switzerland*). Let n be a positive integer and P be a polynomial with integer coefficients such that $u^{2^n} - v^{2^n} | P(u) - P(v)$, for every positive integers u, v. Show that there exists a polynomial Q with integer coefficients such that $P(x) = Q\left(x^{2^n}\right)$.

360. (*Turkey*). For all real numbers x, y, z such that $-2 \leq x, y, z \leq 2$ and $x^2 + y^2 + z^2 + xyz = 4$, determine the least real number K satisfying

$$\frac{z(xz + yz + y)}{xy + y^2 + z^2 + 1} \leq K.$$

361. (*Vietnam*). Find the greatest positive integer k such that the following inequality holds for all positive real numbers a, b, c satisfying $abc = 1$:

$$\frac{1}{a} + \frac{1}{b} + \frac{1}{c} + \frac{k}{a+b+c+1} \geq 3 + \frac{k}{4}.$$

2.43 2014

362. (*Albania*). Prove that for $n \geq 2$ the following inequality holds:

$$\frac{1}{n+1}\left(1 + \frac{1}{3} + \ldots + \frac{1}{2n-1}\right) > \frac{1}{n}\left(\frac{1}{2} + \frac{1}{4} + \ldots + \frac{1}{2n}\right).$$

363. (*Bosnia and Herzegovina*). Let a, b, c be distinct real numbers. Determine the values of

(a) $\frac{1+ab}{a-b} \cdot \frac{1+bc}{b-c} + \frac{1+bc}{b-c} \cdot \frac{1+ca}{c-a} + \frac{1+ca}{c-a} \cdot \frac{1+ab}{a-b}$;
(b) $\frac{1-ab}{a-b} \cdot \frac{1-bc}{b-c} + \frac{1-bc}{b-c} \cdot \frac{1-ca}{c-a} + \frac{1-ca}{c-a} \cdot \frac{1-ab}{a-b}$.
(c) Prove the following inequality:

$$\frac{1+a^2b^2}{(a-b)^2} + \frac{1+b^2c^2}{(b-c)^2} + \frac{1+c^2a^2}{(c-a)^2} \geq \frac{3}{2}.$$

When does the equality hold?

364. (*China*). Let n be a given integer which is >1. Find the greatest constant $\lambda(n)$ such that for any nonzero complex numbers $z_1, z_2, \ldots, z_n$ we have

$$\sum_{k=1}^{n} |z_k|^2 \geq \lambda(n) \min_{1 \leq k \leq n} \left\{ |z_{k+1} - z_k|^2 \right\},$$

where $z_{n+1} = z_1$.

365. (*Estonia*). Let a, b, c be positive real numbers for which $a + b + c = 1$. Prove that

$$\frac{a^2}{b^3 + c^4 + 1} + \frac{b^2}{c^3 + a^4 + 1} + \frac{c^2}{a^3 + b^4 + 1} > \frac{1}{5}.$$

366. (*France, EGMO*). Find all real numbers t such that if a, b, c are the sidelengths of a triangle, then $a^2 + bct$, $b^2 + cat$, $c^2 + abt$ are also the sidelengths of a triangle.

367. (*Hong Kong*). Let (a_n) be a sequence of positive integers. It is given that $a_1 = 1$ and for every $n \geq 1$, a_{n+1} is the smallest positive integer $> a_n$ which satisfies the following condition: for any integers i, j, k, with $1 \leq i, j, k \leq n + 1$, we have

$$a_i + a_j \neq 3a_k.$$

Find a_{2015}.

368. (*India*). Let a, b be positive real numbers. Prove that

$$(1 + a)^8 + (1 + b)^8 \geq 128ab(a + b)^2.$$

369. (*India*). Let x, y be positive rational numbers such that $x^5 + y^5 = 2x^2y^2$. Prove that $1 - xy$ is the square of a rational number.

370. (*Lithuania*). Given positive real numbers a, b, c such that the system of equations

$$\begin{cases} a^2x + b^2y + c^2z = 1 \\ xy + yz + zx = 1 \end{cases}$$

has exactly one solution (x, y, z) in real numbers. Prove that there is a triangle with sidelengths equal to a, b, c.

371. (*Moldova*). Consider $n \geq 2$ numbers $0 < x_1 \leq x_2 \leq \ldots \leq x_n$ such that $x_1 + x_2 + \ldots + x_n = 1$. Prove that if $x_n \leq \frac{2}{3}$, then there exists a positive integer $1 \leq k \leq n$ such that

$$\frac{1}{3} \leq x_1 + x_2 + \ldots + x_k < \frac{2}{3}.$$

372. (*Netherlands*). Let a, b, c be rational numbers for which $a + bc$, $b + ca$, $a + b$ are all nonzero and

$$\frac{1}{a+bc}+\frac{1}{b+ca}=\frac{1}{a+b}.$$

Prove that $\sqrt{(c-3)(c+1)}$ is rational.

373. (*New Zealand*). Determine all pairs (a, b) of real numbers such that the system of equations

$$\begin{cases} x^3+y^3=2 \\ ax+b=y \end{cases}$$

has no solutions (x, y) with x and y real numbers.

374. (*Peru*). Let x, y, z be real numbers such that

$$\begin{cases} x^2+y^2+z^2+(x+y+z)^2=9 \\ xyz\leq \dfrac{15}{32} \end{cases}.$$

Find the maximum possible value of x.

375. (*Saudi Arabia*). Let $a_1, \ldots, a_n$ be a nonincreasing sequence of positive real numbers. Prove that

$$\sqrt{a_1^2+a_2^2+\ldots+a_n^2}\leq a_1+\frac{a_2}{\sqrt{2}+1}+\ldots+\frac{a_n}{\sqrt{n}+\sqrt{n-1}}.$$

When does equality hold?

376. (*Switzerland*). Find all polynomials P with real coefficients such that

$$(x+2014)P(x)=xP(x+1).$$

377. (*Taiwan*). Prove that for positive real numbers a, b, c we have

$$3(a+b+c)\geq 8\sqrt[3]{abc}+\sqrt[3]{\frac{a^3+b^3+c^3}{3}}.$$

378. (*Taiwan*). Let $f(x) = x^n + a_{n-2}x^{n-2} + a_{n-3}x^{n-3} + \ldots + a_1x + a_0$ be a polynomial with real coefficients ($n \geq 2$). Suppose that all roots of f are real. Prove that the absolute value of each root is at most

$$\sqrt{\frac{2(1-n)a_{n-2}}{n}}.$$

379. (*Turkey*). Prove that for all nonnegative real numbers a, b, c with $a^2+b^2+c^2=1$, we have $\sqrt{a+b}+\sqrt{b+c}+\sqrt{c+a}\geq 5abc+2$.

2.44 2015

380. (*Belarus*). Prove that $(a + b + c)^5 \geq 81(a^2 + b^2 + c^2)abc$ for any positive real numbers a, b, c.

381. (*Bosnia and Herzegovina*). Determine the minimum value of the expression

$$\frac{a+1}{a(a+2)}+\frac{b+1}{b(b+2)}+\frac{c+1}{c(c+2)},$$

for positive real numbers a, b, c such that $a + b + c \leq 3$.

382. (*China*). Let $x_1, \ldots, x_n (n \geq 2)$ be a nondecreasing monotonic sequence of positive real numbers such that $x_1, \frac{x_2}{2}, \ldots, \frac{x_n}{n}$ is a nonincreasing monotonic sequence. Prove that

$$\frac{\sum_{i=1}^{n} x_i}{n\left(\prod_{i=1}^{n} x_i\right)^{\frac{1}{n}}} \leq \frac{n+1}{2\sqrt[n]{n!}}.$$

383. (*Estonia*). Let $n \geq 5$ be a positive integer and $a_1, a_2, \ldots, a_n$ real numbers such that all possible sums $a_i + a_j$, $1 \leq i < j \leq n$ form $\frac{n(n-1)}{2}$ consecutive terms of an arithmetic progression when taken in some order. Prove that $a_1 = a_2 = \ldots = a_n$.

384. (*Estonia*). Find all functions $f: \mathbb{R} \to \mathbb{R}$ which satisfy

$$f(f(x)+f(y))=f(x^2)+2x^2f(y)+(f(y))^2, \quad \forall x, y\in\mathbb{R}.$$

385. (*Estonia*). Let $n > 1$ be an integer and $a > b > 0$ real numbers. Prove that

$$(a^n-b^n)\left(\frac{1}{b^{n-1}}-\frac{1}{a^{n-1}}\right)>4n(n-1)\left(\sqrt{a}-\sqrt{b}\right)^2.$$

386. (*Greece*). Determine all functions $f: \mathbb{R} \to \mathbb{R}$ such that for arbitrary $x, y \in \mathbb{R}$ we have

$$f(xy) \leq yf(x) + f(y).$$

387. (*Germany*). Find the least positive integer n such that there is a polynomial

$$P(x) = a_{2n}x^{2n} + a_{2n-1}x^{2n-1} + \ldots + a_1x + a_0$$

with real coefficients that satisfies both of the following properties:

1. $2014 \leq a_i \leq 2015$ for $i = 0, 1, \ldots, 2n$.
2. There is a real number ξ with $P(\xi) = 0$.

388. (*India*). Let f, g be two polynomials with integer coefficients such that the leading coefficients of both polynomials are positive. Suppose that $\deg(f)$ is odd and

$$\{f(a)|a \in \mathbb{Z}\} = \{g(a)|a \in \mathbb{Z}\}.$$

Prove that there exists an integer k such that $g(x) = f(x + k)$.

389. (*Iran*). Find all polynomials $P, Q \in \mathbb{Q}[x]$ such that

$$(P(x))^3 + (Q(x))^3 = x^{12} + 1.$$

390. (*Ireland, EGMO*). Let x, y, z, w be positive real numbers such that $xyzw = 16$. Prove that

$$\frac{x^2}{x+y} + \frac{y^2}{y+z} + \frac{z^2}{z+w} + \frac{w^2}{w+x} \geq 4,$$

with equality only when $x = y = z = w = 2$.

391. (*Moldova*). Let a, b, c be positive real numbers such that $abc = 1$. Prove that

$$a^3 + b^3 + c^3 + \frac{ab}{a^2+b^2} + \frac{bc}{b^2+c^2} + \frac{ca}{c^2+a^2} \geq \frac{9}{2}.$$

392. (*Netherlands*). Determine all polynomials $P(X)$ with real coefficients such that

$$(X+1)P(X-1) - (X-1)P(X)$$

is a constant polynomial.

393. (*Peru*). Find all positive integers n for which there exist real numbers $x_1, x_2, \ldots, x_n$ satisfying the conditions:

(i) $-1 < x_i < 1$ for $1 \leq i \leq n$;
(ii) $x_1 + x_2 + \ldots + x_n = 0$;
(iii) $\sqrt{1-x_1^2} + \sqrt{1-x_2^2} + \ldots + \sqrt{1-x_n^2} = 1$.

394. (*Russia*). Prove that if $n > 4$, then

$$\sum_{k=2}^{n} \sqrt[k]{\frac{k}{k-1}} < n.$$

395. (*Saudi Arabia*). Let a, b, c be positive real numbers satisfying the condition:

$$(a+b+c)\left(\frac{1}{a}+\frac{1}{b}+\frac{1}{c}\right) = 10.$$

Find the greatest value and the least value of

$$T = \left(a^2+b^2+c^2\right)\left(\frac{1}{a^2}+\frac{1}{b^2}+\frac{1}{c^2}\right).$$

396. (*Switzerland*). Find all polynomials P with real coefficients such that

$$P(X)P(X+1) = P\left(X^2+2\right).$$

397. (*Vietnam*). Find the smallest positive integer n such that there exist n real numbers a_1, a_2, …, a_n satisfying the conditions:

(a) $a_1 + a_2 + \ldots + a_n > 0$;
(b) $a_1^3 + a_2^3 + \ldots + a_n^3 < 0$;
(c) $a_1^5 + a_2^5 + \ldots + a_n^5 > 0$.

2.45 2016

398. (*Azerbaijan*). Find all polynomials $P(x)$ with real coefficients such that for all x, y, z satisfying $x + y + z = 0$, the equation below is true:

$$(P(x+y))^3 + (P(y+z))^3 + (P(z+x))^3 = 3P(x+y)P(y+z)P(z+x).$$

399. (*Belarus*). Prove for positive a, b, c that

$$\left(a^2+\frac{b^2}{c^2}\right)\left(b^2+\frac{c^2}{a^2}\right)\left(c^2+\frac{a^2}{b^2}\right) \geq abc\left(a+\frac{1}{a}\right)\left(b+\frac{1}{b}\right)\left(c+\frac{1}{c}\right).$$

400. (*Bosnia and Herzegovina*). Determine all functions $f : \mathbb{Z} \to \mathbb{Z}$ with the property that

$$f(x-f(y))=f(f(x))-f(y)-1$$

holds for all $x, y \in \mathbb{Z}$.

401. (*Brazil*). Determine all functions $f: N \to N$ such that $f(a + b) = f(a) + f(b) + f($-$(a + b) = f(a) + f(b) + f(c) + f(d)$ whenever a, b, c, d are nonnegative integers satisfying $2ab = c^2 + d^2$.

402. (*China*). Find the smallest positive number λ such that for any complex numbers

$$z_1, z_2, z_3 \in \{z \in \mathbb{C} | |z| < 1\},$$

if $z_1 + z_2 + z_3 = 0$, then $|z_1z_2 + z_zz_3 + z_3z_1|^2 + |z_1z_2z_3|^2 < \lambda$.

403. (*Croatia*). Find all functions $f: \mathbb{R} \to \mathbb{R}$ such that for all real x, y:

$$f(x^2) + xf(y) = f(x)f(x + f(y)).$$

404. (*Hong Kong*). Find the largest possible positive integer n, so that there exist n distinct positive real numbers $x_1, x_2, \ldots, x_n$ satisfying the following inequality: for any $1 \le i, j \le n$,

$$(3x_i - x_j)(x_i - 3x_j) \ge (1 - x_ix_j)^2.$$

405. (*Hong Kong*). Let a, b, c be positive real numbers satisfying $abc = 1$. Determine the smallest possible value of

$$\frac{a^3+8}{a^3(b+c)} + \frac{b^3+8}{b^3(c+a)} + \frac{c^3+8}{c^3(a+b)}.$$

406. (*India*). Let a, b, c, d be real numbers satisfying $|a|, |b|, |c|, |d| > 1$ and

$$abc + abd + acd + bcd + a + b + c + d = 0.$$

Prove that

$$\frac{1}{a-1} + \frac{1}{b-1} + \frac{1}{c-1} + \frac{1}{d-1} > 0.$$

407. (*India*). Suppose that a sequence $a_1, a_2, \ldots$ of positive real numbers satisfies

$$a_{k+1} \ge \frac{ka_k}{a_k^2 + k - 1}, \forall k \ge 1.$$

Prove that $a_1 + a_2 + \ldots + a_n \ge n$ for every $n \ge 2$.

408. (*Iran*). Let n be a positive integer. Find the maximum value of

$$\sum_{1\le r<s\le 2n}(s-r-n)x_rx_s,$$

where $-1 \le x_i \le 1$ for all $i = 1, \dots, 2n$.

409. (*Ireland*). Prove that for any positive real numbers a, b and c we have

$$\frac{2a+b}{b+2c}+\frac{2b+c}{c+2a}+\frac{2c+a}{a+2b}\ge 3.$$

410. (*Japan*). Find the smallest positive real number k such that for any positive integer $n \ge 2$ and positive reals $a_0, a_1, a_2, \dots, a_n$,

$$\frac{1}{a_0+a_1}+\frac{1}{a_0+a_1+a_2}+\dots+\frac{1}{a_0+a_1+\dots+a_n}<k\left(\frac{1}{a_0}+\frac{1}{a_1}+\dots+\frac{1}{a_n}\right).$$

411. (*Kosovo*). Solve the equation in real numbers

$$\sqrt{x+\sqrt{4x+\sqrt{16x+\sqrt{\dots+\sqrt{4^nx+3}}}}}-\sqrt{x}=1.$$

412. (*Moldova*). Let $\alpha\in\left(0,\frac{\pi}{2}\right)$. Find the minimum value of the expression

$$P=(1+\cos\alpha)\left(1+\frac{1}{\sin\alpha}\right)+(1+\sin\alpha)\left(1+\frac{1}{\cos\alpha}\right).$$

413. (*Netherlands*). Prove that for all positive reals a, b, c we have

$$a+\sqrt{ab}+\sqrt[3]{abc}\le\frac{4}{3}(a+b+c).$$

414. (*Russia*). Positive real numbers a, b, c are such that $a^2 < 16bc$, $b^2 < 16ca$, $c^2 < 16ab$. Show that $a^2 + b^2 + c^2 < 2(ab + bc + ca)$.

415. (*Serbia*). Define the sequence of polynomials as follows:

$$P_0(x)=x^3-4x,\quad P_{n+1}(x)=P_n(1+x)P_n(1-x)-1, \forall n\ge 0.$$

Prove that $x^{2016}|P_{2016}(x)$.

416. (*Switzerland*). Find all polynomial functions with real coefficients for which

$$(x-2)P(x+2)+(x+2)P(x-2)=2xP(x)$$

for all real x.

417. (*Taiwan*). Suppose that the function $f\colon [0,\infty)\to[0,\infty)$ satisfies:

(a) $f(x)f(y)\le y^2 f\left(\frac{x}{2}\right)+x^2 f\left(\frac{y}{2}\right)$;
(b) $f(x)\le 2016$, $\forall\, x\in[0,1]$.

Prove that $f(x)\le x^2$ for all $x\ge 0$.

2.46 2017

418. (*Belarus*). Find all functions $f\colon \mathbb{R}\to\mathbb{R}$ such that

$$4xyf\left(x^2-y^2\right)=\left(x^2-y^2\right)f(2x)f(2y)$$

for all real x, y.

419. (*Belarus*). Given that x, y, z are positive real numbers satisfying $x+y+z=2$, prove the inequality:

$$\frac{(x-1)^2}{y}+\frac{(y-1)^2}{z}+\frac{(z-1)^2}{x}\ge\frac{1}{4}\left(\frac{x^2+y^2}{x+y}+\frac{y^2+z^2}{y+z}+\frac{z^2+x^2}{z+x}\right).$$

420. (*Bosnia and Herzegovina*). Find the smallest constant $C>0$ for which the following statement holds: among any five positive real numbers a_1, a_2, a_3, a_4, a_5 (not necessarily distinct), one can always choose distinct indices i, j, k, l such that

$$\left|\frac{a_i}{a_j}-\frac{a_k}{a_l}\right|\le C.$$

421. (*China*). Let $x>1$ be a real number and n a positive integer. Prove that

$$\sum_{k=1}^{n}\frac{\{kx\}}{\lfloor kx\rfloor}<\sum_{k=1}^{n}\frac{1}{2k-1}.$$

422. (*Hong Kong*). Let $f(x)$ be a monic cubic polynomial with $f(0)=-64$ and all roots of $f(x)$ are nonnegative real numbers. What is the largest possible value of $f(-1)$?

423. (*Hong Kong*). Given infinite sequences $a_1, a_2, a_3, \ldots$ and $b_1, b_2, b_3, \ldots$ of real numbers satisfying

$$a_{n+1} + b_{n+1} = \frac{a_n + b_n}{2}, a_{n+1}b_{n+1} = \sqrt{a_n b_n}, \forall n \geq 1.$$

Suppose that $b_{2016} = 1$ and $a_1 > 0$. Find all possible values of a_1.

424. (*Hong Kong*). Let $a_0, a_1, a_2, \ldots$ be a sequence of real numbers which satisfies the condition:

$$\sum_{n=0}^{m} (-1)^n \binom{m}{n} a_n = 0$$

for all sufficiently large values of m. Show that there exists a polynomial P such that $a_n = P(n)$ for all $n \geq 0$.

425. (*India*). Let $P_c(x) = x^4 + ax^3 + bx^2 + cx + 1$, $Q_c(x) = x^4 + cx^3 + bx^2 + ax + 1$ with a, b real numbers, $c \in \{1, 2, \ldots, 2017\}$ and $a \neq c$. Define

$$A_c = \{\alpha | P_c(\alpha) = 0\}, B_c = \{\beta | Q_c(\beta) = 0\}.$$

(a) Find the number of unordered pairs of polynomials $P_c(x)$, $Q_c(x)$ with exactly two common roots.
(b) For any $1 \leq c \leq 2017$,find the sum of the elements of $A_c \Delta B_c$.

426. (*India*). Let a, b, c be distinct positive real numbers with $abc = 1$. Prove that

$$\sum_{cyc} \frac{a^6}{(a-b)(a-c)} > 15.$$

427. (*India*). Let f, $g \in \mathbb{R}[x]$ be nonconstant polynomials. Suppose that neither of f, g is the square of a real polynomial, but $f(g(x))$ is. Prove that $g(f(x))$ is not the square of a real polynomial.

428. (*India*). For each $n \geq 2$ define the polynomial $f_n(x) = x^n - x^{n-1} - \ldots - x - 1$. Prove that

(i) For each $n \geq 2$, $f_n(x) = 0$ has a unique positive real root α_n;
(ii) $(\alpha_n)_n$ is a strictly increasing sequence;
(iii) $\lim_{n \to \infty} \alpha_n = 2$.

429. (*Ireland*). Positive real numbers a, b, c satisfy the double inequality

$$\frac{b^2}{a+b} + \frac{c^2}{b+c} + \frac{a^2}{c+a} \geq \frac{c^2}{a+b} + \frac{a^2}{b+c} + \frac{b^2}{c+a} \geq \frac{a^2}{a+b} + \frac{b^2}{b+c} + \frac{c^2}{c+a}.$$

Prove that $a = b = c$.

430. (*Moldova*). Let the sequence $(a_n)_{n \geq 1}$ be defined as:

$$a_n = \sqrt{A^1_{n+2}\sqrt[3]{A^2_{n+3}\sqrt[4]{A^3_{n+4}\sqrt[5]{A^4_{n+5}}}}},$$

where A^k_m are defined by

$$A^k_m = \binom{m}{k} \cdot k!.$$

Prove that

$$a_n < \frac{119n}{120} + \frac{7}{3}.$$

431. (*Moldova*). Let $f(X) = a_nX^n + a_{n-1}X^{n-1} + \ldots + a_1X + a_0$ be a polynomial with real coefficients which satisfies $a_n \geq a_{n-1} \geq \ldots \geq a_1 \geq a_0 > 0$. Prove that for every complex root z of this polynomial, we have $|z| \leq 1$.

432. (*Netherlands*). A function $f : \mathbb{Z}_{>0} \to \mathbb{Z}$ is given, which has the following properties:

(i) $f(p) = 1$ for all prime numbers p,
(ii) $f(xy) = yf(x) + xf(y)$ for all $x, y \in \mathbb{Z}_{>0}$.

Determine the smallest $n \geq 2016$ satisfying $f(n) = n$.

433. (*Netherlands*). Determine all possible values of $\frac{1}{x} + \frac{1}{y}$ if x and y are nonzero real numbers satisfying

$$x^3 + y^3 + 3x^2y^2 = x^3y^3.$$

434. (*New Zealand*). Find all real values for a, b, c such that

(i) $a + b + c = 51$;
(ii) $bc = 4000$;
(iii) $0 < a \leq 10$ and $c \geq 25$.

435. (*Russia*). Define

$$a_n = 2 - \frac{1}{n^2 + \sqrt{n^4 + \frac{1}{4}}}$$

for all positive integers n. Prove that $\sqrt{a_1} + \sqrt{a_2} + \ldots + \sqrt{a_{119}}$ is an integer.

436. (*Serbia*). For a positive integer $n \geq 2$, let $C(n)$ be the smallest positive real constant such that there is a sequence of n real numbers $x_1, x_2, \ldots, x_n$, not all zero, satisfying the following conditions:

(a) $x_1 + x_2 + \ldots + x_n = 0$;
(b) for each $i = 1, 2, \ldots, n$, it holds that $x_i \leq x_{i+1}$ or $x_i \leq x_{i+1} + C(n)x_{i+2}$ (the indices are taken modulo n).

Prove that

(a) $C(n) \geq 2$ for all n;
(b) $C(n) = 2$ if and only if n is even.

437. (*Taiwan*). Given $a, b, c, d > 0$, prove that

$$\sum_{cyc} \frac{c}{a+2b} + \sum_{cyc} \frac{a+2b}{c} \geq 8\left(\frac{(a+b+c+d)^2}{ab+ac+ad+bc+bd+cd} - 1\right).$$

438. (*Taiwan*). Find all polynomials P with real coefficients which satisfy

$$P(x)P(x+1) = P\left(x^2 - x + 3\right).$$

2.47 2018

439. (*Bosnia and Herzegovina*). Let $a_1, a_2, \ldots, a_n$, k and M be positive integers such that

$$\frac{1}{a_1} + \frac{1}{a_2} + \ldots + \frac{1}{a_n} = k, a_1 a_2 \ldots a_n = M.$$

If $M > 1$, prove that the polynomial

$$P(x) = M(x+1)^k - (x+a_1)(x+a_2)\ldots(x+a_n)$$

has no positive roots.

440. (*China*). Prove that there exists a constant $C > 0$ such that

$$H(a_1) + H(a_2) + \ldots + H(a_m) \leq C\sqrt{\sum_{i=1}^{m} i a_i}$$

holds for arbitrary positive integer m and any positive integers $a_1, a_2, \ldots a_m$, where

$$H(n) = \sum_{k=1}^{n} \frac{1}{k}.$$

441. (*Cyprus*). Find all triples (a, b, c) of positive real numbers for which the expression

$$K = \frac{a+3c}{a+2b+c} + \frac{4b}{a+b+2c} - \frac{8c}{a+b+3c}$$

attains its minimum value.

442. (*Estonia*). Find all functions $f: \mathbb{R} \to \mathbb{R}$ that satisfy

$$f(xy + f(xy)) = 2xf(y), \forall x, y \in \mathbb{R}.$$

443. (*Estonia*). A sequence of positive real numbers $a_1, a_2, a_3, \ldots$ satisfies $a_n = a_{n-1} + a_{n-2}$ for all $n \geq 3$. A sequence $b_1, b_2, b_3, \ldots$ is defined by the equations
$b_1 = a_1$, $b_n = a_n + (b_1 + b_3 + \ldots + b_{n-1})$ for even $n > 1$,
$b_n = a_n + (b_2 + b_4 + \ldots + b_{n-1})$ for odd $n > 1$.
Prove that if $n \geq 3$, then

$$\frac{1}{3} < \frac{b_n}{na_n} < 1.$$

444. (*Greece*). Let x, y, z be positive real numbers such that $x + y + z = 9xyz$. Prove that

$$\frac{x}{\sqrt{x^2 + 2yz + 2}} + \frac{y}{\sqrt{y^2 + 2zx + 2}} + \frac{z}{\sqrt{z^2 + 2xy + 2}} \geq 1.$$

Consider when equality applies.

445. (*Hong Kong*). Find all polynomials f such that f has nonnegative integer coefficients, $f(1) = 7$ and $f(2) = 2017$.

446. (*Hong Kong*). Does there exist a polynomial $P(x)$ with integer coefficients such that

$$P\left(1 + \sqrt[3]{2}\right) = 1 + \sqrt[3]{2}, P\left(1 + \sqrt{5}\right) = 2 + 3\sqrt{5}?$$

447. (*India*). Let a_n, b_n be sequences of positive real numbers such that

$$a_{n+1} = a_n + \frac{1}{2b_n}, b_{n+1} = b_n + \frac{1}{2a_n}, \forall n \in \mathbb{N}.$$

Prove that $\max\{a_{2018}, b_{2018}\} > 44$.

448. (*India*). For an integer $n \geq 2$ find all $a_1, a_2, \ldots, a_n, b_1, b_2, \ldots, b_n$ so that

(a) $0 \leq a_1 \leq a_2 \leq \ldots \leq a_n \leq 1 \leq b_1 \leq b_2 \leq \ldots \leq b_n$;

(b) $\sum_{k=1}^{n} (a_k + b_k) = 2n$;

(c) $\sum_{k=1}^{n} \left(a_k^2 + b_k^2\right) = n^2 + 3n$.

449. (*India*). Let $n \geq 2$ be a positive integer. Let $a_1 \leq a_2 \leq \ldots \leq a_n$ be real numbers such that

$$a_1 + a_2 + \ldots + a_n > 0, n\left(a_1^2 + a_2^2 + \ldots + a_n^2\right) = 2(a_1 + a_2 + \ldots + a_n)^2.$$

Show that if $m = \left\lfloor \frac{n}{2} \right\rfloor + 1$, then $a_m > 0$.

450. (*Iran*). Find the maximum value of k for which there exist distinct real numbers $x_1, x_2, \ldots, x_k > 1$ such that for all $1 \leq i, j \leq k$,

$$x_i^{\lfloor x_j \rfloor} = x_j^{\lfloor x_i \rfloor}.$$

451. (*Moldova*). The positive real numbers a, b, c, d satisfy the equality

$$\frac{1}{a+1} + \frac{1}{b+1} + \frac{1}{c+1} + \frac{1}{d+1} = 3.$$

Prove the inequality

$$\sqrt[3]{abc} + \sqrt[3]{bcd} + \sqrt[3]{cda} + \sqrt[3]{dab} \leq \frac{4}{3}.$$

452. (*New Zealand*). Let a, b, c be positive real numbers satisfying

$$\frac{1}{a+2019} + \frac{1}{b+2019} + \frac{1}{c+2019} = \frac{1}{2019}.$$

Prove that $abc \geq 4038^3$.

453. (*Serbia*). Let n be a positive integer and let $x_1, \ldots, x_n$ be positive real numbers. Prove that

$$x_1\left(1 - x_1^2\right) + x_2\left(1 - (x_1 + x_2)^2\right) + x_n\left(1 - (x_1 + \ldots + x_n)^2\right) < \frac{2}{3}.$$

454. (*Slovenia*). Let a, b, c be positive real numbers satisfying $abc = 1$. Prove the inequality:

$$\frac{a+b+c}{3} \geq \frac{a}{a^2b+2} + \frac{b}{b^2c+2} + \frac{c}{c^2a+2}.$$

455. (*Vietnam*). For every positive integer $n \geq 3$, let ϕ_n be the set of all positive integers $<n$ and coprime to n. Consider the polynomial

$$P_n(x) = \sum_{k \in \phi_n} x^{k-1}.$$

(a) Prove that $P_n(x) = (x^{r_n} + 1)Q_n(x)$ for some positive integer r_n and polynomial $Q_n(x) \in \mathbb{Z}[x]$.
(b) Find all n such that $P_n(x)$ is irreducible over $\mathbb{Z}[x]$.

2.48 2019

456. (*Belarus*). Let the sequence (a_n) be constructed in the following way:

$$a_1 = 1, a_2 = 1, a_{n+2} = a_{n+1} + \frac{1}{a_n}, n = 1, 2, \ldots$$

Prove that $a_{180} > 19$.

457. (*Belarus*). Given a quadratic trinomial $p(x)$ with integer coefficients such that $p(x)$ is not divisible by 3 for all integers x. Prove that there exist polynomials $f(x)$ and $h(x)$ with integer coefficients such that

$$p(x) \cdot f(x) + 3h(x) = x^6 + x^4 + x^2 + 1.$$

458. (*Belarus*). Find all functions $f : \mathbb{Z} \to \mathbb{Z}$ satisfying the following conditions:

1. $f(f(x)) = xf(x) - x^2 + 2$ for all $x \in \mathbb{Z}$;
2. f takes all integer values.

459. (*Ecuador*). Determine the value of

$$\left\lfloor 1 + \frac{1}{\sqrt{2}} + \frac{1}{\sqrt{3}} + \frac{1}{\sqrt{4}} + \ldots + \frac{1}{\sqrt{2019}} \right\rfloor.$$

460. (*Ecuador*). Three roots of the equation $x^4 - px^3 + qx^2 - rx + s = 0$ are $\tan A$, $\tan B$, $\tan C$, where A, B, C are the angles of a triangle. Find the fourth root, in function of p, q, r, s.

461. (*Ecuador*). Find all real solutions of the equation

$$(2^x - 4)^3 + (4^x - 2)^3 = (2^x + 4^x - 6)^3.$$

462. (*Ecuador*). Find $a \in \mathbb{R}$ for which the system

$$\begin{cases} x^2 - y^2 = 0 \\ (x-a)^2 + y^2 = 1 \end{cases}$$

has exactly 0, 1, 2, 3, 4, 5 different solutions.

463. (*Ecuador*). Let x, y, z be nonzero real numbers such that

$$x + \frac{1}{x} = y + \frac{1}{y} = z + \frac{1}{z} = 3.$$

Find the sum of all distinct values which the expression $\frac{x}{y} + \frac{y}{z} + \frac{z}{x}$ can take.

464. (*Ecuador*). Find the real solutions of the equation

$$\sqrt{3x^2 + 6x + 7} + \sqrt{5x^2 + 10x + 14} = 4 - 2x - x^2.$$

465. (*Estonia*). Find all functions $f : \mathbb{R} \to \mathbb{R}$ which satisfy

$$f(x^2)f(y^2) + |x|f(-xy^2) = 3|y|f(x^2y), \forall x, y \in \mathbb{R}.$$

466. (*Estonia*). Let us call a real number r *interesting*, if $r = a + b\sqrt{2}$ for some integers a, b. Let $A(x)$, $B(x)$ be polynomial functions with interesting coefficients for which the constant term of $B(x)$ is 1, and $Q(x)$ be a polynomial function with real coefficients such that $A(x) = B(x)Q(x)$. Prove that the coefficients of $Q(x)$ are interesting.

467. (*Hong Kong*). Let a be a real number. Suppose that the function

$$f(x) = \frac{a}{x-1} + \frac{1}{x-2} + \frac{1}{x-6}$$

defined on the interval $3 < x < 5$ attains its maximum at $x = 4$. Find the value of a.

468. (*Hong Kong*). If $57a + 88b + 125c \geq 1148$, where a, b, $c > 0$, what is the minimum value of

$$a^3 + b^3 + c^3 + 5(a^2 + b^2 + c^2)?$$

469. (*Hong Kong*). Determine all functions $f: \mathbb{Q}_> \to \mathbb{Q}_>$ satisfying

$$f\left(x^2(f(y))^2\right) = (f(x))^2 f(y), \forall x, y \in \mathbb{Q}_>.$$

470. (*Hong Kong*). Find the maximal value of $= \sqrt[3]{\frac{a}{b+7}} + \sqrt[3]{\frac{b}{c+7}} + \sqrt[3]{\frac{c}{d+7}} + \sqrt[3]{\frac{d}{a+7}}$, where a, b, c, d are nonnegative real numbers which satisfy $a + b + c + d = 100$.

471. (*India*). Let $n \geq 2$ be an integer. Solve in real numbers

$$|a_1 - a_2| = 2|a_2 - a_3| = 3|a_3 - a_4| = \ldots = n|a_n - a_1|.$$

472. (*Ireland*). Let $0 < x, y, z < 1$. Show that

$$\frac{1}{x(1-y)} + \frac{1}{y(1-z)} + \frac{1}{z(1-x)} \geq 12.$$

473. (*Japan*). Given real numbers x, y such that $x^3 + y, x^2 + y^2, x + y^3$ are all integers. Prove that x, y are integers.

474. (*Moldova*). Let $P(X) = a_{2n+1}X^{2n+1} + a_{2n}X^{2n} + \ldots + a_1X + a_0$ be a polynomial with positive coefficients. Prove that there exists a permutation $(b_{2n+1}, b_{2n}, \ldots, b_1, b_0)$ of $(a_{2n+1}, a_{2n}, \ldots, a_1, a_0)$ such that the polynomial $Q(X) = b_{2n+1}X^{2n+1} + b_{2n}X^{2n} + \ldots + b_1X + b_0$ has exactly one real root.

475. (*Moldova*). Find all polynomials $P(X)$ with real coefficients such that if real numbers x, y, z satisfy $x + y + z = 0$, then the points $(x, P(x))$, $(y, P(y))$, $(z, P(z))$ are collinear.

476. (*SAFEST*). Let $m, n \geq 2$ be integers. Let $f(x_1, \ldots, x_n)$ be a polynomial with real coefficients such that

$$f(x_1, \ldots, x_n) = \left\lfloor \frac{x_1 + \ldots + x_n}{m} \right\rfloor, \forall x_1, \ldots, x_n \in \{0, 1, \ldots, m-1\}.$$

Prove that the total degree of f is at least n.

477. (*SAFEST*). Let $a_1, a_2, \ldots, a_{2019}$ be positive real numbers such that

$$\frac{1}{a_1 + 2019} + \frac{1}{a_2 + 2019} + \ldots + \frac{1}{a_n + 2019} = \frac{1}{2019}.$$

Find the minimum value of $a_1 a_2 \ldots a_{2019}$ and determine for which values of $a_1, a_2, \ldots, a_{2019}$ this minimum occurs.

478. (*Switzerland*). Let S be a nonempty set of positive integers. Prove that at least one of the following two assertions holds:

(i) There exist distinct finite nonempty subsets F and G of S such that

$$\sum_{x\in F}\frac{1}{x}=\sum_{x\in G}\frac{1}{x}.$$

(ii) There exists a positive rational number $r<1$ such that for all finite nonempty subsets F of S it holds that

$$\sum_{x\in F}\frac{1}{x}\neq r.$$

479. (*Switzerland*). Let n be a positive integer. Is there a positive real number ε (depending on n) such that for any positive real numbers it holds:

$$\sqrt[n]{x_1x_2\ldots x_n}\leq(1-\varepsilon)\cdot\frac{x_1+x_2+\ldots+x_n}{n}+\varepsilon\cdot\frac{n}{\frac{1}{x_1}+\frac{1}{x_2}+\ldots+\frac{1}{x_n}}.$$

2.49 2020

480. (*Azerbaijan*). Let x, y, z be positive real numbers such that $x^4+y^4+z^4=1$. Determine with proof the minimum value of

$$\frac{x^3}{1-x^8}+\frac{y^3}{1-y^8}+\frac{z^3}{1-z^8}.$$

481. (*Brazil*). Let (a_n) be a sequence of positive real numbers such that for each positive integer n,

$$\frac{a_1+a_2+\ldots+a_n}{n}\geq\sqrt{\frac{a_1^2+a_2^2+\ldots+a_{n+1}^2}{n+1}}.$$

Prove that the sequence is constant.

482. (*China*). Let ω be a nth primitive root of unity. Given complex numbers a_1, a_2, …, a_n and p of them are nonzero. Let

$$b_k=\sum_{i=1}^{n}a_i\omega^{ki}$$

for $k=1,2,\ldots,n$. Prove that if $p>0$, then at least $\frac{n}{p}$ numbers in b_1, b_2, …, b_n are nonzero.

483. (*Estonia*). Find all functions $f:\mathbb{R}\to\mathbb{R}$ such that

$$f(x^3+y^3)=f(x^3)+3x^3f(x)f(y)+3f(x)(f(y))^2+y^6f(y),\forall x,y\in\mathbb{R}.$$

484. (*France*). Let n be a positive integer. Call a function $f:\mathbb{R}\to\mathbb{R}$ $n-$*positive* if for any real numbers $x_1,\ \ldots,\ x_n$ such that $x_1+\ \ldots\ +x_n=0$, it follows

$$f(x_1)+\ldots+f(x_n)\geq 0.$$

(a) Every function 2020-positive is also 1010-positive?
(b) Every function 1010-positive is also 2020-positive?

485. (*Greece*). Find all functions $f:(0,\infty)\to(0,\infty)$ such that

$$f(xf(y))+f(yf(z))+f(zf(x))=xy+yz+zx,\forall x,y,z\in(0,\ \infty).$$

486. (*Moldova*). Show that for any positive real numbers a, b, c the following inequality holds

$$\frac{a}{b}+\frac{b}{c}+\frac{c}{a}+\frac{a+b+c}{\sqrt{a^2+b^2+c^2}}\geq 3+\sqrt{3}.$$

487. (*Moldova*). Let n be a positive integer. Find all solutions $x\in\left[0,\frac{\pi}{4}\right]$ of the system

$$\begin{cases}\sin x+\cos x=\dfrac{\sqrt{n}}{2}\\ \tan\dfrac{x}{2}=\dfrac{\sqrt{n}-2}{2}\end{cases}.$$

488. (*Moldova*). Find all functions $f:[-1,1]\to\mathbb{R}$ which satisfy $f(\sin x)+f(\cos x)=2020$ for all real x.

489. (*Netherlands*). Find all functions $f:\mathbb{R}\to\mathbb{R}$ satisfying

$$f(x^2y)+2f(y^2)=(x^2+f(y))f(y),\forall x,y\in\mathbb{R}.$$

490. (*Netherlands*). Determine all polynomials $P(x)$ with real coefficients for which

$$P(x^2)+2P(x)=(P(x))^2+2.$$

491. (*Serbia, JBMO*). Given real numbers $a_1,a_2,\ \ldots,\ a_{101}$ from the interval $[-2,10]$ such that their sum is 0. Prove that the sum of their squares is smaller than 2020.

492. (*Switzerland*). Find all polynomials Q with integer coefficients such that every prime number p and any two positive a, b with $p|ab - 1$ satisfy $p|Q(a)Q(b) - 1$.

493. (*Switzerland*). Let a, b, c, d be positive real numbers such that $a + b + c + d = 1$. Prove that

$$\left(\frac{a^2}{a+b}+\frac{b^2}{b+c}+\frac{c^2}{c+d}+\frac{d^2}{d+a}\right)^5 \geq 5^5\left(\frac{ac}{27}\right)^2.$$

494. (*Taiwan*). Let $u_1, u_2, \ldots, u_{2019}$ be real numbers satisfying

$$u_1 + u_2 + \ldots + u_{2019} = 0, \;\; u_1^2 + u_2^2 + \ldots + u_{2019}^2 = 1.$$

Let $a = \min\{u_1, u_2, \ldots, u_{2019}\}$, $b = \max\{u_1, u_2, \ldots, u_{2019}\}$. Prove that

$$ab \leq -\frac{1}{2019}.$$

495. (*Thailand*). Let $n \leq 2$ be a positive integer and $a_1, a_2, \ldots, a_n$ be real numbers such that $a_1 + a_2 + \ldots + a_n = 0$. Define the set A by

$$A = \left\{(i,j) \middle| 1 \leq i < j \leq n, \left|a_i - a_j\right| \geq 1\right\}.$$

Prove that if $A \neq \varnothing$, then

$$\sum_{(i,j)\in A} a_i a_j < 0.$$

496. (*Turkey*). Let x, y, z be real numbers such that $0 < x, y, z < 1$. Find the minimum value of

$$\frac{xyz(x+y+z)+(xy+yz+zx)(1-xyz)}{xyz\sqrt{1-xyz}}.$$

2.50 2021

497. (*China*). Determine the greatest real number C such that for every integer $n \geq 2$, there exist $x_1, x_2, \ldots, x_n \in [-1, 1]$ so that

$$\prod_{1\leq i<j\leq n}(x_i - x_j) \geq C^{\frac{n(n-1)}{2}}.$$

498. (*Estonia*). Find all polynomials $P(x)$ with integer coefficients whose values at points $x = 1, 2, \ldots, 2021$ are numbers $1, 2, \ldots, 2021$ in some order.

499. (*Estonia*). Positive real numbers a, b, c satisfy $abc = 1$. Prove that

$$\frac{a}{1+b} + \frac{b}{1+c} + \frac{c}{1+a} \geq \frac{3}{2}.$$

500. (*Estonia*). Find all polynomials $P(x, y)$ with real coefficients which satisfy

$$P(x+y, x-y) = 2P(x, y), \forall x, y \in \mathbb{R}.$$

501. (*France*). Find all functions $f : \mathbb{N}^* \to \mathbb{N}$ satisfying:

1. $f(xy) = f(x) + f(y)$ for all positive integers x, y;
2. There exist infinitely many positive integers n such that $f(k) = f(n - k)$ for all $k \in \{1, 2, \ldots, n-1\}$.

502. (*Israel*). Which is greater

$$\frac{1^{-3} - 2^{-3}}{1^{-2} - 2^{-2}} - \frac{2^{-3} - 3^{-3}}{2^{-2} - 3^{-2}} + \frac{3^{-3} - 4^{-3}}{3^{-2} - 4^{-2}} - \ldots + \frac{2019^{-3} - 2020^{-3}}{2019^{-2} - 2020^{-2}}$$

or

$$1 - \frac{1}{3} + \frac{1}{5} - \frac{1}{7} + \frac{1}{9} - \ldots + \frac{1}{5781}?$$

503. (*Moldova*). Positive real numbers a, b, c satisfy $a + b + c = 1$. Show that

$$\frac{a+1}{\sqrt{a+bc}} + \frac{b+1}{\sqrt{b+ca}} + \frac{c+1}{\sqrt{c+ab}} \geq \frac{2}{a^2+b^2+c^2}.$$

When does the equality hold?

504. (*Netherlands*). Find all quadruplets (x_1, x_2, x_3, x_4) of real numbers such that the next six equalities apply:

$$\begin{cases} x_1 + x_2 = x_3^2 + x_4^2 + 6x_3x_4 \\ x_1 + x_3 = x_2^2 + x_4^2 + 6x_2x_4 \\ x_1 + x_4 = x_2^2 + x_3^2 + 6x_2x_3 \\ x_2 + x_3 = x_1^2 + x_4^2 + 6x_1x_4 \\ x_2 + x_4 = x_1^2 + x_3^2 + 6x_1x_3 \\ x_3 + x_4 = x_1^2 + x_2^2 + 6x_1x_2 \end{cases}.$$

505. (*Peru*). For any positive integer n, we define

$$S_n = \sum_{k=1}^{n} \frac{2^k}{k^2}.$$

Prove that there are no polynomials P, Q with real coefficients such that for any positive integer n,we have

$$\frac{S_{n+1}}{S_n} = \frac{P(n)}{Q(n)}.$$

506. (*Peru*). Suppose that positive real numbers x, y, z, w satisfy $(x^3 + y^3)^4 = z^3 + w^3$. Prove that $x^4z + y^4w \geq zw$.

507. (*Romania*). Let α be a real number in the interval (0, 1). Prove that there exists a sequence $(\varepsilon_n)_{n \geq 1}$, where each term is either 0 or 1 such that the sequence

$$s_n = \frac{\varepsilon_1}{n(n+1)} + \frac{\varepsilon_2}{(n+1)(n+2)} + \ldots + \frac{\varepsilon_n}{(2n-1)2n}$$

verifies the inequality

$$0 \leq \alpha - 2ns_n \leq \frac{2}{n+1}$$

for any $n \geq 2$.

508. (*Saudi Arabia*). Do there exist polynomials P, Q with integer coefficients such that

(i) Both P and Q have a coefficient with absolute value bigger than 2021,
(ii) All coefficients of PQ by absolute value are at most 1?

509. (*Switzerland*). Find all polynomials P with real coefficients and without multiple roots such that the equations $zP(z) - 1 = 0$ and $P(z - 1)P(z + 1) = 0$ have the same solutions $z \in \mathbb{C}$.

510. (*Taiwan*). Let n, k be positive integers. Prove that for any $a_1, a_2, \ldots, a_n \in [1, 2^k]$ one has

$$\sum_{i=1}^{n} \frac{a_i}{\sqrt{a_1^2 + \ldots + a_i^2}} \leq 4\sqrt{kn}.$$

511. (*Turkey*). For which positive integers n one can find real numbers $x_1, x_2, \ldots, x_n$ such that

$$\frac{x_1^2+x_2^2+\ldots+x_n^2}{(x_1+2x_2+\ldots+nx_n)^2}=\frac{27}{4n(n+1)(2n+1)}$$

and $i \le x_i \le 2i$ for all $i = 1, 2, \ldots, n$?

512. (*Vietnam*). Let a, b, c be nonnegative numbers such that

$$2(a^2+b^2+c^2)+3(ab+bc+ca)=5(a+b+c).$$

Prove that

$$4(a^2+b^2+c^2)+2(ab+bc+ca)+7abc\le 25.$$

2.51 2022

513. (*Brazil*). Prove that if a, b, c are nonnegative real numbers, then

$$a^5+b^5+c^5\ge 5abc(b^2-ac).$$

514. (*China*). Let $C = \{z \in \mathbb{C} : |z| = 1\}$ be the unit circle in the complex plane. Let $z_1, z_2, \ldots, z_{240} \in C$ (not necessarily different) be 240 complex numbers satisfying the conditions:

1. For any open arc Γ of length π on C, there are at most 200 of j ($1 \le j \le 240$) such that $z_j \in \Gamma$.
2. For any open arc γ of length $\frac{\pi}{3}$ on C, there are at most 120 of j ($1 \le j \le 240$) such that $z_j \in \gamma$.

Find the maximum of $|z_1 + z_2 + \ldots + z_{240}|$.

515. (*Cyprus*). Find all pairs (x, y) of real numbers for which

$$\begin{cases} x^2+y^2+xy=133 \\ x+y+\sqrt{xy}=19 \end{cases}.$$

516. (*Estonia*). Find all functions $f : \mathbb{R} \to \mathbb{R}$ that satisfy the following condition:

$$f(x)+f(x+y)\le f(xy)+f(y), \forall x,y\in\mathbb{R}.$$

517. (*France*). Find the smallest possible value of the sum

$$\left\lfloor\frac{a_1}{1}\right\rfloor+\left\lfloor\frac{a_2}{2}\right\rfloor+\ldots+\left\lfloor\frac{a_{2022}}{2022}\right\rfloor,$$

where $(a_1, a_2, \ldots, a_{2022})$ is a permutation of the numbers 1, 2, …, 2022.

(A permutation of a finite set with n elements is an ordered set $(\sigma_1, \sigma_2, \ldots, \sigma_n)$, where each element equals one element from the given set and $\sigma_i \neq \sigma_j$ for $i \neq j$).

518. (*Germany*). Let $n \geq 2$ be an integer and let a_1, a_2, …, a_n be positive real numbers with sum 1. Prove that

$$\sum_{k=1}^{n}\frac{a_k}{1-a_k}(a_1+a_2+\ldots+a_{k-1})^2<\frac{1}{3}.$$

519. (*Greece*). Find the largest constant M such that for any sequence $(a_n)_{n\geq 0}$ of real numbers that satisfies the conditions:

(i) $a_0 = 1, a_1 = 3,$
(ii) $a_0 + a_1 + \ldots + a_{n-1} \geq 3a_n - a_{n+1}$ for any integer $n \geq 1$,

to be true that

$$\frac{a_{n+1}}{a_n}>M$$

for any integer $n \geq 0$.

520. (*Hong Kong*). Let $P(x)$ be a polynomial with integer coefficients. Define a sequence (a_n) by $a_0 = 0$ and $a_n = P(a_{n-1})$ for all $n \geq 1$. Prove that if there exists a positive integer m for which $a_m = 0$, then $a_1 = 0$ or $a_2 = 0$.

521. (*Moldova*). Let $P(X)$ be a polynomial with positive coefficients. Show that for every integer $n \geq 2$ and every n positive numbers x_1, x_2, …, x_n the following inequality is true:

$$\left(P\left(\frac{x_1}{x_2}\right)\right)^2+\left(P\left(\frac{x_2}{x_3}\right)\right)^2+\ldots+\left(P\left(\frac{x_n}{x_1}\right)\right)^2\geq n(P(1))^2.$$

When does the equality hold?

522. (*Netherlands*). For real numbers x, y we define $M(x, y)$ to be the maximum of the three numbers xy, $(x-1)(y-1)$, $x+y-2xy$. Determine the smallest positive value of $M(x, y)$, where x, y range over all real numbers satisfying $0 \leq x, y \leq 1$.

523. (*Romania*). Given positive real numbers such that

$$\sum_{i=1}^{n} \frac{1}{1+x_i^2} = 1.$$

Find the minimal value of the expression

$$\frac{\sum x_i}{\sum \frac{1}{x_i}}$$

and find when it is achieved.

524. (*Saudi Arabia*). Consider the polynomial $f(x) = cx(x-2)$, where c is a positive real number. For any positive integer n, denote by $g_n(x)$ the function composite n times of f and assume that all the 2^n solutions of the equation $g_n(x) = 0$ are real numbers.

1. For $c = 5$, find in terms of n, the sum of all solutions of $g_n(x)$, of which each multiple (if any) is counted only once.
2. Prove that $c \geq 1$.

525. (*Serbia, JBMO*). Prove that for all positive real numbers a, b the following inequality holds:

$$\sqrt{\frac{a^2+b^2}{2}} + \frac{2ab}{a+b} \geq \frac{a+b}{2} + \sqrt{ab}.$$

When does equality hold?

526. (*Switzerland*). Let a, b, c, λ be positive real numbers with $\lambda \geq \frac{1}{4}$. Show that

$$\frac{a}{\sqrt{b^2+\lambda bc+c^2}} + \frac{b}{\sqrt{c^2+\lambda ca+a^2}} + \frac{c}{\sqrt{a^2+\lambda ab+b^2}} \geq \frac{3}{\sqrt{\lambda+2}}.$$

527. (*Switzerland*). Let n be a positive integer. Find all polynomials P with real coefficients such that $P(x^2 + x - n^2) = P(x)^2 + P(x)$.

528. (*Turkey*). What is the minimum value of the expression?

$$xy + yz + zx + \frac{1}{x} + \frac{2}{y} + \frac{5}{z},$$

where x, y, z are positive real numbers.

2.52 2023

529. (*Azerbaijan*). Let $k \geq 2$ be an integer. Find the smallest integer $n \geq k + 1$ with the property that there exists a set of n real numbers such that each of its elements can be written as a sum of k other distinct elements of the set.

530. (*Brazil*). Show that for all positive real numbers a, b, c we have that

$$\frac{a+b+c}{3} - \sqrt[3]{abc} \leq \max\left(\left(\sqrt{a}-\sqrt{b}\right)^2, \left(\sqrt{b}-\sqrt{c}\right)^2, \left(\sqrt{c}-\sqrt{a}\right)^2\right).$$

531. (*China*). Given integer $n \geq 2$. Find the minimum value of λ satisfying that for any real numbers $a_1, a_2, \ldots, a_n, b$

$$\lambda \sum_{i=1}^{n} \sqrt{a_i - b} + \sqrt{n \left| \sum_{i=1}^{n} a_i \right|} \geq \sum_{i=1}^{n} \sqrt{|a_i|}.$$

532. (*Estonia*). For any natural number n and positive integer k, we say that n is – *good* if there exist nonnegative integers $a_1, \ldots, a_k$ such that

$$n = a_1^2 + a_2^4 + a_3^8 + \ldots + a_k^{2^k}.$$

Is there a positive integer k for which every natural number is k-good?

533. (*Germany*). Let P be a polynomial with integer coefficients. Assume that there exists a positive integer n with $P(n^2) = 2022$. Prove that there is no positive rational number r with $P(r^2) = 2024$.

534. (*Hong Kong*). Let x, y, z be real numbers such that $x + y + z \neq 0$. Find the minimum value of

$$\frac{|x| + |x+4y| + |y+7z| + 2|z|}{|x+y+z|}.$$

535. (*Hong Kong*). Suppose that a, b, c are real numbers satisfying $abc = 2$. Prove that at most two of the numbers

$$2a - \frac{1}{b}, 2b - \frac{1}{c}, 2c - \frac{1}{a}$$

are >2.

536. (*India*). Prove that for all integers $k > 2$, there exist k positive integers $a_1, \ldots, a_k$ such that

$$\sum_{1\le i<j\le k} \frac{1}{a_i a_j} = 1.$$

537. (*Malaysia*). Let a_1, a_2, ..., a_n be a sequence of real numbers with $a_1 + a_2 + \ldots + a_n = 0$. Define the score $S(\sigma)$ of a permutation $\sigma = (b_1, \ldots, b_n)$ of $(a_1, \ldots, a_n)$ to be the minima of the sum $(x_1 - b_1)^2 + \ldots + (x_n - b_n)^2$ over all real numbers $x_1 \le \ldots \le x_n$.

Prove that $S(\sigma)$ attains the maxima over all permutations σ if and only if $b_1 + b_2 + \ldots + b_k \ge 0$ for all $1 \le k \le n$.

538. (*Malaysia*). For which $n \ge 3$ does there exist positive integers $a_1 < a_2 < \ldots < a_n$ such that

$$a_n = a_1 + \ldots + a_{n-1}, \ \frac{1}{a_1} = \frac{1}{a_2} + \ldots + \frac{1}{a_n}?$$

539. (*Moldova*). Polynomials $(P_n(X))_{n \in N}$ are defined as

$$P_0(X) = 0, P_1(X) = X + 2, P_n(x) = P_{n-1}(X) + 3P_{n-1}(X)P_{n-2}(X) + P_{n-2}(x), \forall n \ge 2.$$

Show that if k divides m, then $P_k(X)$ divides $P_m(X)$.

540. (*Myanmar*). Find real numbers a, b, c that satisfy

$$2a - b = a^2 b, 2b - c = b^2 c, 2c - a = c^2 a.$$

541. (*North Macedonia*). Let $f : \mathbb{R}^2 \to \mathbb{R}$ be a function satisfying the following property: if A, B, $C \in \mathbb{R}^2$ are the vertices of an equilateral triangle with sides of length 1, then $f(A) + f(B) + f(C) = 0$.

Show that $f(x) = 0$ for all $x \in \mathbb{R}^2$.

542. (*North Macedonia*). Let $(a_n)_{n \ge 1}$ be the sequence defined by

$$a_1 = 1, a_2 = 2, \frac{a_{n+1}^4}{a_n^3} = 2a_{n+2} - a_{n+1}, \forall n \ge 1.$$

Prove that the following inequality holds for every positive integer $n \ge 1$:

$$\sum_{k=1}^{n} \frac{a_k^2}{a_{k+1}} < 3.$$

543. (*SAFEST*). Find all polynomials $P(X)$ with integer coefficients such that for all positive integers m, n,we have $m + n | P^{(m)}(n) - P^{(n)}(n)$.

(Here $P^{(k)}(x)$ denotes the k-fold composition of P, that is $P^{(k)}(x) = \underbrace{P(P(\ldots P(x)\ldots))}_{k \text{ times}}$).

544. (*Serbia*). Let p be a prime number and $P \in \mathbb{R}[X]$ be a polynomial of degree $<p-1$ such that $|P(1)| = |P(2)| = \ldots = |P(p)|$. Prove that P is constant.

545. (*Switzerland*). Find all monic polynomials $P(x) = x^{2023} + a_{2022}x^{2022} + \ldots + a_1x + a_0$ with real coefficients, such that $a_{2022} = 0$, $P(1) = 1$, and all roots of P are real and <1.

546. (*Turkey, EGMO*). Let n be a positive integer and P, Q be polynomials with real coefficients satisfying $P(x) = x^n Q\left(\frac{1}{x}\right)$ and $P(x) \geq Q(x)$ for all real numbers x.

Prove that $P(x) = Q(x)$ for all real numbers x.

2.53 2024

547. (*Belarus*). Prove that for any real numbers $a, b, c, d \geq \frac{1}{3}$, the following inequality holds:

$$\frac{a^6}{b^4+c^3} + \frac{b^6}{c^4+d^3} + \frac{c^6}{d^4+a^3} + \frac{d^6}{a^4+b^3} \geq \left(\frac{a+b+c+d}{4}\right)^2.$$

548. (*China*). Let M be a positive integer and let $f(x) = x^3 + ax^2 + bx + c$ be a polynomial with integer coefficients such that $|a|, |b|, |c| \leq M$. Prove that if x_1, x_2 are distinct roots of f, then

$$|x_1 - x_2| > \frac{1}{M^2 + 3M + 1}.$$

549. (*China*). Find all functions $f : \mathbb{N}_1 \to \mathbb{N}_1$ such that for all positive integers a, b,

$$\sum_{k=0}^{2b} f(a+k) = (2b+1)f(f(a)+b).$$

550. (*China*). Let $m, n \in \mathbb{Z}_{\geq 0}$, $a_0, a_1, \ldots, a_m, b_0, b_1, \ldots, b_n \in \mathbb{R}_{\geq 0}$. For any integer $0 \leq k \leq m+n$ define

$$c_k := \max_{i+j=k} a_i b_j.$$

Prove that

$$\frac{1}{m+n+1}\sum_{k=0}^{m+n} c_k \geq \frac{1}{(m+1)(n+1)}\left(\sum_{i=0}^{m} a_i\right)\left(\sum_{j=0}^{n} b_j\right).$$

551. (*France*). Define the sequence (a_n) by

$$a_1 = 1, a_n = \frac{1}{n} + \frac{1}{a_1 \ldots a_{n-1}}, n \geq 2.$$

Prove that $a_m \leq 1$ for any positive integer $m \geq 3$.

552. (*Hong Kong*). Let $n > 1$ be a positive integer and let $a_0, a_1, \ldots, a_{n-1}$ be integers. It is known that the polynomial $x^n + a_{n-1}x^{n-1} + a_{n-2}x^{n-2} + \ldots + a_1x + a_0$ has n pairwise prime integer roots. Prove that a_0 and a_1 are relatively prime.

553. (*India*). Let $x_1, x_2, \ldots, x_{2024}$ be nonnegative real numbers such that $x_1 \leq x_2 \leq \ldots \leq x_{2024}$ and $x_1^3 + x_2^3 + \ldots + x_{2024}^3 = 2024$. Prove that

$$\sum_{1 \leq i < j \leq 2024} (-1)^{i+j} x_i^2 x_j \geq -1012.$$

554. (*Indonesia*). Given real numbers x, y, z which satisfy

$$|x + y + z| + |xy + yz + zx| + |xyz| \leq 1.$$

Show that $\max\{|x|, |y|, |z|\} \leq 1$.

555. (*Indonesia*). Find all second degree polynomials P such that for all $a \in [1, \infty)$, it holds

$$P(a^2 + a) \geq aP(a+1).$$

556. (*Malaysia*). Find all functions $f: \mathbb{Z} \to \mathbb{Z}$ such that for all integers x, y,

$$f(x - f(y)) = f(f(y)) + f(x - 2y).$$

557. (*Netherlands*). Let a, b, c be real numbers such that $0 \leq a \leq b \leq c$ and $a + b + c = = 1$. Show that

$$ab\sqrt{b-a} + bc\sqrt{c-b} + ac\sqrt{c-a} < \frac{1}{4}.$$

558. (*Serbia*). Does there exist a positive integer n and (a) complex numbers $a_0, a_1, \ldots, a_n$; (b) real numbers $a_0, a_1, \ldots, a_n$ such that $P(x)Q(x) = x^{2024} + 1$, where

$$P(x) = a_n x^n + \ldots + a_1 x + a_0, Q(x) = a_0 x^n + a_1 x^{n-1} + \ldots + a_n?$$

559. (*Switzerland*). Find all monic polynomials P with integer coefficients such that for all integers a, b there exists an integer c such that $P(a)P(b) = P(c)$.

560. (*Taiwan*). Let $\mathbb{R}_+$ be the set of positive real numbers. Find all functions $f : \mathbb{R}_+ \to \mathbb{R}_+$ such that

$$f(xy + x + y) + f\left(\frac{1}{x}\right) f\left(\frac{1}{y}\right) = 1$$

for every $x, y \in \mathbb{R}_+$.

561. (*Turkey, EGMO*). Let $(a_n)_{n \in N^*}$ be a strictly increasing sequence such that the inequality

$$a_n(a_n - 2a_{n-1}) + a_{n-1}(a_{n-1} - 2a_{n-2}) \geq 0$$

holds for all $n \geq 3$. Prove that for all $n \geq 2$ the inequality

$$a_n \geq a_{n-1} + a_{n-2} + \ldots + a_1$$

holds as well.

562. (*Vietnam*). Let $P(x) \in \mathbb{Z}[x]$ be a polynomial. Determine all polynomials $Q(x) \in \mathbb{Z}[x]$ such that for every positive integer n, there exists a polynomial $R_n(x) \in \mathbb{Z}[x]$ that satisfies

$Q(x)^{2n} - 1 = R_n(x)(P(x)^{2n} - 1)$.

Chapter 3
Solutions

In this chapter, the reader will find detailed solutions for each of the problems presented in Chap. 1. The introduction can be used to recap the theory if needed. For topical access, please use the index at the end of the book.

3.1 1968

1. (a) We have

$$S(x,n)=\left(\sum_{p=1}^{n}(x+p)\right)^2-\sum_{p=1}^{n}(x+p)^2=\left(nx+\frac{n(n+1)}{2}\right)^2$$
$$-\left[nx^2+2x\cdot\frac{n(n+1)}{2}+\frac{n(n+1)(2n+1)}{6}\right]$$
$$=n(n-1)\left[x^2+(n+1)x+\frac{(n+1)(3n+2)}{12}\right].$$

(b) Answer: yes.
The equation $S(x,n)=0$ is quadratic, with discriminant

$$\Delta=(n+1)^2-4\cdot\frac{(n+1)(3n+2)}{12}=\frac{n+1}{3}.$$

The roots are integers if and only if $n+1=3a^2$, $a\in\mathbb{N}^*$. In this case,

$$x_{1,2}=\frac{-3a^2\pm a}{2}.$$

C. Mănescu-Avram, *Selection Tests in Algebra for Mathematical Olympiads*,
Problem Books in Mathematics, https://doi.org/10.1007/978-3-032-01032-2_3

3.2 1969

2. Let $b_i = a_i c_i$, $1 \le i \le n$. Then $c_1 c_2 \ldots c_i \ge 1$, $1 \le i \le n$. By the generalized AM–GM inequality,

$$\frac{b_1+b_2+\ldots+b_n}{a_1+a_2+\ldots+a_n}=\frac{a_1c_1+a_2c_2+\ldots+a_nc_n}{a_1+a_2+\ldots+a_n}\ge\left(c_1^{a_1}c_2^{a_2}\ldots c_n^{a_n}\right)^{\frac{1}{a_1+a_2+\ldots+a_n}}$$
$$=\left[c_1^{a_1-a_2}(c_1c_2)^{a_2-a_3}\ldots(c_1c_2\ldots c_{n-1})^{a_{n-1}-a_n}(c_1c_2\ldots c_n)^{a_n}\right]^{\frac{1}{a_1+a_2+\ldots+a_n}}\ge 1.$$

3.3 1972

3. We will use induction on n. For $n = 1$, the statement is obvious. Assume that it holds for k. From the given condition, we have

$$\sum_{j=1}^{k}\left|a_{j1}x_1+\ldots+a_{jk}x_k+a_{j,k+1}\right|+\left|a_{k+1,1}x_1+\ldots+a_{k+1,k}x_k+a_{k+1,k+1}\right|\le M,$$

$$\sum_{j=1}^{k}\left|a_{j1}x_1+\ldots+a_{jk}x_k-a_{j,k+1}\right|+\left|a_{k+1,1}x_1+\ldots+a_{k+1,k}x_k-a_{k+1,k+1}\right|\le M,$$

for each choice of $x_i = \pm 1$. Since $|x + y| + |x - y| \ge 2|x|$ for all x, y, we get

$$2\sum_{j=1}^{k}\left|a_{j1}x_1+\ldots+a_{jk}x_k\right|+2\left|a_{k+1,k+1}\right|\le 2M\Rightarrow\sum_{j=1}^{k}\left|a_{j1}x_1+\ldots+a_{jk}x_k\right|\le M-\left|a_{k+1,k+1}\right|.$$

By the induction hypothesis,

$$\sum_{j=1}^{k}\left|a_{jj}\right|\le M-\left|a_{k+1,k+1}\right|,$$

so the statement holds also for $k + 1$, hence it holds for all n.

3.4 1975

4. Suppose that $\frac{1}{2} \in A$, then there exists a positive integer n for which $a_{n+1} = \frac{1}{2}$, so

$$b_{n+1} = b_n\left(1 - \frac{1}{2}\right) + (1 - b_n)\frac{1}{2} = \frac{1}{2} \in B.$$

Suppose that $\frac{1}{2} \in B$, then there exists a positive integer n for which$b_{n+1} = \frac{1}{2}$, so

$$\frac{1}{2} = b_n(1 - a_{n+1}) + (1 - b_n)a_{n+1} \Rightarrow (2a_{n+1} - 1)(2b_n - 1) = 0.$$

If $b_n \neq \frac{1}{2}$, then $a_{n+1} = \frac{1}{2} \in A$. If $b_n = \frac{1}{2}$, then we write the equality for $n-1$, etc. If there exists n_0 with $b_{n_0+1} = \frac{1}{2}, b_{n_0} \neq \frac{1}{2}$, then $a_{n_0+1} = \frac{1}{2} \in A$. If $b_{n_0} = \frac{1}{2}$ for any $n_0 \leq n$, then $b_1 = \frac{1}{2}$. But $b_1 = a_1$, so $a_1 = \frac{1}{2} \in A$.

5. For $x = y = 0$, we get $f(0) = -1$. For $x = 0$, we obtain $f(y) = f(-y)$, so f is an even function. Choose $x = ky$, $k \in \mathbb{N}$, then

$$f((k+1)y) - f(ky) = f(ky) - f((k-1)y) + 2(f(y) + 1).$$

Adding these equalities for $k = 1, 2, \ldots, p - 1$, we deduce

$$f(py) - f((p-1)y) = (2p - 1)(f(y) + 1).$$

Adding now for $p = 1, 2, \ldots, n$, we have

$$f(ny) = n^2(f(y) + 1) - 1 \Rightarrow f(y) = \frac{f(ny) + 1}{n^2} - 1 \Rightarrow f\left(\frac{1}{n}\right) = \frac{f(1) + 1}{n^2} - 1,$$

$$f\left(\frac{m}{n}\right) = f\left(m \cdot \frac{1}{n}\right) = m^2\left(f\left(\frac{1}{n}\right) + 1\right) - 1 = \left(\frac{m}{n}\right)^2 (f(1) + 1) - 1.$$

The solutions are $f(x) = ax^2 - 1$, with $a \in \mathbb{Q}$ arbitrary.

6. Suppose that

$$\prod_{i=1}^{n} \sin x_i \cos x_i \neq 0.$$

Consider the equation $\sin x_1 \cos x_2 = \sin x_2 \cos x_3$. The following implications hold:

$$x_1 = x_2 \Rightarrow \sin x_1 = \sin x_2 \Rightarrow \cos x_2 = \cos x_3 \Rightarrow x_2 = x_3$$

and we can continue to get $x_1 = x_2 = \ldots = x_n$, whence $x_i = \frac{\pi}{n}, 1 \leq i \leq n$. If $x_1 < x_2$, then $\sin x_1 < \sin x_2 \Rightarrow \cos x_2 > \cos x_3 \Rightarrow x_2 < x_3$. Repeating, we get

$$x_1 < x_2 < x_3 < \ldots < x_n < x_1,$$

contradiction.

3.5 1976

7. Denote $s = x_1 + x_2 + x_3 + x_4$, then the system becomes

$$\begin{cases} x_1^2 - sx_1 + a = 0 \\ x_2^2 - sx_2 + a = 0 \\ x_3^2 - sx_3 + a = 0 \\ x_4^2 - sx_4 + a = 0 \end{cases},$$

that is, x_1, x_2, x_3, x_4 are solutions of the equation $z^2 - sz + a = 0$. These roots are

$$z_{1,2} = \frac{s \pm \sqrt{s^2 - 4a}}{2}.$$

Case 1. $x_1 = x_2 = x_3 = x_4 = \frac{s - \sqrt{s^2 - 4a}}{2}$. Then

$$x_1 + x_2 + x_3 + x_4 = s \Leftrightarrow s = 2\sqrt{s^2 - 4a} \Leftrightarrow 3s^2 = 16a \Leftrightarrow s = 4\sqrt{\frac{a}{3}}.$$

The solution is $(x_1, x_2, x_3, x_4) = \left(\sqrt{\frac{a}{3}}, \sqrt{\frac{a}{3}}, \sqrt{\frac{a}{3}}, \sqrt{\frac{a}{3}}\right)$.

Case 2. $x_1 = x_2 = x_3 = \dfrac{s - \sqrt{s^2 - 4a}}{2}, x_4 = \dfrac{s + \sqrt{s^2 - 4a}}{2}$. Then

$$x_1 + x_2 + x_3 + x_4 = s \Leftrightarrow s = \sqrt{s^2 - 4a} \Leftrightarrow a = 0.$$

If $a \neq 0$, then the system has no solution. If $a = 0$, then we have the solutions

$$(s, 0, 0, 0), (0, s, 0, 0), (0, 0, s, 0), (0, 0, 0, s)$$

with s arbitrary.

Case 3. $x_1 = x_2 = \dfrac{s - \sqrt{s^2 - 4a}}{2}, x_3 = x_4 = \dfrac{s + \sqrt{s^2 - 4a}}{2}$. Then

$$x_1 = x_2 = x_3 = x_4 = s \Leftrightarrow s = 0$$

and we have the solution $(-\sqrt{-a}, -\sqrt{-a}, \sqrt{-a}, \sqrt{-a})$ and all permutations.

Case 4. $x_1 = \dfrac{s-\sqrt{s^2-4a}}{2}, x_2 = x_3 = x_4 = \dfrac{s+\sqrt{s^2-4a}}{2}$. Same solutions as in Case 2.

Case 5. $x_1 = x_2 = x_3 = x_4 = \dfrac{s+\sqrt{s^2-4a}}{2}$. The system has the unique solution

$$\left(-\sqrt{\frac{a}{3}}, -\sqrt{\frac{a}{3}}, -\sqrt{\frac{a}{3}}, -\sqrt{\frac{a}{3}}\right).$$

8. We will prove that $1 \le f(x, y, z, t) \le 2$ for all x, y, z, t as in the enounce. Equality holds on the left only when $x = y = z = t = \frac{1}{2}$, and on the right only when $x = t = 1,\ y = z = 0$.

By the Cauchy–Schwarz inequality,

$$\left(ax^2 + by^2\right)(a+b) \ge (ax+by)^2,$$

consequently,

$$f(x, y, z, t) \ge \frac{ax+by}{a+b} + \frac{az+bt}{a+b} = \frac{a(x+z)+b(y+t)}{a+b} = 1,$$

which proves the first inequality.

Notice that

$$\frac{ax^2+by^2}{ax+by} \le \max\{x, y\},$$

hence

$$f(x, y, z, t) \le \max\{x, y\} + \max\{z, t\} \le \max\{x, y\} + \max\{1-x, 1-y\}.$$

If $x \ge y$, then $f(x, y, z, t) \le x + 1 - y \le 2$, as $x - y \le 1$. The case $x < y$ is similar, which ends the proof.

3.6 1977

9. Let $(z_1, \ldots, z_n)$ be a permutation for which

$$\sum_{i=1}^{n} (x_i - z_i)^2$$

is minimal. Then for any $i < j$ for $x_i \neq x_j$ we have $z_i \geq z_j$, which solves the problem. Indeed, otherwise

$$\begin{aligned}(x_i - z_j)^2 + (x_j - z_i)^2 &= (x_i - z_i)^2 + (x_j - z_j)^2 + 2(x_i z_i + x_j z_j - x_i z_j - x_j z_i)\\ &= (x_i - z_i)^2 + (x_j - z_j)^2 + 2(x_i - x_j)(z_i - z_j) \leq (x_i - z_i)^2 + (x_j - z_j)^2\end{aligned}$$

contradicts the minimality.

10. We will prove that $x + y = 1$ implies $P(x, y) = x - 2y$. Indeed, in this case $P(x, y) = P(x, 1 - x)$ is a polynomial in $z = x - 2y = 3x - 2 : P(x, 1 - x) = F(z)$. Then for $a = b = \frac{x}{2}, c = 1 - x$, we get $P(x, 1 - x) + 2P\left(1 - \frac{x}{2}, \frac{x}{2}\right) = 0$, so that

$$F(z) + 2F\left(-\frac{z}{2}\right) = 0.$$

We have $F(1) = 1$ and $F((-2)^k) = (-2)^k$ for all positive integers k, hence $F(z) = z$, which proves that $P(x, y) = x - 2y$, if $x + y = 1$.

For $x + y \neq 0$, since P is homogenous, we have

$$\begin{aligned}P(x, y) &= (x + y)^n P\left(\frac{x}{x + y}, \frac{y}{x + y}\right) = (x + y)^n \left(\frac{x}{x + y} - 2 \cdot \frac{y}{x + y}\right)\\ &= (x + y)^{n-1}(x - 2y).\end{aligned}$$

The same equality holds also for $x + y = 0$, since P is a polynomial.

11. Consider the function

$$f : [1, \infty) \to \mathbb{R}, f(x) = x + \frac{2}{x},$$

then $f'(x) = 1 - \frac{2}{x^2}$, so f is decreasing on $[1, \sqrt{2}]$ and increasing on $[\sqrt{2}, \infty)$. Since f is continuous, we have

$$f\left(\left[1, \sqrt{2}\right]\right) = \left[f\left(\sqrt{2}\right), f(1)\right] = \left[2\sqrt{2}, 3\right], f\left(\left[\sqrt{2}, 2\right]\right) = \left[f\left(\sqrt{2}\right), f(2)\right] = \left[2\sqrt{2}, 3\right].$$

Moreover, $f((2, \infty)) = (3, \infty)$. Then $f(x) = 3$ if and only if $x \in \{1, 2\}$. By the AM–GM inequality,

$$\sqrt[3]{\left(\sum_{i=1}^{n} x_i\right)\left(\sum_{i=1}^{n} \frac{1}{x_i}\right)^2} \le \frac{1}{3}\left(\sum_{i=1}^{n}\left(x_i + \frac{2}{x_i}\right)\right),$$

with equality only if

$$x_1 + x_2 + \ldots + x_n = \frac{1}{x_1} + \frac{1}{x_2} + \ldots + \frac{1}{x_n}.$$

But $x_i \in [1, 2]$, $1 \le i \le n$ and so

$$\sum_{i=1}^{n} x_i \ge n, \sum_{i=1}^{n} \frac{1}{x_i} \le n,$$

whence $x_i = 1$, $1 \le i \le n$. Now, we have

$$x_i \in [1,2], 1 \le i \le n \Rightarrow x_i + \frac{2}{x_i} \le 3, 1 \le i \le n \Rightarrow \sum_{i=1}^{n}\left(x_i + \frac{2}{x_i}\right)$$

$$\le 3n \Rightarrow \sqrt[3]{\left(\sum_{i=1}^{n} x_i\right)\left(\sum_{i=1}^{n} \frac{1}{x_i}\right)^2} \le n \Rightarrow \left(\sum_{i=1}^{n} x_i\right)\left(\sum_{i=1}^{n} \frac{1}{x_i}\right)^2 \le n^3,$$

with equality only for $x_1 = x_2 = \ldots = x_n = 1$.
Generalization. If $2 < a < b$ and $x_i \in [a, b]$, $1 \le i \le n$, then

$$\left(\sum_{i=1}^{n} x_i\right)\left(\sum_{i=1}^{n} \frac{1}{x_i}\right)^2 \le \frac{bn^3}{a^2}.$$

12. Answer: $\mathbb{R} - \mathbb{Q}$.
   An equivalent proposition is:

$$\forall c > 0, \exists n \in \mathbb{N}^* \text{ such that } \{\alpha n\} \in (0, c) \cup (1 - c, 1).$$

Denote by A the set of numbers α with the required property. If $\alpha \notin \mathbb{Q}$, then the sequence $\{n\alpha\}_{n\ge 1}$ is dense in $(0, 1)$, so $\alpha \in A$. If $\alpha = \frac{m}{n} \in \mathbb{Q}^*$, then we can choose $c = \frac{1}{n}$ and observe that

$$\{k\alpha\} \notin \left(0, \frac{1}{n}\right) \cup \left(1 - \frac{1}{n}, 1\right).$$

If $\alpha = 0$, then $|m| < c$ is clearly false for $c < 1$, so $0 \notin A$.

3.7 1978

13. The system $x + y = a$, $x^2 + y^2 = b$ has positive solutions, if $b < a^2 \leq 2b$, whence it follows that f is constant on the interval $(b, 2b]$, for any $b > 0$. Since the interval $(0, \infty)$ can be covered by such intervals, we deduce that f is constant.
14. Define

$$f(x) = \frac{1}{x - \lfloor x \rfloor}$$

for $x > 0$ not an integer. The continued fraction of $f(x)$ is the continud fraction of x shifted by 1, so if the sequence $x_{n+1} = f(x_n)$ is periodic, then the continued fraction of x_0 is periodic, hence x_0 is a quadratic irrational, that is, x_0 is a root of a quadratic polynomial irreducible over $\mathbb{Z}[X]$.

Conversely, if x_0 is a quadratic irrational, then x_0 has a continued fraction which is periodic from some point, i.e., the sequence $(x_n)_{n \geq 0}$ is eventually periodic.
15. We can take $k = 0$.
16. Let f: $\mathbb{N} \to \mathbb{N}$ defined inductively by

$$f(4k+1) = 2(4k+3), f(4k+3) = 4k+1, f(2k) = 2f(k), k \in \mathbb{N}.$$

Then $f(f(n)) = 2n$ for any $n \in \mathbb{N}$, and we can define

$$F(0) = 0, F(1) = 1, F(n) = p_1^{f(\alpha_1)} \cdots p_k^{f(\alpha_k)}$$

using the canonical representation as product of primes

$$n = p_1^{\alpha_1} \cdots p_k^{\alpha_k}.$$

3.8 1979

17. Using the Viète's formula, we get

$$n! \leq x_1 \ldots x_n = \frac{n+1}{n!} \leq (n+1)!,$$

which is true only for $n = 1$. Then $P_1(x) = x - 2$.
18. For $n = 1, 2$ the inequality is false. For $n = 3$,

$$\sum_{\text{sym}} a_1^2 a_2 \geq \sum_{\text{sym}} a_1 a_2 a_3,$$

which is true. We will prove that the inequality is false for any $n \geq 4$. Take

$$a_1 = \ldots = a_n = a,$$

then $(na^2)(na) \geq 6a^n$ if and only if $n^2 - n \geq 6a^{n-3}$. It is possible to choose a such that RHS is greater than LHS, contradiction.

19. For $|x| \leq 1$, we have the expansion

$$\frac{1}{x-4} = -\frac{1}{4(1-\frac{x}{4})} = -\frac{1}{4}\left(1 + \frac{x}{4} + \frac{x^2}{16} + \frac{x^3}{64} + \cdots\right).$$

Choose

$$P = \frac{1}{4}\left(1 + \frac{1}{4}X + \frac{1}{16}X^2\right),$$

then

$$P(x) + \frac{1}{x-4} = -\frac{1}{4} \cdot \frac{\frac{x^3}{64}}{1-\frac{x}{4}} = \frac{x^3}{64(x-4)}.$$

Since $|x^3| \leq 1$ and $|x-4| \geq 3$, it follows that

$$\left|P(x) + \frac{1}{x-4}\right| \leq \frac{1}{64 \cdot 3} < 0,01$$

for $|x| \leq 1$.

The answer to the second question is negative: suppose that there are real numbers a, b such that

$$\left|ax + b + \frac{1}{x-4}\right| \leq 0,01$$

for $|x| \leq 1$. Choose $x = 0,\ 1,\ -1$, respectively, then

$$\left|b - \frac{1}{4}\right| \leq 0,01 \Leftrightarrow -0,01 + \frac{1}{4} \leq b \leq 0,01 + \frac{1}{4},$$

$$\left|a + b - \frac{1}{3}\right| \leq 0,01,$$

$$\left|-a+b-\frac{1}{5}\right|\leq 0,01,$$

which is impossible.

20. Answer: $P \in \mathbb{R}[X]$, deg $(P) \leq 1$.

Let $P = a_mX^m + a_{m-1}X^{m-1} + \ldots + a_0$, $a_m \neq 0$, the desired polynomial. Then

$$P\left(x+\frac{1}{n}\right)=a_m\left(x+\frac{1}{n}\right)^m+a_{m-1}\left(x+\frac{1}{n}\right)^{m-1}+a_0=a_mx^m+\left(a_m\cdot\frac{1}{n}+a_{m-1}\right)x^{m-1}$$
$$+\left(a_m\cdot\frac{m(m-1)}{2n^2}+a_{m-1}\cdot\frac{m-1}{n}+a_{m-2}\right)x^{m-2}+P_1(x),P\left(x-\frac{1}{n}\right)$$
$$=a_mx^m+\left(-a_m\cdot\frac{1}{n}+a_{m-1}\right)x^{m-1}$$
$$+\left(a_m\cdot\frac{m(m-1)}{2n^2}-a_{m-1}\cdot\frac{m-1}{n}+a_{m-2}\right)x^{m-2}++P_2(x),$$

where P_1, P_2 are polynomials of degree at most $m - 3$. Using the enounce, we have

$$a_mx^m+a_{m-1}x^{m-1}+a_{m-2}x^{m-2}+\frac{a_m\cdot m(m-1)}{2n^2}x^{m-2}+P_3(x)=a_mx^m+a_{m-1}x^{m-1}$$
$$+a_{m-2}x^{m-2}+P_4(x)\Leftrightarrow\frac{a_m\cdot m(m-1)}{n^2}x^{m-2}+P_5(x)=0,\forall x\in\mathbb{Q},$$

where P_3, P_4, P_5 are polynomials of degree at most $m - 3$. The last equality holds for infinitely many values of x, therefore the polynomial from the left is the zero polynomial, hence $m = 0$ or $m = 1$.

21. We will use induction on n.

The base case is obvious. Suppose that the given inequality is true for n and prove it for $n + 1$:

$$a_1^3+\ldots+a_n^3+a_{n+1}^3\geq(a_1+\ldots+a_n+a_{n+1})^2. \quad (1)$$

Without loss of generality, assume that $a_1 < a_2 < \ldots < a_{n+1}$. The inequality (1) is equivalent to

$$a_1^3+\ldots+a_n^3+a_{n+1}^3\geq(a_1+\ldots+a_n)^2+2(a_1+\ldots+a_n)a_{n+1}+a_{n+1}^2.$$

By the induction hypothesis,

$$a_1^3+\ldots+a_n^3\geq(a_1+\ldots+a_n)^2,$$

so it suffices to show that $a_{n+1}^3\geq 2(a_1+\ldots+a_n)a_{n+1}+a_{n+1}^2$, or,

$$a_{n+1}^2 \geq 2(a_1 + \ldots + a_n) + a_{n+1}. \quad (2)$$

Since $a_{n+1} \geq a_n + 1 \geq \ldots \geq a_1 + n \geq n + 1$, the RHS of (2) does not exceed

$$(2n+1)a_{n+1} + n(n+1),$$

which is less than or equal to the LHS of (2), which reduces to

$$(a_{n+1} - n)(a_{n+1} - n - 1) \geq 0,$$

ending the proof.

3.9 1980

22. Denote by S the given sum and consider the product

$$\prod_{i=1}^{n}\left(1 + \frac{1}{i}\right) = \frac{(n+1)!}{n!} = n + 1.$$

Expanding the product, we see that the sum obtained contains all terms of S and 1, so that $S + 1 = n + 1$, whence $S = n$, as desired.

23. By the triangle's inequality,

$$m - 1 \geq \sum_{i=1}^{m-1} \left|a_{(i+1)k} - a_{ik} - a_k\right| \geq |a_{km} - ma_k|, k - 1 \geq |a_{km} - ka_m|,$$

$$k + m - 2 \geq |a_{km} - ma_k| + |a_{km} - ka_m| \geq |ma_k - ka_m|.$$

Dividing by km, we get

$$\left|\frac{a_k}{k} - \frac{a_m}{m}\right| \leq \frac{1}{k} + \frac{1}{m} - \frac{2}{km} < \frac{1}{k} + \frac{1}{m}.$$

24. *b*) For $x = 1$ and $x = -1$, respectively, we get

$$a_0 + a_1 + a_2 + \ldots + a_{2n} = 3^n, a_0 - a_1 + a_2 - \ldots + a_{2n} = 1.$$

Adding and subtracting these equalities, we have

$$a_0 + a_2 + a_4 + \ldots = \frac{3^n + 1}{2}, a_1 + a_3 + a_5 + \ldots = \frac{3^n - 1}{2}.$$

For $x = i$ in the given expansion, we obtain $i^n = S_0 - S_2 + i(S_1 - S_3)$, where

$$S_0 = a_0 + a_2 + a_4 + \ldots, S_1 = a_1 + a_5 + a_9 + \ldots$$
$$S_2 = a_2 + a_6 + a_{10} + \ldots, S_3 = a_3 + a_7 + a_{11} + \ldots.$$

If $4|n$, then $S_1 = S_2 = S_3 = \frac{3^n - 1}{4}, S_0 = \frac{3^n + 3}{4}$.
If $4|n - 1$, then $S_0 = S_1 = S_2 = \frac{3^n + 1}{4}, S_3 = \frac{3^n - 3}{4}$
If $4|n - 2$, then $S_0 = S_1 = S_3 = \frac{3^n - 1}{4}, S_2 = \frac{3^n + 3}{4}$.
If $4|n - 3$, then $S_0 = S_2 = S_3 = \frac{3^n + 1}{4}, S_1 = \frac{3^n - 3}{4}$.

a) We have $a_k = a_{2n-k}, 0 \le k \le 2n$. Indeed, taking $y = \frac{1}{x}$ and comparing the coefficients of the expansions $(1 + x + x^2)^n$ and

$$y^{2n}\left(y^2 + y + 1\right)^n = a_0 y^{2n} + a_1 y^{2n-1} + \ldots + a_{2n},$$

we get the result.

We will prove the assertion by induction. For $n = 2$, we have $a_0 = a_4 = 1$, $a_1 = = a_3 = 2$, $a_2 = 3$ and the property is verified. Suppose that the property is true for $2 \le m \le n$ and prove it for $n + 1$. Consider the expansion

$$\left(1 + x + x^2\right)^{n+1} = b_0 + b_1 x + \ldots + b_{2n+2} x^{2n+2}.$$

Since $(1 + x + x^2)^{n+1} = (1 + x + x^2)(a_0 + a_1 x + \ldots + a_{2n} x^{2n})$, it follows that

$$b_p = a_p + a_{p-1} + a_{p-2}, 0 \le p \le 2n + 2,$$

putting $a_{-2} = a_{-1} = a_{2n+1} = a_{2n+2} = 0$. If $p \le n$ or $p \ge n + 2$, then the property is true, by the induction hypothesis. It remains to prove the inequalities

$$b_{n+1} \ge b_n, b_{n+1} \ge b_{n+2}.$$

We have the equivalences

$$b_{n+1} \ge b_n \Leftrightarrow a_{n+1} + a_n + a_{n-1} \ge a_n + a_{n-1} + a_{n-2} \Leftrightarrow a_{n+1} \ge a_{n-2},$$

which is true, by the induction hypothesis. The second inequality is also true, since

$$b_{n+2} = b_n.$$

25. Answer: $n \equiv 0, 3 \pmod 4$.

Case 1. $n \equiv 0, 3 \pmod 4$. Induction on n. Denote by $S(n)$ the left-hand sum. Then

$$S(3) = 1 + 2 - 3 = 0, S(4) = 1 - 2 - 3 + 4 = 0.$$

Assume that there is a choice of signs for which $S(n) = 0$. The equality

$$S(n+4) - S(n) = (n+1) - (n+2) - (n+3) + (n+4) = 0$$

shows that if the property is valid for n, then it is valid for $n + 4$, which ends the proof by induction.

Case 2. $n \equiv 1 \pmod 4$. Then $n = 4k + 1$ and there are $2k$ even numbers and $2k + 1$ odd numbers, so $S(n) \neq 0$, since it is odd.

Case 3. $n \equiv 2 \pmod 4$. Then $n = 4k + 2$ and there are $2k + 1$ even numbers and $2k + 1$ odd numbers, so $S(n) \neq 0$, since it is odd.

26. For $n = 1, 2, 3$,

$$a + b + c \equiv 0(\text{mod } m), a^2 + 2b + c \equiv 0(\text{mod } m), a^3 + 3b + c \equiv 0(\text{mod } m).$$

whence $b \equiv a - a^2 \equiv a^2 - a^3(\text{mod } m)$, so $a(a - 1)^2 \equiv 0(\text{mod } m)$. But then

$$b^2 \equiv \left(a - a^2\right)^2 \equiv a^2(a-1)^2 \equiv 0(\text{mod } m).$$

For $a = 3$, $b = 6$, $c = 3$, $m = 12$, we have $b \not\equiv 0(\text{mod } m)$.

3.10 1981

27. The multiplicity of ξ is 2 or 4, since the number of non-real roots of a polynomial is always even.

Case 1. ξ is of multiplicity 4. Then $P = (x - \xi)^4$, hence $\xi = -\frac{a_1}{4}$ is rational.

Case 2. ξ is of multiplicity 2. Then the non-real roots of P are simple roots, hence

$$D(x) = \gcd(P(x), P'(x)) = x - \xi.$$

We know that $D(x) \in \mathbb{Q}[x]$, so that $\xi \in \mathbb{Q}$.

28. Induction on n. We can eliminate x_i, x_j, for which $x_i = x_j$, $i \neq j$, choosing $a_i = 1$, $a_j = -1$, so we can assume that $0 < x_1 < x_2 < \ldots < x_n$.

Case 1. $n = 2k + 1$. We will prove that

$$x_1^2 - x_2^2 + x_3^2 - \ldots + x_{2k+1}^2 \geq (x_1 - x_2 + x_3 - \ldots + x_{2k+1})^2.$$

For $k = 0$ this is an equality. Suppose that it is true for k, then

$$(x_1 - x_2 + x_3 - \ldots + x_{2k+1} - x_{2k+2} - x_{2k+3})^2 = (x_1 - x_2)^2$$
$$+ (x_3 - x_4 + \ldots + x_{2k+3})^2 + 2(x_1 - x_2)(x_3 - x_4 + \ldots + x_{2k+3}) \leq (x_1 - x_2)^2$$
$$+ (x_3 - x_4 + \ldots + x_{2k+3})^2 \leq x_1^2 - x_2^2 + x_3^2 - x_4^2 + \ldots + x_{2k+3}^2,$$

where we are used the inequalities $x_1 - x_2 < 0$, $x_3 - x_4 + \ldots + x_{2k+3} > 0$, the base case and the induction hypothesis.

Case 2. $n = 2k$. For $k = 1$, we have $(x_1 - x_2)^2 \leq x_1^2 - x_2^2$. Suppose that the numbers $a_1, a_2, \ldots, a_{2k}$ are chosen, then

$$(a_1x_1 + \ldots + a_{2k}x_{2k} + x_{2k+1} - x_{2k+2})^2 = (a_1x_1 + \ldots + a_{2k}x_{2k})^2 + (x_{2k+1} - x_{2k+2})^2$$
$$+ 2(a_1x_1 + \ldots + a_{2k}x_{2k})(x_{2k+1} - x_{2k+2}) \leq (a_1x_1 + \ldots + a_{2k}x_{2k})^2$$
$$+ (x_{2k+1} - x_{2k+2})^2 \leq a_1x_1^2 + \ldots + a_{2k}x_{2k}^2 + x_{2k+1}^2 - x_{2k+2}^2.$$

3.11 1982

29. We will prove the left inequality by induction. For $n = 3$ it verifies directly. Suppose that it is true for $n - 1$ and prove it for n. We have

$$S_n - S_{n-1} = \sum_{j=1}^{n-1} \frac{1}{\sqrt{j^2 + n^2}} + \sum_{k=1}^{n-1} \frac{1}{\sqrt{n^2 + k^2}} + \frac{1}{\sqrt{2n^2}} > \frac{2(n-1)}{\sqrt{2n^2}} + \frac{1}{\sqrt{2n^2}}$$

for $n \geq 4$, so $S_n \geq n$.

The right inequality follows from

$$S_n - S_{n-1} = \sum_{j=1}^{n-1} \frac{1}{\sqrt{j^2 + n^2}} + \sum_{k=1}^{n-1} \frac{1}{\sqrt{n^2 + k^2}} + \frac{1}{\sqrt{2n^2}}$$
$$= 2\sum_{k=1}^{n} \frac{1}{\sqrt{k^2 + n^2}} - \frac{1}{\sqrt{2n^2}} < \frac{2}{n}\sum_{k=1}^{n} \frac{1}{\sqrt{1 + \left(\frac{k}{n}\right)^2}} < 2\int_0^1 \frac{dx}{\sqrt{1 + x^2}}$$
$$= 2\ln\left(1 + \sqrt{2}\right).$$

The integral can be calculated using the substitution $y=x+\sqrt{1+x^2}$. Now we can choose $C=2\ln\left(1+\sqrt{2}\right)$ and this is the best choice, since the Riemann sum is an approximation of the integral.

30. We will prove that $f(a)=a$ if and only if $a=0$.

If $f(a)=a$, then $f(f(a))=f(a)=a$, hence

$$\frac{a^9}{(a^2+1)(a^6+a^4+2a^2+1)}=a.$$

After calculations, we see that this equality is equivalent to

$$a\left(2a^6+3a^4+3a^2+1\right)=0,$$

which clearly implies $a=0$.

Now, we will show that $f(0)=0$. Denote $f(0)=y$, then $0=f(f(0))=f(y)$ and $y=f(0)=f(f(y))$, and so, as above, we deduce $y=f(0)=0$.

Remark. There are functions f as in the enounce, for example

$$f(x)=\frac{x^3}{x^2+1}, x\in\mathbb{R}.$$

3.12 1983

31. We have

$$a^2p^3+b^2q^3\leq\frac{2a^5+3p^5+2b^5+3q^5}{5}\leq 1.$$

32. Let $a\in[-1,1]$ and $t=\arcsin a$. Then

$$P(a)=P(\sin t)=P(\cos t)=P(\cos(-t))=P(\sin(-t))=P(-\sin t)=P(-a).$$

Now, P is an even polynomial function, so it contains only even powers of X. We will prove the assertion from the enounce by induction on $n=\deg(P)$.

If P is constant, then we can take $Q=P$. Suppose that the assertion is true for polynomials of degree at most $n-1$. From the theorem of division of polynomials, it follows that

$$P(X)=\left(X^4-X^2\right)R(X)+bX^3+cX^2+dX+e.$$

We will show that $R(\sin t) = R(\cos t)$, $t \in \mathbb{R}$, and $b = c = d = 0$. Identifying coefficients, we find $d = 0$. From $P(-1) = P(0) = P(1)$ ($t = 0$ and $t = \pi$), it follows that $b = c = 0$, hence

$$R(\sin t) = \frac{P(\sin t) - e}{(\sin t)^4 - (\sin t)^2} = \frac{P(\cos t) - e}{(\cos t)^4 - (\cos t)^2} = R(\cos t)$$

for every t such that $\sin 2t \neq 0$. But R is a continuous function on $\mathbb{R}$, hence the above equality is true for all $t \in \mathbb{R}$. Since $\deg(R) < \deg(P)$, it follows that there is an $S \in \mathbb{R}[X]$ such that $R(X) = S(X^4 - X^2)$. Then we can take $Q(X) = XR(X) + e$.

3.13 1984

33. Define

$$g_0(x) = \frac{1}{f_0\left(\frac{1}{x}\right)} = \frac{x}{2}, g_1(x) = \frac{1}{f_1\left(\frac{1}{x}\right)} = 1 - x, 0 < x < 1.$$

Then every number $\frac{p}{2^q} \in (0,1), p$ odd, can be obtained from $\frac{1}{2}$ in finitely many steps, each step consisting in applying g_0 or g_1. We will prove this assertion by induction on q: for $q = 1$, there is a unique fraction, namely $\frac{1}{2}$. Assume that the assertion is true for $q - 1 \geq 1$. Take $v = \frac{p}{2^q} \in (0,1)$. One of the numbers v or $1 - v$ equals $\frac{u}{2}, u \in (0,1)$ being a fraction with denominator 2^{q-1}, that is, $v = g_0(u)$ or $v = g_1(g_0(u))$. By hypothesis, u is a finite composition of g_0, g_1 applied to $\frac{1}{2}$, so the same is true for v, which finishes the proof by induction.

Back to the problem, choose $1 < a < b$ and $v = \frac{p}{2^q} \in \left(\frac{1}{b}, \frac{1}{a}\right), p$ odd (this is always possible, since the set $\left\{\frac{p}{2^q}\right\}$ is dense in $\mathbb{R}$). The number v is the image of $\frac{1}{2}$ by a finite composition of g_0, g_1:

$$v = g_{i_k} \circ \ldots \circ g_{i_1}\left(\frac{1}{2}\right).$$

Consider the function $h : (1, \infty) \to (0,1), h(x) = \frac{1}{x}$. We have

$$g_i = h \circ f_i \circ h, i = 0, 1, h \circ h = id,$$

$$v = (h \circ f_{i_k} \circ h) \circ \ldots \circ (h \circ f_{i_1} \circ h)\left(\frac{1}{2}\right) = h \circ (f_{i_k} \circ \ldots \circ f_{i_1}) \circ h\left(\frac{1}{2}\right) = \frac{1}{f_{i_k} \circ \ldots \circ f_{i_1}(2)}.$$

Since $v \in \left(\frac{1}{b}, \frac{1}{a}\right)$, the last denominator belongs to (a, b), as desired.

3.14 1985

34. Multiply the first equation by x, the second by y, then add and subtract them:

$$\begin{cases} x^3+y^3-\frac{9}{4}xy=0 \\ 2(x^4y-xy^4)-x^3+y^3=0 \end{cases}.$$

Denote $s = x + y$, $p = xy$. The the system becomes

$$\begin{cases} 4s(s^2-3p)-9p=0 \\ (x^3-y^3)(2p-1)=0 \end{cases}.$$

Case 1. $x = y$. Then the first equation becomes $8x^3 = 9x^2$, hence

$$(x,y)=(0,0),\left(\frac{9}{8},\frac{9}{8}\right).$$

Case 2. $x \neq y$. Then $p=\frac{1}{2}$, and the first equation becomes

$$8s^3-12s-9=(2s-3)\left(4s^2+6s+3\right)=0,$$

with the unique real solution $s=\frac{3}{2}$. The system

$$x+y=s=\frac{3}{2}, xy=p=\frac{1}{2}$$

has the solutions

$$(x,y)=\left(1,\frac{1}{2}\right),\left(\frac{1}{2},1\right).$$

35. The expression in the parenthesis contains only even functions, so the corresponding equation has an even (possibly 0) number of solutions.

If $a = 0$, then the equation becomes

$$3x^2=\cos\frac{9\pi x}{2}$$

and it has an even number of solutions. In particular, the equation has an even number of solutions in $[-1,5]$, since $|3x^2| \leq 1$ implies $|x| \leq \frac{\sqrt{3}}{3}$.

If $a \neq 0$, then the equation $\sqrt{3-ax}=0$ has the unique solution $x_0=\frac{3}{a}$. If $a>0$, then $x_0 \in [-1,5]$ if and only if $a \geq \frac{3}{5}$. If $a<0$, then $x_0 \in [-1,5]$ if and only if $a \leq -3$.

In conclusion, the given equation has an odd number of solutions if and only if

$$a \in (-\infty, -3] \cup \left[\frac{3}{5}, \infty\right).$$

36. By the Bergström's inequality,

$$\sum_{cyc} \frac{a}{b+c} = \sum_{cyc} \frac{a^2}{ab+ac} \geq \frac{(a+b+c+d)^2}{ab+2ac+bc+2bd+cd+da} \geq 2.$$

The last inequality is equivalent to $(a-c)^2+(b-d)^2 \geq 0$.

3.15 1986

37. Let $-x_1, \ldots, -x_n$ be the roots of the polynomial, then

$$P(x)=\prod_{i=1}^{n}(x+x_i), x_i>0, i=1,\ldots,n.$$

By the Viète's formulae,

$$a_0=\prod_{i=1}^{n} x_i, \frac{a_1}{a_0}=\sum_{i=1}^{n}\frac{1}{x_i}.$$

The AM–GM inequality implies

$$a_1 > n\sqrt[n]{\prod_{i=1}^{n}\frac{a_0}{x_i}}=na_0^{1-\frac{1}{n}}$$

and the inequality is strict, since the x_i are distinct. From the binomial formula,

$$(1+x)^n \geq 1+nx+\frac{n(n-1)}{2}x^2 \geq 2nx, x>0.$$

We deduce

$$P(1) = \prod_{i=1}^{n}(1+x_i) \geq \sqrt[n]{\prod_{i=1}^{n}(2nx_i)} = 2n\sqrt[n]{a_0} > 2n^2 \cdot \frac{a_0}{a_1},$$

which ends the proof.

38. Inequality (i) holds for all $n-$tuples $(x_1, \ldots, x_n)$ if and only if it holds for constant $n-$tuples $(c, \ldots, c)$ and $\left(\underbrace{0, \ldots, 0}_{k}, 1, \ldots, 1\right)$, since any increasing sequence (x_i) can be written as a sum with nonnegative coefficients of such vectors.

For constant vectors, we get

$$\sum_{k=1}^{n} a_k = \sum_{k=1}^{n} b_k.$$

For the vector $\left(\underbrace{0, \ldots, 0}_{k-1}, 1, \ldots, 1\right)$, we get

$$\sum_{i=k}^{n} a_i = \sum_{i=k}^{n} b_i.$$

39. *i*) We will show that $f(A) \leq A$, with equality if and only if

$$A \in \{1, 2, 3, 4, 5, 6, 7, 8, 9, 19\}.$$

If $A = \overline{a_n a_{n-1} \ldots a_0}$, $a_n \neq 0$, then $n = 1$ implies $f(A) - A = a_0 - 9a_1 \leq a_0 - 9 \leq 0$, with equality if and only if $a_0 = 9, a_1 = 1$; $n \geq 2$ implies

$$A \geq 10^n > 25 \cdot 2^n, f(A) < 9(1 + 2 + \ldots + 2^n) = 9\left(2^{n+1} - 1\right).$$

If $A_{k+1} \neq A_k$, then from $A_{k+1} = f(A_k) \leq A_k$, it follows that $A_{k+1} < A_k$. But $A_{A+1} \leq A - A = 0$, contradiction, so there exists k for which $A_{k+1} = A_k$.

ii) We will show that 19 divides A if and only if 19 divides $f(A)$. Indeed,

$$2^n A - f(A) = \sum_{i=0}^{n} 2^{n-i}\left(20^i - 1\right)a_i, 19 | 20^i - 1, 0 \leq i \leq n, \gcd(19, 2^n) = 1.$$

From $19 | A = 19^{86}$, we deduce that $19 | A_k$, so $A_k = 19$.

3.16 1987

40. If $n = \deg(P) = 1$, then the property is obvious. Assume that $n \geq 4$ and let p be a prime number such that $\gcd(n, p-1) = 1$. The number $p = 1987$ satisfies this condition, since $1986 = 2 \cdot 3 \cdot 331$ and prime divisors of n have an even number of digits, by the hypothesis.
Suppose that there are rational numbers $x \neq y, x = \frac{a}{c}, y = \frac{b}{d}$, with a, b, c, d integers, $c, d > 0, \gcd(a, c) = \gcd(b, d) = 1$ and $P(x) = P(y)$. Then

$$d^n\left(a^n - pac^{n-1}\right) = c^n\left(b^n - pbd^{n-1}\right). \qquad (1)$$

The numbers in parentheses are coprime to c, d, respectively, so $c = d$, hence $a \neq b$. Now, we have

$$b^n - a^n = p(b-a)c^{n-1}. \qquad (2)$$

Dividing by $b - a$, we get

$$\sum_{i=0}^{n-1} a^i b^{n-i-1} = pc^{n-1}. \qquad (3)$$

It follows that $p \nmid a$, $p \nmid b$, since $\gcd(a, c) = 1$, hence

$$be \equiv 1 \pmod p \qquad (4)$$

for some $e \in \{1, 2, \ldots, p-1\}$. By (2),

$$(ae)^n = a^n b^n \equiv b^n e^n \equiv 1 \pmod p. \qquad (5)$$

Case 1. $ae \not\equiv 1 \pmod p$. By (5), $ae \not\equiv 0 \pmod p$. The Fermat's theorem implies

$$(ae)^{p-1} \equiv 1 \pmod p. \qquad (6)$$

Let α be the least positive exponent for which $(ae)^\alpha \equiv 1 \pmod p$. Then $\alpha > 1$ and by (5), (6), we have $\alpha | n, \alpha | p-1$, contradiction.
Case 2. $ae \equiv 1 \pmod p$. By (4),

$$a \equiv b \not\equiv 0 \pmod p. \qquad (7)$$

From (3), $na^{n-1} \equiv 0 \pmod p$, hence $n \equiv 0 \pmod p$. But n is a square, therefore

$$n \equiv 0 \pmod{p^2}. \qquad (8)$$

Let $a = kp + r$, $b = lp + r$. By the binomial formula,

$$a^i \equiv ikpr^{i-1} + r^i \pmod{p^2}, b^j \equiv jlpr^{j-1} + r^j \pmod{p^2},$$

hence

$$a^i b^j \equiv (ik + jl)pr^{i+j-1} + r^{i+j} \pmod{p^2}.$$

The LHS in (3) is congruent modulo p^2 to

$$\sum_{i=0}^{n-1} (ik + (n-i-1)l)pr^{n-2} + nr^{n-1} = \frac{n(n-1)}{2}(k+l)pr^{n-2} + nr^{n-1}.$$

By (8), this number is divisible by p^2, so the RHS in (3) must be divisible by p^2, hence $p|c$. Now, (2) becomes

$$b^n - a^n = (b-a)p^n w, w \in \mathbb{Z}. \qquad (9)$$

Assume that

$$b - a = p^m q, m \geq 1, p \nmid q. \qquad (10)$$

We will prove by induction that for every positive integer s not divisible by p and every integer $z \geq 0$, there exists an integer t_z such that

$$b^{sp^z} - a^{sp^z} = p^{m+z} t_z, p \nmid t_z. \qquad (11)$$

For $z = 0$, by (10) and the binomial formula we have

$$b^s - a^s = (a + qp^m)^s - a^s = sa^{s-1}qp^m + Ap^{2m},$$

with A integer, and we can take
$t_0 = sa^{s-1}q + Ap^m \not\equiv 0 \pmod{p}$.
Choose $z \geq 0$ and assume that there exists t_z that satisfies (11). We have

$$b^{sp^{z+1}} - a^{sp^{z+1}} = (b^{sp^z})^p - (a^{sp^z})^p = (a^{sp^z} + p^{m+z}t_z)^p - (a^{sp^z})^p = \sum_{j=1}^{p} \binom{p}{j} a^{sp^z(p-j)} p^{(m+z)j} t_z^j.$$

In this sum, all summands, excepting the first and the last term, are divisible by $p^{1+2(m+z)}$, since the binomial coefficients are divisible by p, therefore

$$b^{sp^{z+1}} - a^{sp^{z+1}} = pa^{sp^z(p-1)}p^{m+z}t_z + Bp^{1+2(m+z)} + p^{(m+z)p}t_z^p,$$

with B integer, and it suffices to take
$t_{z+1} = a^{sp^z(p-1)}t_z + Bp^{m+z} + t_z^p p^{(m+z)(p-1)-1} \not\equiv 0 \pmod p$,
which ends the proof by induction.
Now, write

$$n = sp^z, s \geq 1, z \geq 0, s \not\equiv 0 \pmod p \tag{12}$$

and apply (11) for these s, z and the corresponding t_z. From (9), (10), (11) we get

$$p^{m+z}t_z = p^{m+n}qw,$$

which implies $n \leq z$. But (12) implies $n \geq p^z > z$, contradiction.

This shows that the equality $P(x) = P(y)$ cannot hold for distinct rational numbers x, y.

41. We use the Cauchy–Schwarz inequality

$$\left|a_{n-1}x^{n-1} + \ldots + a_1x\right| \leq \sqrt{a_{n-1}^2 + \ldots + a_1^2}\sqrt{x^{2(n-1)} + \ldots + x^2}.$$

For x real and n even, we have

$$\begin{aligned}P(x) = (a_nx^n + a_0) + \left(a_{n-1}x^{n-1} + \ldots + a_1x\right) \geq (a_nx^n + a_0)\\ - \left|a_{n-1}x^{n-1} + \ldots + a_1x\right| \geq \min\{a_n, a_0\} \cdot (x^n + 1)\\ - \sqrt{a_1^2 + \ldots + a_n^2}\sqrt{x^2 + \ldots + x^{2(n-1)}}.\end{aligned}$$

The condition *b*) implies

$$P(x) \geq \min\{a_n, a_0\} \cdot (x^n + 1) - \frac{2\min\{a_n, a_0\} \cdot \sqrt{x^2 + \ldots + x^{2(n-1)}}}{\sqrt{n-1}}.$$

It suffices to show that

$$x^n + 1 \geq \frac{2\sqrt{x^2 + \ldots + x^{2(n-1)}}}{\sqrt{n-1}}.$$

Squaring, we get

$$(n-1)(x^n+1)^2 \geq 4\sum_{k=1}^{n-1} x^{2k}.$$

As x appears only with even exponents, denote $n = 2N$, $x^2 = y$, to obtain

$$(2N-1)\left(y^N+1\right)^2 - 4\left(y+y^2+\ldots+y^{2N-1}\right) \geq 0,$$

inequality that must be true for $y \geq 0$ and N positive integer. The polynomial

$$Q(Y) = (2N-1)\left(Y^N+1\right)^2 - 4\left(Y^{2N-1}+Y^2+Y\right)$$

has the double root $y_0 = 1$, therefore it is divisible by $(Y-1)^2$. We get the decomposition $Q(Y) = (Y-1)^2 Q_1(Y)$, where

$$Q_1(Y) = (2N-1)Y^{2N-2} + (4N-6)Y^{2N-3} + (6N-15)Y^{2N-4} + \ldots + (2N-1)$$

is a polynomial with only positive coefficients, which proves the required inequality.

42. Let $a_1, a_2, \ldots, a_n$ be arbitrary real numbers and consider the sum

$$S = \sum_{i,j=1}^{n} a_i a_j \cos\left(\alpha_i - \alpha_j\right) = \sum_{i,j=1}^{n} a_i a_j \cos\alpha_i \cos\alpha_j + \sum_{i,j=1}^{n} a_i a_j \sin\alpha_i \sin\alpha_j.$$

The first term of the sum is

$$\sum_{i,j=1}^{n} a_i a_j \cos\alpha_i \cos\alpha_j = \sum_{i=1}^{n} (a_i \cos\alpha_i)^2 + 2\sum_{1\leq i<j\leq n} a_i a_j \cos\alpha_i \cos\alpha_j = \left(\sum_{i=1}^{n} a_i \cos\alpha_i\right)^2.$$

The second term can be expressed analogously, hence

$$S = \left(\sum_{i=1}^{n} a_i \cos\alpha_i\right)^2 + \left(\sum_{i=1}^{n} a_i \sin\alpha_i\right)^2 \geq 0.$$

In particular, for $a_i = i$, we get the required inequality.

43. From

$$f(\tan\alpha) = \tan\left(\alpha + \frac{\pi}{8}\right),$$

it follows that

$$f^{(1987)}(\tan\alpha)=\tan\left(\alpha+\frac{1987\pi}{8}\right)=\tan\left(\alpha+\frac{3\pi}{8}\right)=\tan\left(\frac{\pi}{2}-\left(\frac{\pi}{8}-\alpha\right)\right)=\frac{1}{f(-\tan\alpha)},$$

therefore

$$f^{1987}(x)=\frac{1}{f(-x)}.$$

3.17 1988

44. Let $x-y=u$, $x-z=v$. Then we have

$$Bu^2+(A-B-C)uv+Cv^2\geq 0$$

for any real numbers u, v.

If $B=C=0$, then $Auv\geq 0$, whence $A=0$.

If $BC\neq 0$, by the general theory of the quadratic functions, we must have $B>0$, $C>0$ and similarly, $A\geq 0$. The required condition is satisfied if and only if the discriminant of the quadratic function is negative:

$$\Delta=(A-B-C)^2-4BC=A^2+B^2+C^2-2(AB+BC+CA)\leq 0.$$

Having in mind the Heron's formula for the area of a triangle, this condition can be written as

$$\sqrt{A}\leq\sqrt{B}+\sqrt{C},\sqrt{B}\leq\sqrt{C}+\sqrt{A},\sqrt{C}\leq\sqrt{A}+\sqrt{B},A,B,C\geq 0,$$

which includes also the cases when some of A, B, C are zero.

45. Let $x_1,\ldots,x_n$ be the given numbers and denote $y_i=\frac{1}{x_i}, 1\leq i\leq n$. We have to solve the system of equations

$$y_1+\frac{1}{y_1}=y_2+\frac{1}{y_2}=\ldots=y_n+\frac{1}{y_n}=y_1+y_2+\ldots+y_n.$$

If $n=2$, then the system has only the solutions $(-1,-1)$, $(1,1)$. If $n>2$, then denote

$$s=\sum_{i=1}^{n}y_i.$$

Each component y_i is a solution of the equation $y+\frac{1}{y}=s$, which has real roots only for $|s|\geq 2$. If $|s|=2$, then $n=2$ and we have the above solutions. If $|s|>2$, then the equation $y^2-sy+1=0$ has two distinct real roots

$$\frac{s\pm\sqrt{s^2-4}}{2}.$$

Case 1. $y_1=y_2=\ldots=y_n$. Then all components have the same sign, which is the sign of s and we obtain the solutions

$$y_1=\ldots=y_n=\frac{1}{\sqrt{n-1}};y_1=\ldots=y_n=-\frac{1}{\sqrt{n-1}}.$$

The solutions of the problem are $x_1=\ldots=x_n=\sqrt{n-1}$ and $x_1=\ldots=x_n=-\sqrt{n-1}$. these solutions hold also in the case $n=2$.

Case 2. $y_1=\ldots=y_k=\frac{s+\sqrt{s^2-4}}{2}, y_{k+1}=\ldots=y_n=\frac{s-\sqrt{s^2-4}}{2}$. Then

$$k\cdot\frac{s+\sqrt{s^2-4}}{2}+(n-k)\cdot\frac{s-\sqrt{s^2-4}}{2}=s\Longrightarrow(k+1-n)(k-1)s^2=(n-2k)^2.$$

If $k=1$, then $n=2$, contradiction. If $k>1$, then $k+1>n$, contradiction.

46. The roots of the polynomial $\Phi_p(X)=X^{p-1}+X^{p-2}+\ldots+X+1$ are $z,z^2,\ldots,z^{p-1}$, where $z=\cos\frac{2\pi}{p}+i\sin\frac{2\pi}{p}$. It follows that $z^{p-1}+z^{p-2}+\ldots+z+1=0$, hence

$$\left(z^{p-1}-1\right)+\left(z^{p-2}-1\right)+\ldots+(z-1)=-p,$$

or, equivalently,

$$(z-1)\left(-\frac{1}{p}z^{p-2}-\frac{2}{p}z^{p-3}-\ldots-\frac{p-2}{p}z-\frac{p-1}{p}\right)=1.$$

By the Eisenstein's criterion, the polynomial $\Phi_p(X)$ is irreducible in $\mathbb{Q}[X]$, so that

$$P(X)=(X-1)\left(-\frac{1}{p}X^{p-2}-\frac{2}{p}X^{p-3}-\ldots-\frac{p-1}{p}\right)-1$$

is divisible by $\Phi_p(X)$. As deg $(p)=p-1$, comparing the leading coefficients, we have

$$P(X) = (X-1)\left(-\frac{1}{p}X^{p-2} - \frac{2}{p}X^{p-3} - \ldots - \frac{p-1}{p}\right) - 1$$
$$= -\frac{1}{p}\left(X^{p-1} + X^{p-2} + \ldots + X + 1\right).$$

Finally,

$$p = (X-1)\left(-X^{p-2} - 2X^{p-3} - \ldots - (p-1)\right) + \left(X^{p-1} + X^{p-2} + \ldots + X + 1\right).$$

We will prove that p is the smallest number with this property. Indeed, if $0 < a < p$ and $a = (X - 1)f(X) + \Phi_p(X)g(X)$, then $a = pg(1) \geq p$, contradiction.

47. Denote $x = \frac{b}{a}$, $y = \frac{c}{b}$. By hypothesis, we have $x \geq 1, y \geq 1$, hence Im$f \subset [1, \infty)$. the inequality $c < a + b$ implies $1 + x > xy$ $(*)$.

    Case 1. $x \leq y$. Then from $(*)$ it follows $x^2 - x - 1 < 0$, hence $x < \frac{1+\sqrt{5}}{2}$.

    Case 2. $x > y$. Then $(*)$ implies $y < 1 + \frac{1}{x} < 1 + \frac{1}{y}$, so that $y^2 - y - 1 < 0$, whence $y < \frac{1+\sqrt{5}}{2}$.

    In conclusion, Im$f \subset \left[1, \frac{1+\sqrt{5}}{2}\right)$. Moreover, for any $\lambda \in \left(1, \frac{1+\sqrt{5}}{2}\right)$ there exists a triangle Δ with sidelengths $1, \lambda, 1 + \frac{1}{\lambda}$ for which $f(\Delta) = \lambda$, For $\lambda = 1$ we choose an isosceles triangle. In conclusion, Im$f = \left[1, \frac{1+\sqrt{5}}{2}\right)$.

3.18 1989

48. Answer: $f_{10}(1) > f_9(2)$.

    We will prove by induction that $f_{2k-1}(2) < f_{2k}(1)$ for $k \geq 2$. If $k = 2$, then

$$f_3(2) = 3^{512} < 4^{1000} = 2^{2000} < 2^{6561} = f_4(1).$$

    Assume that the inequality is true for $k \geq 2$; then it also holds for $k + 1$: Denote $f_{2k-1}(2) = a$, $f_{2k}(1) = b$, so $a + 1 \leq b$ by the induction hypothesis. We have

$$f_{2k}(2) = 2^a, f_{2k+1}(1) = 3^b, f_{2k+2}(1) = 2^{3^b} \geq 2^{3^{a+1}} > 8^{2^a} > 3^{2^a} = f_{2k+1}(2),$$

    which ends the proof.

49. Answer: $n = 7$.

    For $n = 7$, factorize $(z + 1)^7 - (z^7 + 1) = 7z(z + 1)(z^2 + z + 1)^2$.

    Now, we will use Gauss's theorem: any root of the derivative of a polynomial is in the convex hull of the roots of the polynomial. Thus, if all roots of a polynomial have modulus at most 1, so does the derivative. Compute the derivative:

$$f(z) = (z+1)^n - z^n - 1, f'(z) = n\left[(z+1)^{n-1} - z^{n-1}\right].$$

Since $f'(0) \neq 0$, the roots of the equation $f'(z) = 0$ are the roots of the equation

$$\left(1 + \frac{1}{z}\right)^{n-1} = 1,$$

that is,

$$1 + \frac{1}{z_k} = \cos\frac{2k\pi}{n-1} + i\sin\frac{2k\pi}{n-1}, 1 \leq k \leq n-2.$$

If $|z_1| \leq 1$, then

$$\left(\cos\frac{2\pi}{n-1} - 1\right)^2 + \left(\sin\frac{2\pi}{n-1}\right)^2 = \left(2\sin\frac{\pi}{n-1}\right)^2 \geq 1 \Leftrightarrow \sin\frac{\pi}{n-1} \geq \frac{1}{2} = \sin\frac{\pi}{6},$$

hence $n \leq 7$.

50. If $\deg(P) = n$, $\deg(Q) = m$, then $mn = 15$.

Case 1. $m = 1$. Then $Q(X) = X$, $P(X) = (X-1)(X-2)\ldots(X-15)$.

Case 2. $m = 3$. Then $Q(X) = X^3 + aX^2 + bX$, $a, b \in \mathbb{Z}$ and $P(X) = (X - z_1)\ldots(X - z_5)$, $z_i \in \mathbb{C}$. By hypothesis,

$$P(Q(X)) = (Q(X) - z_1)\ldots(Q(X) - z_5) = (X-1)\ldots(X-15).$$

Then we have

$$Q(X) - z_i = (X - a_i)(X - b_i)(X - c_i), 1 \leq i \leq 5,$$

with $a_i, b_i, c_i \in \{1, 2, \ldots, 15\}$. Moreover, the sets $M_i = \{a_i, b_i, c_i\}$ are a partition of the set $\{1, 2, \ldots, 15\}$. From the Viète's formulas,

$$a_i + b_i + c_i = -a, a_ib_i + b_ic_i + c_ia_i = b, 1 \leq i \leq 5.$$

It follows

$$a_i^2 + b_i^2 + c_i^2 = a^2 - 2b, 1 \leq i \leq 5.$$

Summing it, we get

$$\sum_{i=1}^{5}\left(a_i^2+b_i^2+c_i^2\right)=5\left(a^2-2b\right)=\sum_{k=1}^{15}k^2=40\cdot 31,$$

hence $a_i^2+b_i^2+c_i^2=a^2-2b=248$. If $a_1=1$, then $b_1^2+c_1^2=247$, which is impossible, since $247\equiv 3(\text{mod } 4)$.

Case 3. $m=5$. Then

$$P(Q(X))=(Q(X)-z_1)(Q(X)-z_2)(Q(X)-z_3)=(X-1)(X-2)\ldots(X-15).$$

As above, we get a partition of the set $M=\{1,2,\ldots,15\}$ in subsets $M_i=\{a_i,b_i,c_i,d_i,e_i\}$, $1\le i\le 3$, such that

$$a_i^2+b_i^2+c_i^2+d_i^2+e_i^2=\frac{1}{3}\sum_{k=1}^{15}k^2=\frac{40\cdot 31}{3},$$

which is impossible.

Case 4. $m=15$. Then $P(X)=X-15!$, $Q(X)=(X-1)(X-2)\ldots(X-15)+15!$.

51. Consider the double sequence $(a_{kn})_{k\ge 1,\, n\ge 0}$ defined by

$$a_{kn}=\sqrt{k+\sqrt{k+1+\ldots+\sqrt{k+n}}}.$$

We will prove by induction on n that for any $k\ge 1$ holds: $a_{kn}<k+1$.

The base case $n=0$ is obvious: $\sqrt{k}<k+1$. Suppose that the inequality is true for$n$ and prove it for $n+1$. We have

$$a_{k,n+1}<k+1\Leftrightarrow a_{k,n+1}^2<k^2+2k+1\Leftrightarrow k+a_{k+1,n}<k^2+2k+1\Leftrightarrow a_{k+1,n}<k^2+k+1.$$

The last inequality is a consequence of $a_{k+1,n}<k+2<k^2+k+1$.

The inequality from the enounce is $a_{1n}<2$.

3.19 1990

52. *a*) Write ab for $a\circ b$. From $(aa)a=a(aa)$, we deduce $aa=a$ for all a. Then

$$a(a(ba))=(aa)(ba)=a(ba),\ (a(ba))a=(ab)(aa)=(ab)a=a(ba).$$

From *ii*) it follows that $a(ba)=a$ for all $a,b\in S$. Now,

$$(ac)(a(bc)) = (a(ca))(bc) = a(bc), (a(bc))(ac) = (ab)(c(ac)) = (ab)c,$$

hence $a(bc) = ac$.

$b)$ Define $ab = a$ for all $a, b \in S$.

53. Answer: $n = 11$.
Choose $a_i = i^3$, $1 \le i \le n \le 10$. Then the inequality cannot hold, since

$$0 < i^3 - j^3 \Rightarrow i - j \ge 1 \Rightarrow i^3 - j^3 = (i-j)^3 + 3ij(i-j) \ge 1 + 3ij.$$

Now, consider the partition

$$[1, 1000] = \{1\} \cup \left(\bigcup_{k=1}^{9} \left(k^3, (k+1)^3 \right] \right).$$

Among any 11 distinct numbers $a_1, a_2, \ldots, a_{11} \in [1, 1000]$, by the pigeonhole principle, there exist $a_i > a_j$ in the same interval. Then

$$0 < \sqrt[3]{a_i} - \sqrt[3]{a_j} < 1 \Rightarrow 0 < a_i - a_j = \left(\sqrt[3]{a_i} - \sqrt[3]{a_j} \right)^3 + 3\sqrt[3]{a_i a_j} < 1 + 3\sqrt[3]{a_i a_j}.$$

54. We have

$$x = y \Rightarrow f = g; y = 0 \Rightarrow xh(x) = f(x) - f(0).$$

Denote $l(x) = f(x) - f(0)$, then

$$(x-y)l(x+y) = (x+y)(l(x) - l(y)).$$

Now,

$$y = 1 \Rightarrow l(x+1) = \frac{x+1}{x-1}(l(x) - l(1)),$$

$$x \mapsto x+1 \Rightarrow l(x+2) = \frac{x+2}{x}\left[\frac{x+1}{x-1}(l(x) - l(1)) - l(1)\right] = \frac{x+2}{x-2}(l(x) - l(2)),$$

whence it follows $l(x) = ax + b$, with a, b constants. But $l(0) = 0$, so

$$l(x) = ax, f(x) = g(x) = ax + b, h(x) = a.$$

55. The hypothesis implies that there exists a positive integer N_0 such that $f(k) > k$ for all $k \ge N_0$. Assume that there is a function g: $\mathbb{N} \to \mathbb{N}$ for which the set

$$B = \{k | g(f(k)) \geq g(k)\}$$

is finite. It follows that there exists a positive integer N_1 such that $g(f(k)) < g(k)$ for all $k \geq N_1$. If a positive integer $n_1 \geq \max\{N_0, N_1\}$, then

$$n_2 = f(n_1) > n_1 \text{ and } g(f(n_1)) < g(n_1).$$

Since $n_2 = f(n_1) > \max\{N_0, N_1\}$, we have

$$f(n_2) > n_2 \text{ and } g(f(n_2)) < g(n_2) < g(n_1).$$

Repeating this process, we get an infinite decreasing sequence of positive integers

$$g(n_1) > g(n_2) > g(n_3) > \ldots, \text{where } n_{k+1} = f(n_k) > n_k \text{ for all } k,$$

contradiction.

56. If $x = 1$, then $(P(1))^2 - 2P(1) - 2 = 0$, hence $P(1) = 1 \pm \sqrt{3}$. We will show that P is one of the constant polynomials $P_1 = 1 - \sqrt{3}$ or $P_2 = 1 + \sqrt{3}$.
Consider the case $P(1) = 1 + \sqrt{3}$. By division, we get

$$P(X) = (X - 1)Q(X) + 1 + \sqrt{3}.$$

The condition from the enounce implies

$$(X-1)\left[4(X+1)Q(2X^2-1) - (X-1)(Q(X))^2 - 2\left(1+\sqrt{3}\right)Q(X)\right]$$
$$= 0 \Leftrightarrow 4(X+1)Q(2X^2-1) = (X-1)(Q(X))^2 + 2\left(1+\sqrt{3}\right)Q(X).$$

Taking $x = 1$, we get $X - 1 | Q(X)$. Inductively, it follows that $(X-1)^n | Q(X)$ for all $n \geq 1$, hence $Q(X) = 0$ and $P(X) = 1 + \sqrt{3}$.

57. If $n = 3$, then $x_1x_2 + x_2x_3 + x_3x_1 = x_1x_2 - (x_1 + x_2)^2 = -\left(x_1^2 + x_1x_2 + x_2^2\right) \leq 0$.
If $n = 4$, then

$$x_1x_2 + x_2x_3 + x_3x_4 + x_4x_1 = (x_1 + x_3)(x_2 + x_4) = -(x_1 + x_3)^2 \leq 0.$$

For $n \geq 5$, we can choose the sequence $x_1 = x_2 = 1$, $x_4 = -2$, $x_i = 0$, $i \notin \{1, 2, 4\}$, as a counterexample.

3.20 1991

58. The equality $f(x) = f\left(\frac{1}{x}\right)$ suggests that $f(1) = \frac{2n-1}{4}$ is a good candidate. Fix $x > 0$ and an integer k, $1 \leq k \leq n-1$ and choose

$$a_1 = \ldots = a_k = x^n, b_1 = \ldots = b_{n-k} = 1.$$

By the AM–GM inequality, we have

$$x^k = \sqrt[n]{x^{nk}} \leq \frac{kx^n + (n-k)}{n}.$$

Then

$$\sum_{k=1}^{n-1} x^k \leq \frac{n(n-1)}{2} \cdot \frac{x^n + 1}{n} = \frac{(n-1)(x^n+1)}{2},$$

hence

$$\sum_{k=1}^{2n-1} x^k = x^n + (1+x^n) \sum_{k=1}^{n-1} x^k \leq x^n + \frac{(n-1)(1+x^n)^2}{2} \leq \frac{(1+x^n)^2}{4} + \frac{(n-1)(1+x^n)^2}{2} = \frac{(2n-1)(1+x^n)^2}{4}.$$

We conclude that

$$f(x) \leq \frac{2n-1}{4}, \forall x > 0,$$

with equality only for $x = 1$.

59. Denote $x - b_i = a_i$. Then we have to prove that

$$(1+a_1)\ldots(1+a_n)\left(\frac{1}{a_1} + \ldots + \frac{1}{a_n}\right) \geq 2n^2.$$

By the Huygens inequality, we find that

$$LHS \geq \frac{n\left(1 + \sqrt[n]{a_1 \ldots a_n}\right)^n}{\sqrt[n]{a_1 \ldots a_n}}.$$

The function $g(x)=\dfrac{(1+x)^n}{x}, x>0$, has the derivative $g'(x)=\dfrac{(1+x)^{n-1}[(n-1)x-1]}{x^2}$, hence it attains the minimum value at $x=\dfrac{1}{n-1}$ and

$$g\left(\frac{1}{n-1}\right)=(n-1)\left(\frac{n}{n-1}\right)^n>2n.$$

60. Denote

$$f(n)=\sum_{j+h=n, j\geq h}(-1)^h\frac{1}{j}\binom{j}{h}2^{j-h}, g(n)=\sum_{j+h=n, j\geq h}(-1)^h\binom{j}{h}2^{j-h}.$$

Then

$$\begin{aligned}nf(n)=&\sum_{j+h=n,j\geq h}(-1)^h\frac{j+h}{j}\binom{j}{h}2^{j-h}=\sum_{j+h=n,j\geq h}(-1)^h\binom{j}{h}2^{j-h}\\&+\sum_{j+h=n,j\geq h}(-1)^h\frac{h}{j}\binom{j}{h}2^{j-h}=\sum_{j+h=n,j\geq h}(-1)^h\binom{j}{h}2^{j-h}\\&+\sum_{j+h=n,j\geq h}(-1)^h\binom{j-1}{h-1}2^{j-1-(h-1)}=g(n)-g(n-2),\end{aligned}$$

$$\begin{aligned}g(n)=&\sum_{j+h=n,j\geq h}(-1)^h\binom{j}{h}2^{j-h}=\sum_{j+h=n,j\geq h}(-1)^h\left[\binom{j-1}{h}+\binom{j-1}{h-1}\right]2^{j-h}\\=&2\sum_{j+h=n,j\geq h}(-1)^h\binom{j-1}{h}2^{j-1-h}-\sum_{j+h=n,j\geq h}(-1)^{h-1}\binom{j-1}{h-1}2^{j-1-(h-1)}\\=&2g(n-1)-g(n-2).\end{aligned}$$

By induction it follows that $g(n)=n+1$, hence $f(n)=\frac{2}{n}$.

61. Denote by S the given sum. Applying the AM–GM inequality, we deduce

$$\begin{aligned}S\leq&\left(x_1^2+x_{n+1}^2\right)\left(x_2^2+x_{n+2}^2\right)\ldots\left(x_n^2+x_{2n}^2\right)\\&<(x_1+x_{n+1})^2(x_2+x_{n+2})^2\ldots(x_n+x_{2n})^2\leq\left(\frac{x_1+x_2+\ldots+x_{2n}}{n}\right)^{2n}=\frac{1}{n^{2n}}.\end{aligned}$$

3.21 1992

62. Clearly, $1 \le m < n$ and $(x_k - x_m)(1 - x_k) \ge 0$ for $k \le m$, so

$$x_k^2 \le x_k - x_m + x_k x_m.$$

Adding, we obtain

$$\sum_{k=1}^{m} x_k^2 \le \sum_{k=1}^{m} x_k - m x_m + x_m \sum_{k=1}^{m} x_k,$$

$$\sum_{k=m+1}^{n} x_k^2 = 1 - \sum_{k=1}^{m} x_k^2 \ge 1 - \sum_{k=1}^{m} x_k + m x_m - x_m \sum_{k=1}^{m} x_k. \quad (1)$$

Using the inequality

$$x_m \sum_{k=m+1}^{n} x_k \ge \sum_{k=m+1}^{n} x_k^2$$

in (1), we get

$$1 - \sum_{k=1}^{m} x_k + m x_m - x_m \sum_{k=1}^{n} x_k \le 0. \quad (2)$$

From (2) and

$$x_m \left(m - \sum_{k=1}^{n} x_k \right) \ge 0,$$

we deduce the required inequality.

63. Lemma. If $0 < x_1 y_1 < x_2 y_2 < \ldots < x_k y_k$ and $x_1 + \ldots + x_i \ge y_1 + \ldots + y_i$, $1 \le i \le k$, then

$$\frac{1}{x_1} + \frac{1}{x_2} + \ldots + \frac{1}{x_k} \le \frac{1}{y_1} + \frac{1}{y_2} + \ldots + \frac{1}{y_k},$$

with equality only if $x_i = y_i$, $1 \le i \le k$.
Proof. Define

$$s_i=(x_1-y_1)+\ldots+(x_i-y_i), z_i=\frac{1}{x_iy_i}, 1\leq i\leq k, s_0=0, z_{k+1}=0.$$

By hypothesis, $s_i \geq 0$, $z_i \geq z_{i+1}$, $1 \leq i \leq k$. Changing the order of summation, we get

$$\sum_{i=1}^{k}\left(\frac{1}{y_i}-\frac{1}{x_i}\right)=\sum_{i=1}^{k}\frac{x_i-y_i}{x_iy_i}=\sum_{i=1}^{k}(s_i-s_{i-1})z_i=\sum_{i=1}^{k}s_i(z_i-z_{i-1})\geq 0,$$

with equality if and only if $s_i = 0$, $1 \leq i \leq k$, that is, if and only if $x_i = y_i$, $1 \leq i \leq k$.

Back to the problem, we can suppose that $1 \leq a_1 < a_2 < \ldots < a_k$. By the pigeonhole principle, $a_1 + a_2 + \ldots + a_i \geq 1 + 2 + \ldots + 2^{i-1} = 2^i - 1, 1 \leq i \leq k$. We have to prove

$$\frac{1}{a_1}+\frac{1}{a_2}+\ldots+\frac{1}{a_k}\leq 1+\frac{1}{2}+\ldots+\frac{1}{2^{k-1}}.$$

Since $a_i < a_{i+1}$ implies $2^{i-1}a_i < 2^ia_{i+1}$, $1 \leq i \leq k$, we can use the above lemma for $x_i = a_i$, $y_i = 2^{i-1}$, $1 \leq i \leq k$, to obtain the required result.

64. *a*) If $z = x^2 - xy + y^2$, then $x^2 + y^2 = z + xy$ and

$$x^4+y^4=z^2+2xyz-(xy)^2=2z^2-(z-xy)^2\leq 2z^2\leq 8,$$

since $1\leq z\leq 2$. Equality holds for $z=xy=2$, that is, $x=y=\pm\sqrt{2}$.

For the first inequality, denote $x + y = 2a$, $x - y = 2b$, $x = a + b$, $y = a - b$, $z = = a^2 + 3b^2 \in [1, 2]$. Then

$$x^4+y^4=2\left[a^4+b^4+6(ab)^2\right]=2\left[a^4+\frac{(z-a^2)^2}{9}+6a^2\cdot\frac{z-a^2}{3}\right]=\frac{2}{9}$$
$$\times\left(z^2-8a^4+16a^2z\right).$$

We see that $-8a^4 + 16a^2z \geq - 8a^4 + 16a^4 = 8a^4$, since $z = a^2 + 3b^2 \geq a^2$. Hence

$$x^4+y^4\geq\frac{2}{9},$$

with equality only for $z=1, a=0$, that is $x, y\in\left\{\pm\frac{1}{\sqrt{3}}\right\}$.

b) Define the polynomial

$$P(z,a)=\sum_{k=0}^{n}\binom{2n}{2k}(3a^2)^{n-k}(z-a^2)^k-z^n.$$

Then

$$x^{2n}+y^{2n}=(a+b)^{2n}+(a-b)^{2n}=2\sum_{k=0}^{n}\binom{2n}{2k}a^{2n-2k}b^{2k}=\frac{2}{3^n}(z^n+P(z,a)).$$

For $n=4$, we have

$$P(z,a)=\sum_{k=0}^{4}\binom{8}{2k}(3a^2)^{4-k}(z-a^2)^k-z^4=16a^2(-8a^6-16a^4z+24a^2z^2+5z^3).$$

But $z\geq a^2$, hence $5z^3-5a^6\geq 0$ and

$$16a^2z^2-16a^4z=16a^2z(z-a^2)\geq 0, 8a^2z^2-3a^6=5a^2z^2+3a^2(z^2-a^4)\geq 0.$$

It follows that $-8a^6-16a^4z+24a^2z^2+5z^3\geq 0$, hence $P(z,a)\geq 0$ and

$$x^8+y^8\geq\frac{2}{3^4}z^4\geq\frac{2}{3^4},$$

with equality only for $a=0$, $z=1$, that is, x, $y\in\left\{\pm\frac{1}{\sqrt{3}}\right\}$. Now we will show that $x^{2n}+y^{2n}\geq\frac{2}{3^n}$ for all $n\geq 3$, with equality only for $x,y\in\left\{\pm\frac{1}{\sqrt{3}}\right\}$. Indeed,

$$\binom{2n}{2k}\geq\binom{n}{k}$$

for all $k\in\{0,1,\ldots,n\}$, hence

$$\sum_{k=0}^{n}\binom{2n}{2k}(3a^2)^{n-k}(z-a^2)^k\geq\sum_{k=0}^{n}\binom{n}{k}(3a^2)^{n-k}(z-a^2)^k=(z+2a^2)^n.$$

It follows that $P(z,a)\geq(z+2a^2)^n-z^n\geq 0$, whence

$$x^{2n}+y^{2n}\geq\frac{2}{3^n}z^n\geq\frac{2}{3^n},$$

with equality only for $a=0, z=1$, that is, $x,y\in\left\{\pm\frac{1}{\sqrt{3}}\right\}$.

65. Since

$$\frac{x_1}{1+x_1} = \frac{1}{1+x_2} + \ldots + \frac{1}{1+x_{n+1}} \geq n\sqrt[n]{\prod_{i\neq 1}\frac{1}{1+x_i}},$$

multiplying $n + 1$ analogous inequalities leads to

$$\prod_{i=1}^{n+1}\frac{x_i}{1+x_i} \geq n^{n+1}\prod_{i=1}^{n+1}\sqrt[n]{\prod_{j\neq i}\frac{1}{1+x_j}} = n^{n+1}\prod_{i=1}^{n+1}\frac{1}{1+x_i},$$

hence $x_1 \ldots x_{n+1} \geq n^{n+1}$.
Note. The result is a consequence of the inequality

$$\sum_{i=1}^{n+1}\frac{1}{1+x_i} \geq \frac{n+1}{1+\sqrt[n+1]{x_1\ldots x_{n+1}}}.$$

66. Consider the sequence of polynomials $P_n(x) = a_n + b_nx + c_nx^2 + d_nx^3$. By the hypothesis, $\{P_n(x)\}$ is periodic. Moreover, it satisfies the recurrence relation

$$\begin{aligned} P_{n+1}(x) &= a_{n+1} + b_{n+1}x + c_{n+1}x^2 + d_{n+1}x^3 \\ &= (a_n + b_n) + (b_n + c_n)x + (c_n + d_n)x^2 + (d_n + a_n)x^3 \\ &= P_n(x) + \frac{1}{x}P_n(x) + \frac{(x^4-1)a_n}{x}. \end{aligned}$$

For $x = \pm 1, \pm i$, we have

$$P_{n+1}(-1) = 0, P_{n+1}(1) = 2P_n(1), P_{n+1}(i) = (1-i)P_n(i), P_{n+1}(-i) = (1+i)P_n(-i),$$

and finally

$$P_n(1) = 2^{n-1}P_1(1), P_n(i) = (1-i)^{n-1}P_1(i), P_n(-i) = (1+i)^{n-1}P_1(-i)$$

for all positive integers n. The sequences $\{P_n(1)\}$, $\{P_n(i)\}$, $\{P_n(-i)\}$ are periodic, but the geometric progressions $\{2^{n-1}\}$, $\{|1 - i|^{n-1}\}$, $\{|1 + i|^{n-1}\}$ are unbounded, so that

$$P_1(1) = P_1(i) = P_1(-i) = 0,$$

which implies

$$P_n(1)=P_n(i)=P_n(-i)=0$$

for all positive integers n. From $P_2(1) = P_2(-1) = P_2(i) = P_2(-i) = 0$, it follows that

$$\begin{cases} a_2+b_2+c_2+d_2=0 \\ a_2-b_2+c_2-d_2=0 \\ a_2+ib_2-c_2-id_2=0 \\ a_2-ib_2-c_2+id_2=0 \end{cases}.$$

It is easy to see, adding and subtracting the equations, that this linear homogenous system has only the trivial solution $a_2 = b_2 = c_2 = d_2 = 0$.

3.22 1993

67. If $b = 0$, then the inequalities become

$$\frac{a}{4} \le \frac{a}{4} \le \frac{a}{3} \le \frac{a}{2\sqrt{2}},$$

with equality for every nonnegative a in the first inequality and only for $a=0$ in the second and the third inequalities.

If $b > 0$, then we can divide by it. In order to avoid radicals, we will consider a sixth power, so we will take $a=x^6$, $b=1$, with $x \ge 1$. Then we have to prove

$$\left(\frac{x^3+1}{2}\right)^2 \le \frac{x^6+x^4+x^2+1}{4} \le \frac{x^6+x^3+1}{3} \le \sqrt{\left(\frac{x^4+1}{2}\right)^3}.$$

These inequalities are equivalent, respectively, to

$$(x^2-x)^2 \ge 0,$$
$$(x-1)^2(x^4+2x^3+2x+1) \ge 0,$$
$$(x-1)^4\left[(x^4-1)^2+4(x^7+x)+10(x^6+x^2)+4(x^5+x^3)\right] \ge 0,$$

which are true, with equality only for $x = 1$.

In conclusion in all cases equality holds only for $a = b$, with the exception of $ab = 0$, when the first inequality becomes an equality.

68. Apply the Jensen's inequality for the convex function $f(x) = \frac{1}{\sqrt{1-x}}$. Then

$$\sum_{i=1}^{n} \frac{1}{\sqrt{1-x_i}} \geq \frac{n}{\sqrt{1-\frac{\sum_{i=1}^{n} x_i}{n}}} = \frac{n}{\sqrt{1-\frac{1}{n}}} = n\sqrt{\frac{n}{n-1}}.$$

69. Answer: $x = 6$.

For $1 \leq n \leq x-1$, define $A_n = \left\{ y \in \mathbb{N} \middle| n^2 \leq y < (n+1)^2 \right\}$. Then $\lfloor \sqrt{y} \rfloor = n$ for any $y \in A_n$. Since A_n has $(n+1)^2 - n^2 = 2n+1$ elements, the equation becomes

$$\sum_{n=1}^{x-1} n|A_n| = \sum_{n=1}^{x-1} n(2n+1) = \frac{x(x-1)(4x+1)}{6} = 125,$$

whence $x = 6$.

70. Using repeatedly the inequality

$$\frac{(1-a)(1-b)}{(1+a)(1+b)} - \frac{1-(a+b)}{1+(a+b)} = \frac{2ab(a+b)}{(1+a)(1+b)(1+a+b)} \geq 0,$$

for nonnegative real numbers a, b, we have

$$\frac{(1-x_1)(1-x_2)\dots(1-x_n)}{(1+x_1)(1+x_2)\dots(1+x_n)} \geq \frac{(1-(x_1+x_2))(1-x_3)\dots(1-x_n)}{(1+(x_1+x_2))(1+x_3)\dots(1+x_n)} \geq \dots$$
$$\geq \frac{1-(x_1+x_2+\dots+x_n)}{1+(x_1+x_2+\dots+x_n)} = \frac{1}{3},$$

with equality only when one x_i is $\frac{1}{2}$ and the others are 0.

71. We have

$$z_n = \prod_{k=1}^{n} \sqrt{k^2+1}(\cos\alpha_k + i\sin\alpha_k),$$

where

$$\cos\alpha_k = \frac{k}{\sqrt{k^2+1}}, \ \sin\alpha_k = \frac{1}{\sqrt{k^2+1}}.$$

It follows that $\alpha_k \in \left(0, \frac{\pi}{2}\right)$ for all $k \geq 1$. Moreover,

$$\frac{1}{k+1} < \sin \alpha_k = \frac{1}{\sqrt{k^2+1}} < \frac{1}{k}.$$

The sign of $\mathrm{Im}(z_n)$ is given by $\sin\left(\sum_{k=1}^{n} \alpha_k\right)$, so it suffices to show that there are infinitely many positive integers n for which $\beta_n = \sum_{k=1}^{n} \alpha_k$ belong to the set

$$A = \bigcup_{m \geq 0} (2m\pi, (2m+1)\pi)$$

and infinitely many n for which β_n belong to the set

$$B = \bigcup_{m \geq 0} ((2m+1)\pi, (2m+2)\pi).$$

As $\sin\alpha_k < \alpha_k$, we have

$$\beta_n = \sum_{k=1}^{n} \alpha_k > \sum_{k=1}^{n} \sin \alpha_k > \sum_{k=1}^{n} \frac{1}{k+1} = \frac{1}{2} + \frac{1}{3} + \ldots + \frac{1}{n+1}.$$

Then we have $\lim_{n\to\infty} \beta_n = \infty$ and for any $\varepsilon > 0$ there exists $N(\varepsilon)$ such that

$$\beta_{n+1} - \beta_n = \alpha_{n+1}, \forall n \geq N(\varepsilon).$$

From this we deduce that $\beta_n \in A$ for infinitely many n and $\beta_n \in B$ for infinitely many n.

72. Induction on n. If $n = 1$, then we can choose $A = \varnothing$, $B = \{1\}$. In this case f is constant and $A\Delta B = \{1\}$.

Suppose that $n \geq 2$ and the assertion holds for $n-1$. Let $f : \mathcal{P}(X) \to X$ be a function.

Case 1. $f(A) \neq n$ for any $A \subset \{1, 2, \ldots, n-1\}$. Denote $Y = X - \{n\}$, $g : \mathcal{P}(Y) \to Y$ the restriction of f to $\mathcal{P}(Y)$. By induction, we obtain the required result.

Case 2. There is $A \subset Y$ such that $f(A) = n$. Define $\mathbb{F} = \{B | B \subset X, n \in B\}$. Clearly $|\mathbb{F}| = 2^{n-1}$ and $n \in A\Delta B$ for all $B \in \mathbb{F}$. If there exists $B \subset \mathbb{F}$ such that $f(B) = n$, then

$$f(A) = f(B) = n = \max\{k | k \in A\Delta B\}.$$

Otherwise, for any $B \in \mathbb{F}$, we have $f(B) < n$. But $B \in \mathbb{F}$ implies $B - \{n\} \subset Y$. The restriction of f to $\mathcal{P}(Y)$ defines a function $g : \mathcal{P}(Y) \to Y, g(C) = f(C \cup \{n\})$. Applying to g the induction hypothesis, we find $C, D \subset Y$ such that

$$g(C)=g(D)=\max\{k|k\in C\Delta D\}.$$

Then $f(C\cup\{n\})=g(C\cup\{n\})$.

73. Let $f=X^{m+n}-X^{m+1}-X+1$, $g=X^{m+n}+X^{n+1}-X+1$. From $g(1)=2$ we deduce that $X-1$ is not a common divisor. From

$$f(-1)=(-1)^{m+n}+(-1)^{m}+2,\quad g(-1)=(-1)^{m+n}+(-1)^{n+1}+2$$

we deduce that $X+1$ is a common divisor if and only if m is odd and n is even.

Let $z\in\mathbb{C}$ be a common root of f and g; subtracting the equalities $f(z)=0$, $g(z)=0$ we obtain $z^n(z^{m-n}+1)=0$. But $z\neq 0$, hence $z^{m-n}+1=0$. From

$$f(z)=z^m(z^n-z+1)-(z^m+z-1)=0,$$

$$g(z)=z^n-z+1+z^n(z^m+z-1)=0,$$

we see that $z^n-z+1=0$ if and only if $z^m+z-1=0$.

Case 1. $z^n-z+1=0$. Then $z^n=z-1$, $z^{m-n}=-1$, hence $|z-1|=|z|=1$. The complex numbers satisfying these equalities are

$$\omega=\cos\frac{\pi}{6}+i\sin\frac{\pi}{6},\quad \overline{\omega}=\cos\frac{5\pi}{6}+i\sin\frac{5\pi}{6}.$$

These are the roots of the polynomial X^2-X+1, hence of the polynomial X^3+1. The equalities $z^{m-n}=z^3=-1$ imply $m-n=6k+3$, $k=0,1,2,\ldots$.

In conclusion: $\gcd(f,g)=X^3+1$ if $m\equiv 1(\text{mod }2)$, $n\equiv 0(\text{mod }2)$, $m-n\equiv 3$ (mod 6)

$\gcd(f,g)=X^2-X+1$, if $m-n\equiv 3(\text{mod }6)$, m is even or n is odd.

Case 2. $z^n-z+1\neq 0$. Then the determinant of the system $f(z)=g(z)=0$ is zero:

$$\begin{vmatrix} z^m & -1 \\ 1 & z^n \end{vmatrix}=0,$$

whence it follows $z^{m+n}+1=0$. From $z^{m+n}=z^{m-n}=-1$, we deduce $z^{2m}=z^{2n}=1$, hence z is a root of the polynomial $\gcd(X^{2m}-1, X^{2n}-1)=X^{\gcd(2m,2n)}-1=X^2-1$, so that $z=-1$.

If m is odd, n is even and $m-n\not\equiv 3(\text{mod }6)$, then $\gcd(f,g)=X+1$.

3.23 1994

74. We will prove by induction on m, where $m > 1$ is an odd integer, that if x_1, x_2, ..., x_m are nonnegative integers that satisfy the system containing the first m equations, then $x_1 = x_m$, or there is j with $1 \leq j \leq m-1$ such that $x_j = x_{j+1}$. By convention, the indices are taken modulo m.
Observe that for each j, x_{j-1} and x_{j+1} are solutions of the equation

$$x^2 - 2(x_j - 1)x + (x_j + 1)^2 - n^2 = 0, \ (E_j)$$

whose discriminant must be nonnegative, $\Delta_j = 4(n^2 - 4x_j) \geq 0$, in order to have real roots. In fact, $\Delta_j \geq 4$, because n is odd. The equation (E_j) has integral coefficients and at least one integer root, so both roots are integers and are distinct, since $\Delta_j > 0$. Let $\alpha_j < \beta_j$ be these roots. We have $\{x_{j-1}, x_{j+1}\} \subset \{\alpha_j, \beta_j\}$ and $\beta_j = x_j$ or $\alpha_j < x_j < \beta_j$. Indeed, if $\alpha_j > x_j$, then $\alpha_j + \beta_j > 2x_j - 2$, and if $\beta_j < x_j$, then $\alpha_j + \beta_j \leq 2x_j - 3$.
Let j be such that $x_j = \max\{x_1, \ldots, x_m\}$. Since x_{j-1}, x_{j+1} are the roots of (E_j), we have

$$x_{j-1} = x_j \text{ or } x_j = x_{j+1} \text{ or } x_{j-1} = x_{j+1} \text{ or } x_{j-1} < x_j < x_{j+1} \text{ or } x_{j+1} < x_j < x_{j-1}.$$

The last two cases are impossible, due to the choice of j, so $x_j \neq x_{j-1}$, x_{j+1} implies $x_{j-1} = x_{j+1}$. If $m = 3$, then we are nothing to prove, by cyclicity. If $m > 3$, then removing x_{j-1}, x_j, we have a solution with $m-2$ components, by the induction hypothesis.

75. Let $X = (x_1, x_2, x_3, x_4, x_5)$,

$$F(X) = \ln \frac{x_1x_2x_3x_4x_5 + 1}{x_1x_2x_3x_4x_5 - 1}.$$

Then F is a convex function on $(0, \infty)^5$, since the Hessian

$$H(X) = \left(\frac{\partial^2 F}{\partial x_i \partial x_j}\right), 1 \leq i, j \leq 5$$

is nonnegative defined in every point $(x_1, x_2, x_3, x_4, x_5)$. Denote $p = x_1x_2x_3x_4x_5$, then

$$\frac{\partial^2 F}{\partial x_i \partial x_j} = \frac{2p(p^2+1)}{x_ix_j(p^2-1)^2}, i \neq j, \quad \frac{\partial^2 F}{\partial x_i^2} = \frac{4p^3}{x_i^2(p^2-1)^2}.$$

Now, it remains to show that the matrix with non-diagonal entries $p^2 + 1$ and diagonal entries $2p^2$ is not negative defined.

Apply the Jensen's inequality to obtain the required result.

76. If f has an integer root, then it is reducible in $\mathbb{Z}[x]$. Conversely, suppose that f is reducible in $\mathbb{Z}[x]$. Since $\widehat{f}(x)=x^{n-1}(x+\widehat{a})$ in $\mathbb{F}_p[x]$, from $f=gh, g, h\in\mathbb{Z}[x]$, it follows that

$$g(x)=x^k+pg_1(x), h(x)=x^{n-k-1}(x+a)+ph_1(x), p_1, h_1\in\mathbb{Z}[x].$$

Multiplying, we get

$$f(x)=x^{n-1}(x+a)+pg_1(x)x^{n-k-1}(x+a)+ph_1(x)x^k+p^2g_1(x)h_1(x).$$

If $0<k<n-1$, we have $pq=f(0)=p^2g_1(0)h_1(0)$, contradiction. As $k\geq 1$, we deduce $k=n-1$, hence f has an integer root.

Taking $x\in\{\pm1,\pm p,\pm q\}$ in the equality $f(x)=0$, we find the required values of a, since these are the only possible integer roots x of f.

If $n=2$, then $a=\pm(p+q), \pm(1+pq)$.

If $n>2$, then $\pm p$, $\pm q$ cannot be roots, so $x=\pm 1$ and $a=-(1+pq)$, $1+(-1)^n pq$.

77. We have

$$\sum_{cyc}x(1-y^2)(1-z^2)=\sum_{cyc}x-\sum_{cyc}x(y^2+z^2)+xyz\sum_{cyc}yz=\sum_{cyc}x(1-xy-zx)+xyz=4xyz.$$

By the AM–GM inequality,

$$\sqrt{\frac{xy+yz+zx}{3}}\geq\sqrt[3]{xyz},$$

hence

$$xyz\leq\frac{\sqrt[3]{3}}{9}, \sum_{cyc}x(1-y^2)(1-z^2)=4xyz\leq\frac{4\sqrt{3}}{9}.$$

3.24 1995

78. By the Lucas' theorem, the binomial coefficient $\binom{n}{k}$ is odd if and only if the set of binary digits of k is a subset of the set of binary digits of n. Since a finite set with d elements has 2^d subsets, the result follows.

79. By the weighted AM–GM inequality,

$$a^p b^q \le \left(\frac{ap+bq}{p+q}\right)^{p+q} = ap + bq,$$

$$(1-a)^p(1-b)^q \le (1-a)p + (1-b)q.$$

Adding these inequalities, it follows the desired result.

80. If $f(x + 1) = f(x) + 1$, $\forall\, x \in \mathbb{R}$, then the given conditions are satisfied with equality.

Assume that f is a solution; then

$$f(x+24k) \le f(x) + 24k, f(x+77k) \ge f(x) + 77k,$$

for any nonnegative integer k, by an easy induction. We deduce

$$f(x+1) + 384 \ge f(x + 24 \cdot 16 + 1) = f(x+385) = f(x + 77 \cdot 5) \ge f(x) + 385,$$

hence $f(x + 1) \ge f(x) + 1$. Suppose that there exists x such that $f(x + 1) > f(x) + 1$. Then $f(x + k) > f(x) + k$ for all positive integers k, which implies $f(x + 24) > f(x) + 24$, contradiction. It follows that $f(x + 1) = f(x) + 1$ for all x.

81. Induction on n.

For $n = 1$, we have $a_1^5 + a_1^7 - 2(a_1^3)^2 = a_1^5(a_1 - 1)^2 \ge 0$, which is true, with equality only when $a_1 = 1$. Assume that the inequality is true for $n - 1$ and denote

$$S = a_1^3 + a_2^3 + \ldots + a_{n-1}^3,$$

thus,

$$\left(a_1^5 + a_2^5 + \ldots + a_{n-1}^5\right) + \left(a_1^7 + a_2^7 + \ldots + a_{n-1}^7\right) \ge 2S^2.$$

Adding $a_n^5 + a_n^7$ to both sides of the inequality, we get

$$\left(a_1^5 + a_2^5 + \ldots + a_n^5\right) + \left(a_1^7 + a_2^7 + \ldots + a_n^7\right) \ge a_n^5 + a_n^7 + 2S^2.$$

Now, it suffices to prove that

$$a_n^5 + a_n^7 + 2S^2 \ge 2\left(S + a_n^3\right)^2.$$

Squaring, we obtain after calculations

$$S \le \left[\frac{a_n(a_n - 1)}{2}\right]^2 = 1^3 + 2^3 + \ldots + (a_n - 1)^3,$$

which is true, since each term of LHS is also a term of RHS.
Equality holds only when $a_k = k$, $1 \le k \le n$.

82. From $x_i = (x_{i-1} + a)^2 + b^2 - a^2 \ge 0$, we deduce that $\min_i x_i \ge 0$. If $x_1 = x_2$, then

$$x_1^2 + 2ax_1 + b^2 = x_2^2 + 2ax_2 + b^2,$$

so $x_2 = x_3$ and finally, $x_1 = x_2 = \ldots = x_n$. If $x_1 > x_2$, then $x_1 > x_2 > \ldots > x_n > x_1$, contradiction. Similarly, if $x_1 < x_2$, then $x_1 < x_2 < \ldots < x_n < x_1$, contradiction. It follows that all x_i are equal. The solutions of the quadratic equation

$$x^2 + (2a - 1)x + b^2 = 0$$

are

$$x = \frac{1 - 2a \pm \sqrt{(1 - 2a)^2 - 4b^2}}{2}.$$

Both solutions are admissible, since $\Delta = (1 - 2a)^2 - 4b^2 \ge 0$ implies $a + b \le \frac{1}{2}$, hence $2a \le \frac{1}{2}$, so $1 - 2a \ge \frac{1}{2}$.

83. Answer: $k \in \{-3071, -819, 821\}$.

By the Eisenstein's criterion for $p = 5$, we deduce that P has an irreducible factor of degree at least n. So, if P is reducible over the rationals, then P must have a rational root. But then P has an integer root, which is a divisor d of 1995. If $|d| > 1$, then

$$|d^{n+1} + kd^n| > |870d^2 - 1945d - 1995|,$$

hence $P(d) \ne 0$. It follows that $d = \pm 1$.
For $d = 1$, we get $1 + k - 870 + 1945 + 1995 = 0$, hence $k = -3071$.
For $d = -1$, we get

$$1 - k - 870 - 1945 + 1995 = 0 \text{ or } -1 + k - 870 - 1945 + 1995 = 0,$$

hence $k = -819$ or 821.

3.25 1996

84. *a*) For $m = 1$, the given inequality is an equality. If $m > 1$, then the function $x \mapsto x^m$ is convex, so by the Jensen's inequality,

$$\frac{a^m + b^m}{2} \geq \left(\frac{a+b}{2}\right)^m.$$

Adding with two similar inequalities, we obtain the desired result.

b) 1. Answer: no. Fix b, c and consider the polynomial function

$$P(a) = 2^m(a^m + b^m + c^m) - (a+b)^m - (b+c)^m - (c+a)^m.$$

Take $m \geq 3$ an odd integer. Since the coefficient of a^m is $2^m - 2$, we have

$$\lim_{a \to -\infty} P(a) = -\infty .$$

2. Answer: no. Choose $a, b, c > 0$, $c \to 0$ and $m = -1$. Then

$$\frac{1}{a+b} \leq -\frac{a+b}{2ab},$$

contradiction.

85. From

$$\begin{aligned} f(x) &= x^2 f\left(\frac{1}{x}\right) = x^2 f\left(1 + \frac{1-x}{x}\right) = x^2 + x^2 f\left(\frac{1-x}{x}\right) = x^2 + x^2 \cdot \left(\frac{1-x}{x}\right)^2 f\left(\frac{x}{1-x}\right) \\ &= x^2 + (1-x)^2 f\left(\frac{1}{1-x} - 1\right) = x^2 + (1-x)^2 f\left(\frac{1}{1-x}\right) - (1-x)^2 \\ &= x^2 - (1-x)^2 + f(1-x) = 2x + f(-x) = 2x - f(x), \end{aligned}$$

we deduce that $f(x) = x$, $x \neq 0, 1$. This function extends to $f(x) = x$ for all x, which is the only solution.

86. Denote

$$F_n(x) = f(x) f(x^2)^{\frac{1}{2}} \ldots f(x^n)^{\frac{1}{n}}.$$

Induction on n we will prove that $F_n(x) \geq f(x^n)$ for every positive real number x. The base case is obvious. For the inductive step, suppose that $F_n(x) \geq f(x^n)$ for $1 \leq n \leq m$, then

$$\begin{aligned} F_{m+1}(x)^{m+1} &= F_m(x)^{m+1}f(x^{m+1}) = F_{m-1}(x)^m f(x^m)f(x^{m+1})F_m(x) = \ldots \\ &= f(x)f(x^2)\ldots f(x^{m+1})F_m(x)F_{m-1}(x)\ldots F_1(x) \\ &\geq (f(x)f(x^m))(f(x^2)f(x^{m-1}))\ldots(f(x^m f(x)))f(x^{m+1}) \geq f(x^{m+1})^{m+1}. \end{aligned}$$

87. The polynomial function $P(x)$ is even and $P(0) = 1$, so $P(x) = ax^4 + bx^2 + 1$. Clearly, $x_2 = -x_1$, so $x_1 = \pm 1$. From $P'(x_1) = P'(x_2) = 0$, it follows that $b = -2a$, hence $P(x) = ax^4 - 2ax^2 + 1$. We must have $a > 0$, otherwise $P(x) < 0$ for large x. Also, $P(1) \geq 0$ implies $a \leq 1$. These polynomials satisfy the required conditions.
88. For any additive function f: $\mathbb{R} \to \mathbb{R}$ one has $f(m) = cm$, $\forall\, m \in \mathbb{Z}$, where $c = f(1)$. Then

$$\prod_{i=1}^{n} f_i(1+mx) = \prod_{i=1}^{n} [c_i + mf_i(x)] = a(1+mx)^n, c_i = f_i(1), m \in \mathbb{Z}, x \in \mathbb{R}.$$

Consider the polynomials

$$P_x(T) = \prod_{i=1}^{n} [c_i + f_i(x)T], Q_x(T) = a(1+xT)^n.$$

Case 1. $a \neq 0$. Then

$$a = \prod_{i=1}^{n} f_i(1) = \prod_{i=1}^{n} c_i \neq 0,$$

hence $c_i \neq 0$, $1 \leq i \leq n$. In this case, $P_x(T) = Q_x(T) \neq 0$, so there exist real numbers b_i, $1 \leq i \leq n$, such that $c_i + f_i(x)T = b_i(1 + xT)$, therefore $b_i = c_i$, $f_i(x) = c_i x$, $1 \leq i \leq n$, $x \in \mathbb{R}$.

Case 2. $a = 0$. We will prove by induction on n that there exists i such that $f_i(x) = 0$ for all $x \in \mathbb{R}$. For $n = 1$ we are nothing to prove. Assume that the assertion is valid for n and let $f_1, f_2, \ldots, f_{n+1}$ be additive functions such that

$$\prod_{i=1}^{n+1} f_i(x) = 0$$

and $f_{n+1}(x) \neq 0$. Then there exists $x_0 \in \mathbb{R}$ for which $f_{n+1}(x_0) \neq 0$. Consider the product

$$0 = \prod_{i=1}^{n+1} f_i(x_0 + my) = \prod_{i=1}^{n+1} [f_i(x_0) + mf(y)], m \in \mathbb{Z}, y \in \mathbb{R}.$$

We conclude that

$$P_{x_0,y}(T)=\prod_{i=1}^{n+1}[f_i(x_0)+f_i(y)T]=0.$$

Since $f_{n+1}(x_0)+f(y)T\neq 0$, it follows that

$$Q_{x_0,y}(T)=\prod_{i=1}^{n}[f_i(x_0)+f_i(y)T]=0,$$

hence

$$\prod_{i=1}^{n}f_i(y)=0, \forall y\in\mathbb{R}$$

and the induction hypothesis applies.

89. We have

$$\sum_{k=1}^{n-1}k^2x_k=\sum_{k=1}^{n-1}[(k-1)(k+1)+1]x_k=\sum_{k=1}^{n-1}x_k$$
$$+\sum_{k=1}^{n-1}(k-1)(k+1)x_k\le n+\sum_{k=1}^{n-1}(k-1)nx_k=n$$
$$+n\sum_{k=1}^{n-1}(k-1)x_k=n+n(2n-2-n)=n^2-n,$$

hence

$$\sum_{k=1}^{n-1}k(2n-k)x_k=2n(2n-2)-\sum_{k=1}^{n-1}k^2x_k\ge 2n(2n-2)-\left(n^2-n\right)=3n(n-1).$$

Equality holds when $x_1=n-1, x_2=\ldots=x_{n-2}=0, x_{n-1}=1$.

90. $(i)\Longrightarrow(ii)$: By the Cauchy–Schwarz inequality,

$$a^2x+b^2y+c^2z\ge\left(a^2x+b^2y+c^2z\right)\left(\frac{1}{x}+\frac{1}{y}+\frac{1}{z}\right)$$
$$\ge\left(a\sqrt{x}\cdot\frac{1}{\sqrt{x}}+b\sqrt{y}\cdot\frac{1}{\sqrt{y}}+c\sqrt{z}\cdot\frac{1}{\sqrt{z}}\right)^2=(a+b+c)^2>d^2$$

$(ii)\Longrightarrow(i)$: If $x\le 0$, then taking $a=d=n, b=c=1$, we get $y+z>n^2(1-x)$, which is impossible for large n. Therefore, $x>0$, and similarly, $y, z>0$.

Choose now a quadrilateral with sidelengths

$$a=\frac{1}{x}, b=\frac{1}{y}, c=\frac{1}{z}, d=\frac{1}{x}+\frac{1}{y}+\frac{1}{z}-\frac{1}{n}$$

with n sufficiently large. Then we have

$$\frac{1}{x^2}\cdot x+\frac{1}{y^2}\cdot y+\frac{1}{z^2}\cdot z=\frac{1}{x}+\frac{1}{y}+\frac{1}{z}>\left(\frac{1}{x}+\frac{1}{y}+\frac{1}{z}-\frac{1}{n}\right)^2,$$

so that

$$\frac{1}{x}+\frac{1}{y}+\frac{1}{z}\geq\left(\frac{1}{x}+\frac{1}{y}+\frac{1}{z}\right)^2,$$

whence

$$\frac{1}{x}+\frac{1}{y}+\frac{1}{z}\leq 1.$$

91. From

$$\begin{aligned}&\frac{7}{4}\left((a+b)^4+(b+c)^4+(c+a)^4\right)-(a^4+b^4+c^4)\\&=\frac{3}{4}\sum_{cyc}\left((a+b)^2-(a+c)^2\right)^2+(a+b+c)^4\geq 0,\end{aligned}$$

it follows that the minimum value is 0, attained for $a=b=c=0$.

3.26 1997

92. Denote by $P(x,y)$ the given functional equation and let $a=f(0)$. Then

$$\begin{aligned}P(0,0)&\Rightarrow f(a)=a,\\P(a,a)&\Rightarrow f(2a)=0,\\P(0,2a)&\Rightarrow a=0\Rightarrow f(0)=0,\\P(0,y)&\Rightarrow f(f(y))=-y.\end{aligned}$$

It follows that $-f(y)=f(f(f(y)))=f(-y)$, that is, f is an odd function. We deduce

$$f(u+v)=f(u+f(-v))=f(u)+f(v)$$

for all integers u, v. By induction, we have $f(n) = nf(1)$ for all integers. If $n = f(1)$, then $f(f(1)) = (f(1))^2$, but $f(f(1)) = -1$, contradiction.

93. *a*), *b*) We can suppose that $p, q \geq 0$. Then

$$p^2+q^2+1 \geq p^2+2\left(\frac{q+1}{2}\right)^2 \geq 2\sqrt{\frac{p^2(q+1)^2}{2}}=\sqrt{2}p(q+1),$$

with equality for $(p,q)=(\sqrt{2},1)$, hence $b=\sqrt{2}$.

c) If $(p,q) = (1,1)$, then

$$\frac{p^2+q^2+1}{p(q+1)}=\frac{3}{2},$$

hence $c \leq \frac{3}{2}$. In fact, $c=\frac{3}{2}$, since

$$q=0: p^2+1 \geq \frac{3}{2}p;$$

$$q=1: p^2+2 \geq 3p \Leftrightarrow (p-1)(p-2) \geq 0;$$

$$q=2: p^2+5 \geq \frac{9}{2}p \Leftrightarrow (p-2)\left(p-\frac{5}{2}\right) \geq 0;$$

$$q \geq 3: p(3-q)-2<0 \leq 2(p-q)^2 \Rightarrow p^2+q^2+1 \geq \frac{3}{2}p(q+1).$$

94. Answer: $f(x) = a_0(x^2 + \alpha^2)^n$, $a_0 > 0$, $\alpha \in \mathbb{R}^*$.

Denote $[n] = \{1, 2, \ldots, n\}$ and let Z be the set of roots of f. Then $0 \notin Z$ and $\alpha i \in Z$ implies $-\alpha i \in Z$, $\alpha \in \mathbb{R}^*$, hence

$$Z=\{\alpha_1 i, \ldots, \alpha_n i, -\alpha_1 i, \ldots, -\alpha_n i\}, \alpha_1, \ldots, \alpha_n \in \mathbb{R}^*.$$

We can assume that $\alpha_1, \ldots, \alpha_n > 0$. Define $Z' = \{\alpha_1 i, \ldots, \alpha_n i\}$, $A_k = \{S \subset Z \mid |S| = k\}$. By the Viète's formulas,

$$\frac{a_{2k}}{a_0}=\sum_{S \in A_{2k}}\left(\prod_{a \in S} a\right).$$

We will show that

$$\frac{a_{2k}}{a_0} = \sum_{|S|=k,\, S\subset Z'} \left(\prod_{a\in S} a\right)^2, k\in[n].$$

For $k = 1$,

$$\frac{a_2}{a_0} = \sum_{S\in A_2}\left(\prod_{a\in S} a\right) = \sum_{j=1}^{n}\alpha_j^2 - \sum_{k,l\in[n]}(\alpha_k\alpha_l + (-\alpha_k)(-\alpha_l) + (-\alpha_k)\alpha_l + \alpha_k(-\alpha_l)) = \sum_{j=1}^{n}\alpha_j^2,$$

so the assertion is true.

95. (*Generalization*). Let $a_1, a_2, \ldots, a_n$ be real numbers such that $a_1^2 \le 2a_2$ and at least one of the numbers $a_3, \ldots, a_n$ is nonzero. Prove that the polynomial

$$P(x) = x^n + a_1x^{n-1} + a_2x^{n-2} + \ldots + a_{n-1}x + a_n$$

cannot have only real roots.

Solution. Let $x_1, \ldots, x_n$ be the roots of the polynomial. By the Viète's relations,

$$\sum_{i=1}^{n} x_i^2 = \left(\sum_{i=1}^{n} x_i\right)^2 - 2\sum_{1\le i<j\le n} x_ix_j = (-a_1)^2 - 2a_2 \le 0.$$

If $a_1^2 < 2a_2$, then the roots cannot be all real, since the sum of squares of real numbers is nonnegative. If $a_1^2 = 2a_2$ and the roots are all real, then $x_1 = \ldots = x_n = 0$, contradiction, since at least one of the numbers $a_3, \ldots, a_n$ is nonzero.

96. If $y = z$, then from the first equality we deduce $x = y$. It follows x, y, z that are distinct. For $t = 1$ there are infinitely many triples (x, y, z) satisfying the conditions:

$$\left(x, \frac{1}{1-x}, \frac{x-1}{x}\right),\ x\neq 0, 1.$$

For $t = -1$ there are infinitely many triples (x, y, z) satisfying the conditions:

$$\left(x, -\frac{1}{x+1}, -\frac{x+1}{x}\right),\ x\neq 0, -1.$$

For $t \neq \pm 1$ there are no such triples: subtract each pair of equations, multiply the obtained equalities, simplify by $(x - y)(y - z)(z - x)$ and get $(xyz)^2 = 1$, whence $t = -xyz = \pm 1$, contradiction.

97. Answer: no.

If f is such a function, then the first equality implies that it is bijective. Putting $x = 0$, respectively, $x = 1$ in the given equalities, we obtain

$$f(f(0)) = 0 = f(f(1) + 1),$$

hence $f(0) = f(1) + 1$. Putting $x = 1$, respectively, $x = 0$, we obtain

$$f(f(1)) = 1 = f(f(0) + 1),$$

hence $f(1) = f(0) + 1$, which contradicts the above equality.

98. Using the Cauchy–Schwarz inequality,

$$\begin{aligned}
&\left(\frac{1}{a+b} + \frac{1}{a+2b} + \dots + \frac{1}{a+nb}\right)^2 \\
&\quad \le n\left[\frac{1}{(a+b)^2} + \frac{1}{(a+2b)^2} + \dots + \frac{1}{(a+nb)^2}\right] \\
&\quad < n\left[\frac{1}{a(a+b)} + \frac{1}{(a+b)(a+2b)} + \dots + \frac{1}{(a+(n-1)b)(a+nb)}\right] \\
&\quad = \frac{n^2}{a(a+nb)},
\end{aligned}$$

whence the conclusion.

99. Denote $\alpha + \beta = r \in \mathbb{Q}$, then $p(\alpha) = p(r - \beta) = 0$. It follows that β is a root of the polynomial $p(r - X)$. The polynomials $p(r - X)$ and $q(X)$ have rational coefficients, are irreducible, and have the common root β, therefore they differ by a rational multiplicative constant: $p(r - X) = cq(X)$. Taking into account that p, q are monic, we deduce $c = \pm 1$. From $p(r - X) = \pm q(X)$, we obtain $p^2(r - X) = q^2(X)$. Then

$$p^2\left(r - \frac{r}{2}\right) = q^2\left(\frac{r}{2}\right),$$

whence it follows that $\frac{r}{2}$ is a root of the polynomial $p^2 - q^2$.

100. We can take $v = 1$, due to symmetry. Then $u = e^{\alpha i}$, with $a = \frac{2k\pi}{1997}$, $0 \le k \le 1996$,

$$|u + v|^2 = (1 + \cos\alpha)^2 + (\sin\alpha)^2 = 2(1 + \cos\alpha).$$

Therefore,

$$P\left(|u+v|\geq\sqrt{2+\sqrt{3}}\right)=P\left(\cos\alpha\geq\frac{\sqrt{3}}{2}\right)=P\left(\alpha\in\left[-\frac{\pi}{6},\frac{\pi}{6}\right]\right).$$

Now, $\alpha=\frac{2k\pi}{1997}$, $-998\leq k\leq 998$,

$$P\left(\alpha\in\left[-\frac{\pi}{6},\frac{\pi}{6}\right]\right)=P\left(-\frac{\pi}{6}\leq\frac{2k\pi}{1997}\leq\frac{\pi}{6}\right)=P(-166\leq k\leq 166)=\frac{333}{1997}.$$

101. Answer: 1999.

We will prove that the pairs having the desired property are $(0,0)$ and

$$\left(\cos\left(\frac{\pi}{3996}+\frac{2k\pi}{1998}\right),\ \sin\left(\frac{\pi}{3996}+\frac{2k\pi}{1998}\right)\right), 0\leq k\leq 1997.$$

Indeed, $a_{n+1}+ib_{n+1}=(\alpha+i\beta)(a_n+ib_n)$, hence $a_n+ib_n=(\alpha+i\beta)^n$ for all n. Then $(a_{1997},b_{1997})=(b_1,a_1)$ if and only if $(\alpha+i\beta)^{1997}=i(\alpha-i\beta)$, that is, $z^{1997}=i\bar{z}$, with $z=\alpha+i\beta$. A solution of this equation is $z=0$. Clearly, a solution $z\neq 0$ satisfies $|z|=1$, i. e., $z\bar{z}=1$. The solutions of the equation $z^{1998}=i$ are those listed above by components, whence the conclusion.

102. Answer: $\frac{1}{n(n-1)}$.

More generally, we will find the minimal value of

$$\sum_{i=1}^{n}\frac{x_i^p}{S-x_i^r},$$

subject to $x_1^s+x_2^s+\ldots+x_n^s=1$, where $S=x_1^r+x_2^r+\ldots+x_n^r$ and p,r,s are positive real numbers such that $r\leq s\leq\frac{p}{2}$. By the Cauchy–Schwarz inequality,

$$\left(\sum_{i=1}^{n}(S-x_i^r)\right)\left(\sum_{i=1}^{n}\frac{x_i^p}{S-x_i^r}\right)\geq\left(\sum_{i=1}^{n}x_i^{\frac{p}{2}}\right)^2,$$

with equality only when there is a positive real constant λ such that

$$S-x_i^r=\lambda x_i^{\frac{p}{2}}, 1\leq i\leq n.$$

Applying the power mean inequality, we get

$$\left(\sum_{i=1}^{n} x_i^{\frac{p}{2}}\right)^2 = n^2\left(\sum_{i=1}^{n} \frac{x_i^{\frac{p}{2}}}{n}\right)^2 \geq n^2\left(\sum_{i=1}^{n} \frac{x_i^s}{n}\right)^{\frac{p}{s}} = n^{2-\frac{p}{s}},$$

with equality only when $p = 2s$ or $x_1 = x_2 = \ldots = x_n = \left(\frac{1}{n}\right)^{\frac{1}{s}}$. Again, the power mean inequality implies

$$\sum_{i=1}^{n}(S - x_i^r) = (n-1)\sum_{i=1}^{n} x_i^r = n(n-1)\sum_{i=1}^{n}\frac{x_i^r}{n} \leq n(n-1)\left(\sum_{i=1}^{n}\frac{x_i^s}{n}\right)^{\frac{r}{s}} = \frac{n(n-1)}{n^{\frac{r}{s}}},$$

with equality only when $r = s$ or $x_1 = x_2 = \quad = x_n = \left(\frac{1}{n}\right)^{\frac{1}{s}}$. Finally,

$$\sum_{i=1}^{n}\frac{x_i^p}{S - x_i^r} \geq \frac{\left(\sum_{i=1}^{n} x_i^{\frac{p}{2}}\right)^2}{\sum_{i=1}^{n}(S - x_i^r)} \geq \frac{\frac{n^2}{n^{\frac{p}{s}}}}{\frac{n(n-1)}{n^{\frac{r}{s}}}} = \frac{n^{\frac{r+s-p}{s}}}{n-1}.$$

In the given case, $p = 5$, $r = 1$, $s = 2$.

103. The set of solutions of the given equation is

$$S = \{0\} \cup \left\{-\frac{1}{4n^2} \,\middle|\, n \in \mathbb{N}^*\right\}.$$

Indeed, we have the equivalences

$$4n^2x = \log_2(2n^2x + 1) \Leftrightarrow 4^{2n^2x} = 2n^2x + 1 \Leftrightarrow 2n^2x \in \left\{-\frac{1}{2}, 0\right\} \Leftrightarrow x \in S.$$

We will prove that $a^x + b^x \geq 2 + 3x$ for all $x \in S$ if and only if $ab \leq e^3$. We have

$$\lim_{x\to 0}\frac{a^x + b^x - 2}{x} - \ln(ab).$$

If $a^x + b^x \geq 2 + 3x$ for all $x \in S$, then

$$\frac{a^{-\frac{1}{4n^2}} + b^{-\frac{1}{4n^2}} - 2}{-\frac{1}{4n^2}} \leq 3$$

for all positive integers n. Taking $n \to \infty$, we obtain $\ln(ab) \leq 3$, hence $ab \leq e^3$.

Conversely, if $ab \leq e^3$, then the power mean inequality implies

$$\left(\frac{a^x+b^x}{2}\right)^{\frac{1}{x}} \leq \sqrt{ab} \leq \sqrt{e^3}, x<0.$$

It follows that

$$\frac{a^x+b^x}{2} \geq \sqrt{e^{3x}} \geq 1+\frac{3x}{2}.$$

This inequality is true also for $x = 0$, hence it holds for all $x \in S$.

3.27 1998

104. *a*) Let $P(x) = x^n + a_1x^{n-1} + \dots + a_{n-1}x + a_n$ and $x_1, x_2, \dots, x_n$ be its roots, with $|x_i| = 1$, $1 \leq i \leq n$. By the Viète's formulas,

$$|a_k| = \left|(-1)^k \sum_{1 \leq i_1 < \dots < i_k \leq n} x_{i_1}x_{i_2}\dots x_{i_k}\right| \leq \binom{n}{k}.$$

Since the coefficients are integers, it follows that they have only finitely many values, so the number of the polynomials $P(x)$ is finite.

b) Let r be a positive integer and consider the polynomial

$$Q_r(x) = \left(x - x_1^r\right)\left(x - x_2^r\right)\dots\left(x - x_n^r\right).$$

Then the coefficients of $Q_r(x)$ are symmetric polynomials of $x_1, x_2, \dots, x_n$, so can be expressed as polynomials with integer coefficients of the elementary symmetric polynomials, hence as polynomials with integer coefficients of the coefficients of $P(x)$. Therefore, $Q_r(x)$ is a monic polynomial with integer coefficients having all roots on the unit circle.

Let $x_k = e^{i\alpha_k}$. It suffices to show that

$$\alpha_k = \frac{2\pi p_k}{q_k}, p_k, q_k \in \mathbb{Z}, q_k > 0, 1 \leq k \leq n.$$

Assume that there is k for which α_k is not a rational multiple of 2π. Then the set

$$\{x_k^r | r \in \mathbb{N}\}$$

is infinite, which contradicts *a*).

105. We will prove a more general assertion: if n is a positive integer, $k > 1$ and $x_0, x_1, \dots, x_n$ are strictly positive real numbers, then

$$\sum_{i=0}^{n} \frac{x_i}{x_i + k \sum\limits_{j=0, j\neq i}^{n} x_j} \geq \frac{n+1}{nk+1}.$$

Denote

$$S = \sum_{i=0}^{n} x_i, a_i = x_i + k \sum_{\substack{j=0 \\ j\neq i}}^{n} x_j,$$

then

$$x_i = \frac{kS - a_i}{k-1}$$

for all i and

$$\frac{x_i}{a_i} \leq \frac{x_j}{a_j} \Leftrightarrow \frac{kS - a_i}{a_i} \leq \frac{kS - a_j}{a_j} \Leftrightarrow a_j \leq a_i$$

for all $i, j \in \{0, 1, \ldots, n\}$. It follows that

$$\sum_{i=0}^{n} \sum_{j=0}^{n} \left(\frac{x_i}{a_i} - \frac{x_j}{a_j}\right)(a_j - a_i) \geq 0 \Leftrightarrow \left(\sum_{i=0}^{n} \frac{x_i}{a_i}\right)\left(\sum_{i=0}^{n} a_i\right) \geq (n+1)S = \frac{n+1}{nk+1} \sum_{i=0}^{n} a_i,$$

hence

$$\sum_{i=0}^{n} \frac{x_i}{a_i} \geq \frac{n+1}{nk+1}.$$

106. Let $ABCDA'B'C'D'$ be the parallelepiped, $a = AB$, $b = AA'$, $c = AD$, $S_1 = S[ABB'A']$, $S_2 = S[ABCD]$, $S_3 = S[ADD'A']$ and V be the volume of the parallelepiped. Then

$$V^2 \leq S_1 S_2 S_3,$$

with equality if and only if the parallelepiped is rectangular. Indeed, let $A'M \perp$

$\perp (ABCD), A'N \perp AB$. Then

$$V = A'M \cdot S_2 \leq A'N \cdot S_2 = \frac{S_1}{a} \cdot S_2$$

and similarly

$$V \leq \frac{S_1 S_3}{b}, V \leq \frac{S_2 S_3}{c},$$

whence

$$V^3 \leq \frac{(S_1 S_2 S_3)^2}{abc}.$$

Since $V = S_2 \cdot A^{'}M \leq ac \cdot AA^{'} = abc$, it follows $V^4 \leq abcV^3 \leq (S_1S_2S_3)^2$, so

$$V^2 \leq S_1 S_2 S_3,$$

with equality only if $AB \perp AD \perp AA^{'} \perp AB$.
On the other hand,

$$S_1 S_2 S_3 \leq \left(\frac{S_1 + S_2 + S_3}{3}\right)^3 = 36^3 = 216^2 = V^2,$$

whence $S_1 = S_2 = S_3$ and the rectangular parallelepiped is a cube.

107. For $n = 0$, the conclusion is obvious. Assume $n \geq 1$. To each polynomial

$$g = a_0 + a_1 X + \ldots + a_n X^n \in \mathbb{Z}[X]$$

we will associate the polynomial

$$\widehat{g} = \widehat{a_0} + \widehat{a_1} X + \ldots + \widehat{a_n} X^n \in \mathbb{Z}_2[X].$$

Since

$$\left(X^2 + X + \widehat{1}\right)^{2^n} = \left[(X^2 + X)^2 + \widehat{1}\right]^{2^{n-1}} = \left[(x^2 + X)^{2^2} + \widehat{1}\right]^{2^{n-2}} = \ldots = (X^2 + X)^{2^n} + \widehat{1},$$

it follows that $\widehat{f} = \left(X^2 + X + \widehat{1}\right)^{2^n}$.

If $f = gh$, with $g, h \in \mathbb{Z}[X]$, then $\widehat{f} = \widehat{g}\widehat{h}$. The polynomial $X^2 + X + \widehat{1}$ is irreducible over $\mathbb{Z}_2$, hence $\widehat{g} = \left(X^2 + X + \widehat{1}\right)^p, \widehat{h} = \left(X^2 + X + \widehat{1}\right)^{2^n - p}$, $1 \leq p \leq 2^n - 1$. We deduce

$$g = \left(X^2 + X + 1\right)^p + 2u(X), h = \left(X^2 + X + 1\right)^{2^n - p} + 2v(X), u, v \in \mathbb{Z}[X].$$

Let ε be a root of the polynomial $X^2 + X + 1$. Replacing X by ε in the equality

$$f(X) = \left(X^2 + X\right)^{2^n} + 1 = \left[\left(X^2 + X + 1\right)^p + 2u(X)\right]\left[\left(X^2 + X + 1\right)^{2^n - p} + 2v(X)\right]$$

we get $2 = 2u(\varepsilon) \cdot 2v(\varepsilon)$, therefore $u(\varepsilon)v(\varepsilon) = \frac{1}{2}$. But uv is an integer polynomial and $\varepsilon^2 = -\varepsilon - 1$, whence we deduce that $u(\varepsilon)v(\varepsilon) = a + b\varepsilon$, with a, b integers, contradiction.

108. Denote $b_k = \sqrt{a_k} - \sqrt{a_{k+1}}, 1 \le k \le n$. Then $a_i = (b_i + \ldots + b_n)^2$ and after squaring, the inequality becomes

$$\sum_{k=1}^{n} kb_k^2 + 2 \sum_{1 \le k < l \le n} kb_k b_l \le \sum_{k=1}^{n} kb_k^2 + 2 \sum_{1 \le k < l \le n} \sqrt{kl} b_k b_l,$$

which is obvious.

109. The minimum value of the function is attained when the two radicals coincide, that is,

$$a^2 + x^2 = (b - x)^2 + c^2,$$

whence it follows

$$x = \frac{b^2 + c^2 - a^2}{2b}, f_{min} = f\left(\frac{b^2 + c^2 - a^2}{2b}\right) = \frac{\sqrt{4a^2b^2 + \left(b^2 + c^2 - a^2\right)^2}}{b}.$$

3.28 1999

110. If $\cos\theta = \frac{1}{x}$, then $\sin\theta = \frac{1}{4 - \sqrt{3}x}$, hence $\sqrt{3}\sec\theta + \csc\theta = 4$. Conversely, if θ is a solution of this equation, then $x = \sec\theta$ is a solution of the given equation. Now, we have the equivalences

$$\sqrt{3}\sec\theta + \csc\theta = 4 \Leftrightarrow 2\sin\theta\cos\theta = \frac{\sqrt{3}}{2}\sin\theta + \frac{1}{2}\cos\theta \Leftrightarrow \sin 2\theta$$
$$= \sin\left(\theta + \frac{\pi}{6}\right) \Leftrightarrow \theta \in \left\{\frac{\pi}{6} + 2k\pi | k \in \mathbb{Z}\right\} \cup \left\{\frac{5\pi}{18} + \frac{2k\pi}{3} | k \in \mathbb{Z}\right\}.$$

Distinct values of x are

$$\sec\frac{\pi}{6}=\frac{2\sqrt{3}}{3},\ \sec\frac{5\pi}{18},\ \sec\frac{17\pi}{18},\ \sec\frac{29\pi}{18}.$$

111. In fact, the statement holds for all positive real numbers: if, say, $c \leq a$, $c \leq b$, then the discriminant of the second equation

$$\Delta = 4\left(c^2 - 2ab\right)$$

is negative.

112. *a*) Let x_k, $1 \leq k \leq 4$, be the roots of P. Since $P \in \mathbb{R}[x], x_1 \notin \mathbb{R}$ and $|x_1| = 1$, it follows that $x_2 = \overline{x_1}$ and $x_1x_2 = x_1\overline{x_1} = |x_1|^2 = 1$. By the Viète's relations,

$$x_1 + x_2 + x_3 + x_4 = x_1x_2(x_3 + x_4) + x_3x_4(x_1 + x_2) = -3,$$

hence $x_1 + x_2 = x_3x_4(x_1 + x_2)$. But $x_1 + x_2 = x_1 + \overline{x_1} = 2\,Re(x_1) \neq 0$, so $x_3x_4 = 1$ and

$$p = x_1x_2 \cdot x_3x_4 = 1.$$

b) The polynomial $P = x^4 + 3x^3 + 3x + 1$ is reciprocal, so we can use the substitution $x + \frac{1}{x} = t$, where t are the roots of the quadratic function $t^2 - 2 + 3t$, so

$$t_{1,2} = \frac{-3 \pm \sqrt{17}}{2}.$$

Then

$$P = \left(x^2 + \frac{3 - \sqrt{17}}{2} \cdot x + 1\right)\left(x^2 + \frac{3 + \sqrt{17}}{2} \cdot x + 1\right) = P_1P_2.$$

Here x_1, x_2 are the roots of P_1 and x_3, x_4 are the roots of P_2.

c) We need to show that if

$$2\cos\alpha = \frac{\sqrt{17} - 3}{2},$$

then α is not a rational multiple of π. The polynomial P is the minimal polynomial of x_1 over $\mathbb{Q}$ and it is not a cyclotomic polynomial, so x_1 is not a root of unity.

113. Denote

$$P(x_1, \ldots, x_n) = \sum_{j=1}^{n} \left(x_j^4 - x_j^5\right).$$

Fix x_3, …, x_n and note that

$$\left(x_1^4 - x_1^5\right) + \left(x_2^4 - x_2^5\right) \leq (x_1 + x_2)^4 - (x_1 + x_2)^5 \Leftrightarrow$$

$$\Leftrightarrow 5(x_1 + x_2)\left(x_1^2 + x_1x_2 + x_2^2\right) \leq 4x_1^2 + 6x_1x_2 + 4x_2^2,$$

which is true for $x_1 + x_2 \leq \frac{4}{5}$. Then we have

$$P(x_1, x_2, \ldots, x_n) \leq P(0, x_1 + x_2, x_3, \ldots, x_n) \leq P(0, \ldots, 0, x, y).$$

Choose $x = \frac{1}{2} + t, y = \frac{1}{2} - t,$ and so

$$P(0, \ldots, 0, x, y) = -3\left(t^2 - \frac{1}{12}\right)^2 + \frac{1}{12} \leq \frac{1}{12}.$$

114. *a*) Answer: $f(x) = x$ for all x; $f(x) = -x$ for all x.

Denote $f^2 = f \circ f$. Let f be a function satisfying the required properties. From $x = y = 0$ we get $f^2(0) = f(0)$, hence $f(0) = 0$, because f is injective.

If f is strictly increasing, then $f^2(a) = a$ implies $f(a) = a$: indeed, $f(a) < a$ yields $a = f^2(a) < f(a)$, contradiction, and similarly, $f(a) > a$ leads to a contradiction. For $a = x + f(y)$, we have

$$f^2(a) = f(y + f(x)) = f(y) + x = a,$$

hence $f(a) = a$. But then $f(x) + y = x + f(y)$, therefore the function $x \mapsto f(x) - x$ is constant, and this constant is 0, since $f(0) = 0$. Finally, $f(x) = x$ for all x.

If f is strictly decreasing, then for all x, y,

$$f\left(x + f^2(y)\right) = f(x) + f(y) \Rightarrow f^2\left(x + f^2(y)\right) = f(f(x) + f(y)) \Rightarrow f^2(x) = x,$$

since f^2 is strictly increasing. We obtain the Cauchy functional equation

$$f(x + y) = f(x) + f(y),$$

with solution $f(x) = xf(1)$. Then $f^2(1) = (f(1))^2$. Taking $(x, y) = (0, 1)$ in the given functional equation, we obtain $f^2(1) = f(0) + 1 = 1$, so $(f(1))^2 = 1$. But $f(1) < 0$, because f is strictly decreasing and $f(0) = 0$. Finally, $f(x) = -x$ for all x.

Clearly, these functions satisfy the required conditions.

b) If f is strictly monotone and satisfies

$$f(x+f(y))=f(x)+y^n$$

for all x, y and a fixed integer $n>1$, then $f(0)=0$ and $f^2(y)=y^n$ for all y. Hence

$$f(f^2(y))=f^2(f(y))=(f(y))^n$$

for all y. For all x, y, we have

$$f^2(x+f(y))=(x+f(y))^n$$

and also

$$f^2(x+f(y))=f(f(x)+y^n)=f(y^n)+x^n=f(f^2(y))+x^n=(f(y))^n+x^n.$$

It follows that

$$(x+f(y))^n=x^n+(f(y))^n$$

for all x, y. Choose $x=1$ and $b:=f(y)>0$ to get $(1+b)^n=1+b^n$, contradiction.

115. Let (x,y,z) be a solution in nonzero real numbers of the given system. The system has the equivalent form

$$\frac{p}{y^2}+\frac{p}{z^2}=a,\quad \frac{p}{z^2}+\frac{p}{x^2}=b,\quad \frac{p}{x^2}+\frac{p}{y^2}=c,$$

where $p=xyz$. We deduce $p>0$ and, after calculations,

$$\frac{1}{x^2}=\frac{-a+b+c}{2p},\quad \frac{1}{y^2}=\frac{a-b+c}{2p},\quad \frac{1}{z^2}=\frac{a+b-c}{2p}. \qquad (1)$$

It follows that all numerators are positive, hence a, b, c are the sidelengths of a triangle.

Conversely, if a, b, c are the sidelengths of a triangle, then all numerators in (1) are positive, hence $p>0$. Multiplying the equalities in (1), we get

$$\frac{p^3}{x^2y^2z^2}=p=\frac{(-a+b+c)(a-b+c)(a+b-c)}{8}.$$

Returning to (1), we can find x, y, z, so the given system has a real solution.

116. Denote $x_k=a_k^n, k=1,2,\ldots,n$. Then $a_1a_2\ldots a_n=1$ and

$$\sum_{k=1}^{n}\frac{1}{n-1+x_k}=\sum_{k=1}^{n}\frac{1}{n-1+\frac{a_k^{n-1}}{a_1\dots a_{k-1}a_{k+1}\dots a_n}}\leq\sum_{k=1}^{n}\frac{1}{n-1+\frac{(n-1)a_k^{n-1}}{a_1^{n-1}+\dots+a_{k-1}^{n-1}+a_{k+1}^{n-1}+\dots+a_n^{n-1}}}$$

$$=\frac{1}{n-1}\sum_{k=1}^{n}\frac{a_1^{n-1}+\dots+a_{k-1}^{n-1}+a_{k+1}^{n-1}+\dots+a_n^{n-1}}{a_1^{n-1}+\dots+a_n^{n-1}}$$

$$=\frac{(n-1)\sum_{k=1}^{n}a_k^{n-1}}{(n-1)\sum_{k=1}^{n}a_k^{n-1}}=1.$$

117. We will use the following

Lemma. Let a, b be positive real numbers and k be a positive integer. Then

$$\frac{a}{b}<\frac{a-x^2+(x+k+1)^2}{b+k+1}$$

if and only if

$$\frac{a}{b}<2x+k+1.$$

Proof. Obvious.

We have to show that the lower bound of the function

$$f(x_1,\dots,x_n)=\frac{x_1^2+x_2^2+\dots+x_n^2}{x_1+x_2+\dots+x_n}$$

over the set of ordered systems of integers $0<x_1<\dots<x_n$, is $\frac{2n+1}{3}$. Observe that

$$f(1,2,\dots,n)=\frac{2n+1}{3}\quad(*).$$

Now apply the lemma for an ordered system of the form

$$x_1<x_2<\dots<x_p<x+1<\dots<x+k+1,x\geq x_p+2,$$

and taking

$$a=x_1^2+\dots+x_p^2+(x+1)^2+\dots+(x+k+1)^2,$$

$$b=x_1+\dots+x_p+(x+1)+\dots+(x+k+1),$$

we deduce $f(x_1,\ldots,x_p,x,\ldots,x+k) < f(x_1,\ldots,x_p,x+1,\ldots,x+k+1)$. By induction, it follows that the lower bound is realized on a sequence of the form $x+1, \ldots, x+n$. Applying the lemma, we obtain $f(x,x+1,\ldots,x+n-1) < f(x+1,x+2,\ldots,x+n)$. Taking account of $(*)$, the result follows.

Second solution. We may suppose that $0 < x_1 < x_2 < \ldots < x_n$, then $1 \le x_1$, $2 \le \le x_2, \ldots, n \le x_n$. Let $x_k = k + a_k$, $1 \le k \le n$. Since $x_{k+1} - x_k \ge 1$, it follows that $a_{k+1} \ge a_k$. We can rewrite the proposed inequality as

$$3\sum_{k=1}^{n}(k+a_k)^2 \ge (2n+1)\sum_{k=1}^{n}k + (2n+1)\sum_{k=1}^{n}a_k.$$

If $a_k = 0$, $1 \le k \le n$, this inequality becomes an equality. Otherwise, we obtain

$$3\sum_{k=1}^{n}a_k^2 + 6\sum_{k=1}^{n}ka_k \ge (2n+1)\sum_{k=1}^{n}a_k.$$

By the Chebyshev's inequality,

$$6\sum_{k=1}^{n}ka_k \ge 6\cdot\frac{\left(\sum_{k=1}^{n}k\right)\left(\sum_{k=1}^{n}a_k\right)}{n} = \frac{6n(n+1)}{2n}\sum_{k=1}^{n}a_k > (2n+1)\sum_{k=1}^{n}a_k,$$

hence

$$3\sum_{k=1}^{n}a_k^2 + 6\sum_{k=1}^{n}ka_k > (2n+1)\sum_{k=1}^{n}a_k.$$

The inequality is verified, with equality if and only if $\{a_1, a_2, \ldots, a_n\} = \{1, 2, \ldots, n\}$.

118. Define

$$B_k = \left(1 + \frac{1}{m}\right)^k$$

and observe that $B_k > 1 + \frac{k}{m}, k \ge 2$. Then

$$B_k \le 1 + \frac{k}{m} + \frac{1}{m}\left[\frac{n(n-1)}{2!m} + \frac{n(n-1)(n-2)}{3!m^2} + \ldots + \frac{n(n-1)\ldots(n-k+1)}{k!m^{k-1}}\right], k \le n.$$

For $m > n^2$, we get

$$B_k < 1 + \frac{k}{m} + \frac{1}{m}\left(\frac{1}{2!} + \frac{1}{3!} + \ldots + \frac{1}{k!}\right) < 1 + \frac{k+1}{m}.$$

Define

$$A_k = 1 + \frac{k+1}{m}, 1 \le k \le n,$$

then $B_1 < A_1 < B_2 < A_2 < \ldots < B_n < A_n$. Clearly, the sequencens

$$a_k = m^n A_k, b_k = m^n B_k$$

satisfy the required conditions.

For $n = 5$, choose a geometric progression s^4, s^3t, s^2t^2, st^3, t^4, with $0 < s < t$ and an arithmetic progression a, $a + r$, $a + 2r$, $a + 3r$, $a + 4r$. Then

$$r = a_2 - a_1 < b_3 - b_1 = s^2t^2 - s^4, 3r = a_5 - a_2 > b_5 - b_3 = t^4 - s^2t^2,$$

hence $3s^2(t^2 - s^2) > 3r > t^2(t^2 - s^2)$. It follows that $3s^2 > t^2 \ge (s + 1)^2$, so $s \ge 2$. Choose $s = 2$, $t = 3$, $a = r = 17$, then the progressions

$$16 < 17 < 24 < 34 < 36 < 51 < 54 < 68 < 81 < 85$$

are as desired.

119. By induction we will prove that

$$x_{n+1} > \sum_{k=1}^{n} kx_k.$$

For $n = 1$ we have $x_2 \ge 3x_1 > x_1$ by definition. Suppose that the inequality is true for n, then

$$x_{n+2} \ge (n+3)x_{n+1} - \sum_{k=1}^{n} kx_k = (n+1)x_{n+1} + 2x_{n+1} - \sum_{k=1}^{n} kx_k > (n+1)x_{n+1}$$

$$+ 2\sum_{k=1}^{n} kx_k - \sum_{k=1}^{n} kx_k = \sum_{k=1}^{n+1} kx_k.$$

We deduce

$$x_{n+1} > \sum_{k=1}^{n} kx_k > nx_n.$$

By induction, $x_{n+1} > nx_n > n(n-1)x_{n-1} > \ldots > n!\,a$. For sufficiently large n we have $n!\,a > 1999!$.

120. If z is a root of f, then $|z| > 1$, otherwise

$$p = |z^n + az| = |z|\left|z^{n-1} + a\right| \leq 1 + |a|,$$

contradiction. Let $f = gh$ be a decomposition of the polynomial in $\mathbb{Z}[x]$. Then $p = g(0)h(0)$, hence $|g(0)| = 1$ or $|h(0)| = 1$. Suppose that $|g(0)| = 1$ and let $z_1, \ldots, z_k$ be the roots of g. These are also the roots of f hence

$$1 = |g(0)| = |z_1 \ldots z_k| > 1,$$

contradiction.

121. Define

$$L_k(x) = \prod_{\substack{j=0 \\ j \neq k}}^{2n} (x - j), 0 \leq k \leq 2n.$$

The polynomial

$$\sum_{k=0}^{2n} \frac{P(k)}{L_k(k)} L_k(x)$$

has degree at most $2n$ and takes the same values as P at $0, 1, \ldots, 2n$, therefore

$$P(x) = \sum_{k=0}^{2n} \frac{P(k)}{L_k(k)} L_k(x).$$

If $P(x) = a_{2n}x^{2n} + a_{2n-1}x^{2n-1} + \ldots + a_0$, then

$$a_{2n} = \sum_{k=0}^{2n} \frac{P(k)}{L_k(k)} = \sum_{k=0}^{2n} \frac{P(k)}{(-1)^k k!(2n-k)!} = \frac{1}{(2n)!} \sum_{k=0}^{2n} (-1)^k \binom{2n}{k} P(k)$$
$$= \frac{1}{(2n)!} \cdot \frac{1 + (1-2)^{2n}}{2} = \frac{1}{(2n)!}.$$

The polynomial $Q(x) = 2P(x) - P(x+1)$ has the same degree and the same leading coefficient as $P(x)$, so

$$Q(x)=a_{2n}(x-1)(x-2)\dots(x-(2n-1))(x-r)$$

for some real number r. From $Q(0) = 1$, we deduce

$$1=Q(0)=a_{2n}\cdot(-1)^{2n}\cdot r\cdot(2n-1)!=\frac{r}{2n},$$

hence $r = 2n$ and

$$2P(2n+1)-P(2n+2)=Q(2n+1)=\frac{(2n)!}{(2n)!}=1.$$

122. By the Cauchy–Schwarz inequality,

$$\left(\sum_{k=1}^{n}a_k\right)\left(\sum_{k=1}^{n}a_k^3\right)\geq\left(\sum_{k=1}^{n}a_k^2\right)^2,$$

with equality only for $a_k^3=\lambda a_k, 1\leq k\leq n$. Then $a_k=a>0, 1\leq k\leq n$ and

$$an=96,\quad a^2n=144, a^3n=216,$$

whence $a=\frac{3}{2}, n=64$.

123. Answer: $f(x) = x$ for all x.

Let f be a function which satisfies the given conditions and define

$$Y=\{y=x-f(x)|x\in\mathbb{R}\}.$$

Then

$$f(y-1)=y-1, 2y=(y-1)-f(y-1),$$

hence $2y \in Y$. By induction, $2^n y \in Y$ for all $y \in Y$ and nonnegative integers n. The set

$$\left\{\frac{f(2^n a-1)}{2^n a-1}|n\in\mathbb{N}\right\}$$

is finite for all fixed $a \in Y$, $a \neq 2^{-n}$. Then we may choose positive integers m, n, $m \neq n$ such that

$$\frac{f(2^n a-1)}{2^n a-1}=\frac{f(2^m a-1)}{2^m a-1}.$$

This implies

$$\frac{-2^n a-1}{2^n a-1}=\frac{-2^m a-1}{2^m a-1},$$

whence $a = 0$ and so, $Y = \{0\}$. We deduce that f is the identic function. Conversely, the identic function satisfies the given conditions.

3.29 2000

124. Answer: constant polynomials.
If P is constant, then we can take $n = 1$.
Suppose that P is nonconstant, deg $(P) = d \geq 1$. We can suppose P monic. Then

$$\sum_{1\leq i\leq n} i(P(x+2i)-P(x+2i+1))=0$$

for infinitely many x. It follows that the polynomial from the left is the zero polynomial. But the coefficient of x^{d-1} is $2id - (2i + 1)d = -d \neq 0$, contradiction.

125. Let $a \in \left\{-1, \frac{1\pm\sqrt{5}}{2}\right\}$. Then $x_i = a, 1 \leq i \leq n,$ is the only solution. Indeed, let

$$x_i = \max\{x_j | 1 \leq j \leq n\}.$$

Then $x_i^3 \geq x_{i-1}^3$, hence $x_{i+2} = x_i$. Repeating this, we find $x_i = x_{i+2} = x_{i+4} = \ldots$. From $x_{i-1}^3 = x_i + x_{i+1} + 1 = x_{i+2} + x_{i+1} + 1 = x_i^3$, we deduce $x_{i-1} = x_i$, hence all x_i must be equal. Putting $x_i = a$, we find $a^3 - 2a - 1 = (a-1)(a^2 - a - 1) = 0$, whence the conclusion.

126. Answer: $c > b$.
We will prove that $2b - a > c > b$. Indeed, the given expression can be written as

$$2(c-b)^3+3(b-a)(c-b)^2-2(b-a)^3=0.$$

The polynomial function $P(x) = 2x^3 + 3x^2 - 2$ has only one real root $t \in (0, 1)$, since the derivative $P'(x) = 6x(x + 1)$ has the roots 0 and 1, and is negative on $(-1, 0)$, hence P is decreasing on $(-1, 0)$ and increasing outside this interval. From

$$P(-1) = -1, P(0) = -2, P(1) = 3,$$

we deduce $t \in (0, 1)$. Putting $c - b = t(b - a)$, we get

$$0 < \frac{c-b}{b-a} < 1,$$

as desired.

127. *a*) The diophantine eqs. $3x^2 - y^2 = 0, 1$ have no solutions (the first, since $\sqrt{3}$ is irrational, the second since -1 is a quadratic non-residue mod 3), hence

$$3n^2 - \left\lfloor n\sqrt{3} \right\rfloor^2 \geq 2,$$

therefore

$$n\sqrt{3} \geq \sqrt{\left\lfloor n\sqrt{3} \right\rfloor^2 + 2}, \left\{n\sqrt{3}\right\} \geq \sqrt{\left\lfloor n\sqrt{3} \right\rfloor^2 + 2} - \left\lfloor n\sqrt{3} \right\rfloor$$
$$= \frac{2}{\sqrt{\left\lfloor n\sqrt{3} \right\rfloor^2 + 2} + \left\lfloor n\sqrt{3} \right\rfloor} > \frac{1}{\sqrt{\left\lfloor n\sqrt{3} \right\rfloor^2 + 2}} \geq \frac{1}{n\sqrt{3}}.$$

b) Since $3x^2 - y^2 = 3(2x + y)^2 - (3x + 2y)^2$ and $3 \cdot 1^2 - 1^2 = 2$, there are infinitely many solutions in positive integers of the eq. $3x^2 - y^2 = 2$. For these solutions, $y = \lfloor x\sqrt{3} \rfloor$. We need to find the minimum of $f(n) = n\sqrt{3}\{n\sqrt{3}\}$, where n runs over the positive integers. Let n' be the smallest positive integer which satisfies

$$3n'^2 - \left\lfloor n'\sqrt{3} \right\rfloor^2 = 2.$$

Then $f(n) \geq f(n')$. Indeed,

$$f(n) = n\sqrt{3}\left\{n\sqrt{3}\right\} = \frac{1}{2}\left(1 + \frac{3n^2 - \lfloor n\sqrt{3} \rfloor^2}{(n\sqrt{3} + \lfloor n\sqrt{3} \rfloor)^2}\right)\left(3n^2 - \left\lfloor n\sqrt{3} \right\rfloor^2\right).$$

Then $3n^2 - \lfloor n\sqrt{3} \rfloor^2 \geq 2 = 3n'^2 - \lfloor n'\sqrt{3} \rfloor^2$ and

$$f(n) \geq \frac{1}{2}\left(1 + \frac{3n'^2 - \lfloor n'\sqrt{3} \rfloor^2}{(n'\sqrt{3} + \lfloor n'\sqrt{3} \rfloor)^2}\right)\left(3n'^2 - \left\lfloor n'\sqrt{3} \right\rfloor^2\right) = f(n').$$

Now, we need to find the minimum of $f(n)$, where n runs over those positive integers for which $3n^2 - \lfloor n\sqrt{3} \rfloor^2 = 2$. Then $\lfloor n\sqrt{3} \rfloor = \sqrt{3n^2 - 2}$ and

$$f(n) = n\sqrt{3}\left(n\sqrt{3} - \left\lfloor n\sqrt{3} \right\rfloor\right) = n\sqrt{3}\left(n\sqrt{3} - \sqrt{3n^2 - 2}\right) = 3n^2 - \sqrt{(3n^2 - 1)^2 - 1}$$
$$= 1 + \frac{1}{3n^2 - 1 + \sqrt{(3n^2 - 1)^2 - 1}}.$$

Taking $n \to \infty$, we find that the minimum of f is 1.

128. From $x^5 - 55x + 21 = (x^2 - 3x + 1)(x^3 + 3x^2 + 8x + 21)$, we deduce

$$x_{1,2} = \frac{3 \pm \sqrt{5}}{2},$$

since the cubic factor has no roots with the product 1.

129. Use Cauchy–Schwarz for the second inequality:

$$\frac{a^2}{a^2 + 2bc} + \frac{b^2}{b^2 + 2ca} + \frac{c^2}{c^2 + 2ab} \geq \frac{(a + b + c)^2}{a^2 + b^2 + c^2 + 2ab + 2bc + 2ca} = 1.$$

For the first inequality, multiply by 2, take out 1 from each fraction, then use the second inequality:

$$2\left(\frac{bc}{a^2+2bc} + \frac{ca}{b^2+2ca} + \frac{ab}{c^2+2ab}\right) = 1 - \frac{a^2}{a^2+2bc} + 1 - \frac{b^2}{b^2+2ca} + 1 - \frac{c^2}{c^2+2ab} \leq 3 - 1 = 2.$$

Dividing by 2, we get the desired inequality.

130. Set $a = y + z$, $b = z + x$, $c = x + y$, then $a + b + c = 2(x + y + z)$, so

$$x = \frac{-a + b + c}{2}, y = \frac{a - b + c}{2}, z = \frac{a + b - c}{2}.$$

Since a, b, c are the sidelengths of a triangle, we deduce that x, y, z are positive. The given equation becomes

$$x(x - z)^2 + y(y - x)^2 + z(z - y)^2 = 0,$$

whence it follows $x = y = z$, therefore $a = b = c$.

A counterexample is $(a, b, c) = (10, 3, 24)$.

131. The given sum is 0. We will use a fact from the theory of finite differences: If n is a positive integer and P is a polynomial of degree $< n$, then

$$\sum_{i=0}^{n}(-1)^i\binom{n}{i}P(i)=0.$$

Since $1 \le k+n-m \le n$, the polynomial

$$P(x)=\frac{1}{n!}\prod_{\substack{1\le i\le n\\ i\ne k+n-m}}(x+i+m)$$

has degree $n-1<n$, the above result applies. We have

$$P(x)=\frac{1}{x+n+k}\binom{x+n+m}{n},$$

whence

$$\sum_{i=0}^{n}(-1)^i\binom{n}{i}\frac{1}{n+k+i}\binom{i+n+m}{n}=0.$$

Now,

$$0=\sum_{i=0}^{n}(-1)^i\binom{n}{i}\frac{1}{n+k+i}\binom{i+n+m}{n}=\sum_{i=0}^{n}(-1)^i\frac{n!}{i!(n-i)!}\cdot\frac{1}{n+k+i}$$
$$\cdot\frac{(i+n+m)!}{n!(m+i)!}=\sum_{i=0}^{n}\frac{(-1)^i}{n+k+i}\cdot\frac{(m+n+i)!}{i!(n-i)!(m+i)!},$$

which is the given expression.

132. We have successively,

$$f(2p)=f(2)f(p)\Rightarrow f(2)=2; f(12)=f(4)f(3)=f(5)+f(7)=2f(3)+6\Rightarrow f(3)$$
$$=3, f(4)=4; f(1999)=f(3)f(667)-2, f(667)=f(3)[f(25)f(9)-2]$$
$$-2; f(9)=f(7)+2=9, f(25)=\frac{f(3)f(11)+f(17)}{2}; f(21)=4+f(17)$$
$$=f(3)f(7)\Rightarrow f(21)=21; f(22)=2f(11)=f(13)+f(9)=f(11)$$
$$+11\Rightarrow f(11)=11.$$

133. We can use the inequality

$$\frac{a^2}{a+b} \geq \frac{3}{4}a - \frac{1}{4}b.$$

Add it with three similar inequalities and apply the given condition.

134. We have

$$(a_{n+1}+b_{n+1})^2 = (a_n+b_n)^2 + \left(\frac{1}{a_n}+\frac{1}{b_n}\right)^2 + 2\left(2+\frac{a_n}{b_n}+\frac{b_n}{a_n}\right) > (a_n+b_n)^2 + 8.$$

Since $(a_2+b_2)^2 \geq 16$, it follows that $a_n + b_n > 2\sqrt{2n}$, for any $n \geq 2$.

135. If $x_1, x_2, \ldots, x_n$ have the same sign, then the inequality is obvious. Let N be the set of negative numbers among them and P be the set of positive numbers and suppose, without loss of generality, that P has at most $\left\lfloor \frac{n}{2} \right\rfloor$ elements. Then

$$\sum_{k=1}^{n} |x_k| - \left|\sum_{k=1}^{n} x_k\right| = 2\min\left\{-\sum_{x_k \in N} x_k, \sum_{x_k \in P} x_k\right\} \leq 2\sum_{x_k \in P} x_k.$$

If $P = \{x_{k_1}, x_{k_2}, \ldots, x_{k_l}\}$ in increasing order, then $x_{k_1} \leq 1, x_{k_2} \leq 2, \ldots, x_{k_l} \leq l$, hence

$$2\sum_{x_k \in P} x_k \leq 2\left(1+2+\ldots+\left\lfloor \frac{n}{2} \right\rfloor\right) \leq \frac{n^2-1}{4}.$$

136. Clearly, P and Q have the same degree. Let

$$P = X^n + a_{n-1}X^{n-1} + \ldots + a_1X + a_0,$$

$$Q = X^n + b_{n-1}X^{n-1} + \ldots + b_1X + b_0.$$

If $P \neq Q$, then let $k = \max\{p | a_p \neq b_p\}$. The coefficient of $X^{n(n-1)+k}$ in $P(P(X))$ is the coefficient C of $X^{n(n-1)+k}$ in $P^n(x)$ if $k > 0$ and is equal to $C + a_{n-1}$ if $k = 0$.

In the case $k > 0$ the coefficient C is obtained by adding the results of the multiplications of X^n taken $n-1$ times, a_kX^k taken once and of terms of degree higher than k. This gives $C = na_k + C'$ is a combination of the coefficients $a_{k+1}, \ldots, a_{n-1}$. Similarly, the coefficient of $X^{n(n-1)+k}$ in $Q(Q(X))$ is $nb_k + C'$, which implies $a_k = b_k$, contradiction.

If $k = 0$, then $(n+1)a_0 + C' = (n+1)b_0 + C'$, hence $a_0 = b_0$, contradiction.

137. Denote $S = a_1 + \ldots + a_{15}, T = a_1^2 + \ldots + a_{15}^2$. Then $S = 100 - a_{16}, T = 1000 - a_{16}^2$. By the AM – QM inequality, $S^2 \leq 15T$, so that

$$(100 - a_{16})^2 - 15(1000 - a_{16}) \leq 0,$$

that is,

$$16a_{16}^2 - 200a_{16} - 5000 = 8(2a_{16} + 25)(a_{16} - 25) \leq 0,$$

whence $a_{16} \leq 25$. Equality holds in the AM $-$ QM inequality only for $a_1 = \ldots = a_{15}$, then $a_{16} = 25$, $a_1 = \ldots = a_{15} = 5$.

138. Define $a_m = f(mx), x \in \mathbb{R}$. Then $|a_{mn} - na_m| \leq n - 1$. The sequence $\{\frac{a_m}{m}\}$ is Cauchy, because

$$\left|\frac{a_m}{m} - \frac{a_n}{n}\right| \leq \frac{1}{m} + \frac{1}{n},$$

hence it is convergent. Let

$$g(x) = \lim_{m \to \infty} \frac{a_m}{m}.$$

We have

$$g(x+y) - g(x) - g(y) = \lim_{n \to \infty} \frac{f(n(x+y)) - f(nx) - f(ny)}{n} = 0,$$

so g is additive. Moreover,

$$|f(x) - g(x)| = \lim_{n \to \infty} \frac{|a_n - na_1|}{n} \leq \frac{n-1}{n} < 1,$$

therefore g satisfies the required conditions.

3.30 2001

139. Adding the given equations, we obtain

$$2\sum_{k=1}^{n} x_k = \sum_{k=1}^{n} x_k^2 := s.$$

By the AM–QM inequality,

$$\frac{s}{2n} \leq \sqrt{\frac{s}{n}},$$

whence $s \leq 4n$, with equality only when all numbers are equal. On the other hand,

$$\sum_{k=1}^{n} (x_k - 1)^2 = s - s + n = n.$$

Case 1. There exists a such that $x_a \leq 1$. Then

$$x_a^2 = x_{a-1} + x_{a-2} \leq 1 \Rightarrow x_{a-1} \leq 1 \Rightarrow \ldots \Rightarrow x_i \leq 1, 1 \leq i \leq n \Rightarrow -1 \leq x_i - 1 \leq 0, 1 \leq i \leq n \Rightarrow \sum_{i=1}^{n} (x_i - 1)^2 \leq n,$$

with equality if and only if $x_1 = x_2 = \ldots = x_n = 0$.

Case 2. $x_i > 1, 1 \leq i \leq n$. Let x_a be the maximal value, then $(x_a - 1)^2 \geq 1$, so $x_a \geq 2$. But $x_a^2 = x_{a-1} + x_{a-2} \leq 2x_a$, hence $x_a \leq 2$, therefore $x_a = 2$. Repeating this argument, we find $x_1 = x_2 = \ldots = x_n = 2$.

140. The additivity, with $f(x^k) = f^k(x)$ for some $k > 1$ implies that $f(x) = x$ for all x. Indeed, for $n \in \mathbb{Z}$, we have $f(x + n) = f(x) + n$, so

$$\sum_i \binom{k}{i} f^i(x) n^{l-i} = (f(x) + n)^k = f^k(x+n) = f\left((x+n)^k\right) = \sum_i \binom{k}{i} f(x^i) n^{k-i}.$$

For x fixed, we may view this equality as a polynomial identity with respect to n. Comparing the coefficients of n^{k-1}, we see that $f(x^2) = (f(x))^2 \geq 0$, so $f(x) \geq 0$ for $x \geq 0$, which implies that f is increasing, hence $f(x) = x$ for all x.

141. We know that $(a + b + c)^2 \geq 3(ab + bc + ca)$. Let $a = xy$, $b = yz$, $c = zx$, then

$$(xyz)^2 \geq (xy + yz + zx)^2 \geq 3xyz(x + y + z),$$

whence the result.

142. Define the sequences $a_1 = 1, a_{n+1} = 1 + \frac{1}{a_n}, b_1 = 3, b_{n+1} = 1 + \frac{2}{b_n}$. By induction,

$$a_n x < f(x) < b_n x.$$

From

$$\lim_{n\to\infty} a_n = \lim_{n\to\infty} b_n = 2,$$

it follows that $f(x) = 2x$.

143. We will prove the assertion by induction on n.

For $n = 0$, we have $P = c$ is constant. The system of inequalities

$$|c-1| < 1, |c-a| < 1, a \geq 3,$$

is incompatible, therefore $\max\{|c-1|, |c-a|\} \geq 1$.

Suppose that the assertion is true for all $i = 0, 1, \ldots, n-1$, and define

$$Q(x) = \frac{P(x+1) - P(x)}{a-1}.$$

Then

$$\left|Q(i) - a^i\right| = \frac{\left|-P(i+1) + a^{i+1} + P(i) - a^i\right|}{a-1} \leq \frac{\left|P(i+1) - a^{i+1}\right|}{a-1} + \frac{\left|P(i) - a^i\right|}{a-1}$$
$$\leq \frac{2}{a-1} \max_{0\leq i\leq n+1} \left|P(i) - a^i\right| \leq \max_{0\leq i\leq n+1} \left|P(i) - a^i\right|.$$

By the induction hypothesis,

$$\max_{0\leq i\leq n} \left|Q(i) - a^i\right| \geq 1,$$

therefore

$$\max_{0\leq i\leq n+1} \left|P(i) - a^i\right| \geq 1,$$

which finishes the proof.

144. The Hölder's inequality implies

$$\frac{a}{4} + \frac{3b}{4} \geq a^{\frac{1}{4}} b^{\frac{3}{4}}, \quad \frac{b}{5} + \frac{4c}{5} \geq b^{\frac{1}{5}} c^{\frac{4}{5}}, \quad \frac{c}{3} + \frac{2a}{3} \geq c^{\frac{1}{3}} a^{\frac{2}{3}}.$$

Then,

$$\left(\frac{a+3b}{4} \cdot \frac{b+4c}{5} \cdot \frac{c+2a}{3}\right)^{60} \geq \left(a^{\frac{1}{4}+\frac{2}{3}} b^{\frac{3}{4}+\frac{1}{5}} c^{\frac{4}{5}+\frac{1}{3}}\right)^{60} = a^{55} b^{57} c^{68}.$$

It remains to show that $a^{55} b^{57} c^{68} \geq (abc)^{60}$, which is true, since $c^8 \geq a^5 b^3$.

145. Denote $p = a + b + c$, $q = ab + bc + ca$, $r = abc$. Then a, b, c are the roots of the polynomial $P(x) = x^3 - px^2 + qx - r$. The given equalities are equivalent to

$$pa + bc = b, pb + ca = c, pc + ab = a.$$

Adding these equalities, we get $q = p - p^2$. Multiplying these equalities by a, b, c and adding them, we have, after a short computation, $3r = -3p^3 + p^2 + p$. Write the given equalities as

$$(p-c)(p-b) = b, (p-a)(p-c) = c, (p-b)(p-a) = a.$$

By multiplication, it follows that

$$[(p-a)(p-b)(p-c)]^2 = abc.$$

Using the above equalities, we have after simple calculations,

$$p(4p-3)(p+1)^2 = 0.$$

If $p = 0$, then $P(x) = x^3$, so $a = b = c = 0$.

If $p = -1$, then $P(x) = x^3 + x^2 - 2x - 1$. This polynomial has three real roots in each interval $(-2, -1), (-1, 0), (1, 2)$. In fact, these roots are $2\cos\frac{2\pi}{7}, 2\cos\frac{4\pi}{7}, 2\cos\frac{6\pi}{7}$.

If $p = \frac{3}{4}$, then $P(x) = x^3 - \frac{3}{4}x^2 + \frac{3}{16}x - \frac{1}{64}$, so $a = b = c = \frac{1}{4}$.

146. If

$$\frac{4n-2}{n+5} = \frac{a^2}{b^2}, \gcd(a,b) = 1,$$

then

$$n = \frac{2b^2 + 5a^2}{4b^2 - a^2} = -5 + \frac{22b^2}{4b^2 - a^2}.$$

Since $\gcd(b^2, 4b^2 - a^2) = 1$, it follows that $4b^2 - a^2 | 22$.

Observe that $4b^2 - a^2 \equiv 0, 3 \pmod 4$, hence $4b^2 - a^2 \in \{-1, 11\}$. In the first case, we get $b = 0$, contradiction. In the second case, $2b - a = 1$, $2b + a = 11$, whence $a = 5$, $b = 3$, so that $n = 13$.

147. If $n \geq 6$, we choose $x_i = 0, 1 \leq i \leq n-6, x_{n-5} = \ldots = x_{n-1} = \frac{1}{2}, x_n = 1$. Then

$$\frac{9}{4} = \text{LHS} < \text{RHS} = \frac{5}{2},$$

so the inequality is not valid.

The inequality holds for $n = 2, 3, 4, 5$. The cases $n = 2, 3$ are easy. The case $n = 4$ is similar to the case $n = 5$. Consider the case $n = 5$: the inequality is equivalent to

$$x_5^2 - (x_1 + x_2 + x_3 + x_4)x_5 + x_1^2 + x_2^2 + x_3^2 + x_4^2 \geq 0.$$

The discriminant of the quadratic function in x_5 is

$$\begin{aligned}\Delta &= (x_1 + x_2 + x_3 + x_4)^2 - 4\left(x_1^2 + x_2^2 + x_3^2 + x_4^2\right) \\ &= -\left[(x_1 - x_2)^2 + (x_1 - x_3)^2 + (x_1 - x_4)^2 + (x_2 - x_3)^2 + (x_2 - x_4)^2 + (x_3 - x_4)^2\right] \leq 0,\end{aligned}$$

whence the conclusion.

148. Denote $a = \frac{x}{3}, b = \frac{4y}{5}, c = \frac{3z}{2}$. Then

$$15 \geq 7xy + 3yz + 5zx \geq 15x^{\frac{12}{15}}y^{\frac{10}{12}}z^{\frac{8}{12}} \Longrightarrow 1 \geq x^6y^5z^4.$$

Hence

$$2P(x, y, z) = \frac{6}{x} + \frac{5}{y} + \frac{4}{z} \geq 15, P_{min} = \frac{15}{2}.$$

3.31 2002

149. From the inequality $|\cos a - \cos b| \leq |a - b|$, it follows that

$$|x_1 - x_2| \geq |\cos x_1 - \cos x_2| = |x_2 - x_3|,$$

hence $|x_1 - x_2| \geq |x_2 - x_3| \geq \ldots \geq |x_{n-1} - x_n| \geq |x_n - x_1| \geq |x_1 - x_2|$, therefore

$$x_1 = x_2 = \ldots = x_n.$$

The equation $\cos x = x$ has a unique solution, so that the given system has a unique solution.

150. Let $s = y + z, p = yz$. Then $s = 3 - x, p = x^2 - 3x + a$, so y, z are solutions of

$$t^2 + (x-3)t + x^2 - 3x + a = 0.$$

A solution exists if and only if

$$(x-3)^2 - 4\left(x^2 - 3x + a\right) \geq 0 \Leftrightarrow 3x^2 - 6x + 4a - 9 \leq 0,$$

whose discriminant is 16(9 – 3a), so a solution exists if and only if $a \leq 3$. Then

$$\max(x) = 1 + \frac{2}{3}\sqrt{9-3a}, \quad \min(x) = 1 - \frac{2}{3}\sqrt{9-3a},$$

and $\frac{4}{3}\sqrt{9-3a} = 8$ implies $a = -9$.

151. By the AM–GM inequality,

$$\frac{a^2}{1+2bc} + \frac{b^2}{1+2ca} + \frac{c^2}{1+2ab} \geq \frac{a^2}{1+b^2+c^2} + \frac{b^2}{1+c^2+a^2} + \frac{c^2}{1+a^2+b^2}$$
$$= \frac{a^2}{2-a^2} + \frac{b^2}{2-b^2} + \frac{c^2}{2-c^2}.$$

Consider the real function defined on (0, 1) by

$$f(x) = \frac{x}{2-x}.$$

Then

$$f'(x) = \frac{2}{(2-x)^2}, f''(x) = \frac{4}{(2-x)^3} > 0,$$

hence f is a convex function. The Jensen's inequality get us

$$\frac{a^2}{2-a^2} + \frac{b^2}{2-b^2} + \frac{c^2}{2-c^2} \geq 3 \cdot \frac{\frac{1}{3}}{2-\frac{1}{3}} = \frac{3}{5},$$

with equality only when $a^2 = b^2 = c^2 = \frac{1}{3}$.

152. We have

$$f(x_1, x_2, x_3) = (x_1 + x_2 - 2x_3)(x_1 - 2x_2 + x_3)(-2x_1 + x_2 + x_3).$$

Then

$$|f(r, r+2, x) + s| = |2y^3 - 18y + s|,$$

whence $y = r - x + 1$. Consequently,

$$g(r, s, t) = \max_{r-t-1 \le y \le r-t+1} |2y^3 - 18y + s|.$$

Let $h(x) = 2x^3 - 18x$. Therefore

$$\min_{r,s,t \in \mathbb{R}} g(r, s, t) = \frac{1}{2} \min_{a \in \mathbb{R}} \left(\max_{a \le x \le a+2} h(x) - \min_{a \le x \le a+2} h(x) \right) = 6\sqrt{3} - \frac{20\sqrt{6}}{9}.$$

153. Suppose that the assertion is false, i.e.,

$$\sin(x_1 - \alpha) + \sin(x_2 - \alpha) + \ldots + \sin(x_n - \alpha) < 0,$$

therefore

$$\cos x_1 + \cos x_2 + \ldots + \cos x_n > \frac{\cos \alpha}{\sin \alpha} (\sin x_1 + \sin x_2 + \ldots + \sin x_n) \ge n \cos \alpha,$$

hence

$$(\sin x_1 + \sin x_2 + \ldots + \sin x_n)^2 + (\cos x_1 + \cos x_2 + \ldots + \cos x_n)^2 > n^2.$$

But

$$(\sin x_1 + \ldots + \sin x_n)^2 + (\cos x_1 + \ldots + \cos x_n)^2 = \sum_{i=1}^{n} \sum_{j=1}^{n} \cos(x_i - x_j) \le n^2,$$

contradiction.

154. Define $A = \{n+1, n+2, \ldots, 2n\}$, $B = \{1, 2, \ldots, n\}$. Call a pair (a_i, a_{i+1}) descent, if $a_i > a_{i+1}$ and ascent, otherwise. Observe that

1. $\sum$descents $- \sum$ ascents $= a_1 - a_{2n}$.
2. The set of differences of consecutive elements is $\{1, 2, \ldots, 2n - 1\}$.
3. $\sum$descents $+ \sum$ ascents $= 1 + 2 + \ldots + (2n - 1)$.

" $\Leftarrow$ " By 1) and 3), it follows that

$$\sum \text{descents} = n^2, \sum \text{ascents} = n^2 - n.$$

Let (a_i, a_{i+1}), $i \in I$, be the descents, then

$$\sum \text{descents} = \sum_{i \in I} a_i - \sum_{i \in I} a_{i+1}.$$

This value is at most $2n + (2n - 1) + \ldots + (n + 1) - (1 + 2 + \ldots + n) = n^2$ with equality only if each element from A appears as a left element and never as a right element of a descent. Because $a_1 \in A$, $a_{2n} \in B$, all elements from A must have odd indices.

" $\Rightarrow$ " The elements from A have odd indices, hence the sum of descents is n^2. Since

$$\text{ascents} + \text{descents} = 1 + 2 + \ldots + 2n - 1,$$

it follows that the sum of ascents is $n^2 - n$. Then

$$a_1 - a_{2n} = \text{descents} - \text{ascents} = n.$$

155. Answer: $a \in \{-2, -1, 0, 1, 2\}$.

If $x = y = 0$, then $a = 0$. Suppose that $a \neq 0$ and denote $p := xy$. We have

$$a = x^3 + y^3 = (x + y)^3 - 3xy(x + y) = a^3 - 3ap,$$
$$a = x^5 + y^5 = (x + y)^5 - 5xy(x^3 + y^3) - 10x^2y^2(x + y) = a^5 - 5ap - 10ap^2.$$

and so,

$$a^2 = 3p + 1, a^4 = 10p^2 + 5p + 1.$$

Eliminating a, we obtain the eq. $10p^2 + 5p + 1 = (3p + 1)^2$, hence $p \in \{0, 1\}$. If $p = 0$, then $a^2 = 1$, so $a = \pm 1$. If $p = 1$, then $a^2 = 4$, so $a = \pm 2$. Each of these values for a can be obtained:

$$(x, y, a) : (-1, -1, -2), (-1, 0, -1), (0, 0, 0), (1, 0, 1), (1, 1, 2).$$

156. If x_1, x_2, x_3 are real numbers and y_1, y_2, y_3 are positive real numbers, then the Cauchy–Schwarz inequality yields

$$\left(\sum_{i=1}^{3} x_i\right)^2 = \left(\sum_{i=1}^{3} \sqrt{y_i} \cdot \frac{x_i}{\sqrt{y_i}}\right)^2 \leq \left(\sum_{i=1}^{3} y_i\right)\left(\sum_{i=1}^{3} \frac{x_i^2}{y_i}\right).$$

For $x_1 = a\sqrt{a}, x_2 = b\sqrt{b}, x_3 = c\sqrt{c}, y_1 = a^2(b^2 + 1), y_2 = b^2(c^2 + 1), y_3 = c^2(a^2 + 1)$, it becomes

$$\left(a\sqrt{a}+b\sqrt{b}+c\sqrt{c}\right)^2 \le \left(a^2b^2+b^2c^2+c^2a^2+a^2+b^2+c^2\right)$$
$$\times\left(\frac{a}{b^2+1}+\frac{b}{c^2+1}+\frac{c}{a^2+1}\right).$$

Using the inequality

$$\sqrt{\frac{uv+vw+wu}{3}} \le \frac{u+v+w}{3},$$

which is true for positive real numbers u, v, w, we obtain

$$a^2b^2+b^2c^2+c^2a^2 \le \frac{\left(a^2+b^2+c^2\right)^2}{3} = \frac{1}{3},$$

by the condition $a^2 + b^2 + c^2 = 1$, whence it follows the proposed inequality.

157. Eliminating denominators, we have

$$(a+b+c)\left(a^3+b^3+c^3\right) \ge (3abc)^2 = \left(a^2+b^2+c^2\right)^2,$$

which is known.

158. Using the AM–QM inequality, it suffices to prove that

$$\frac{x_1^2}{\left(1+x_1^2\right)^2}+\frac{x_2^2}{\left(1+x_1^2+x_2^2\right)^2}+\ldots+\frac{x_n^2}{\left(1+x_1^2+\ldots+x_n^2\right)^2}<1.$$

For $k \ge 2$, we have

$$\frac{x_k^2}{\left(1+x_1^2+\ldots+x_k^2\right)^2} \le \frac{1}{1+x_1^2+\ldots+x_{k-1}^2} - \frac{1}{1+x_1^2+\ldots+x_k^2}.$$

If $k = 1$, then

$$\frac{x_1^2}{\left(1+x_1^2\right)^2} \le 1 - \frac{1}{1+x_1^2}.$$

Summing these inequalities, we get

$$\frac{x_1^2}{(1+x_1^2)^2}+\ldots+\frac{x_n^2}{(1+x_1^2+\ldots+x_n^2)^2}\leq 1-\frac{1}{1+x_1^2+\ldots+x_n^2}<1.$$

159. Denote $\frac{a}{b}=x$, $\frac{b}{c}=y$, $\frac{c}{a}=z$, then $xyz=1$ and the proposed inequality becomes

$$x+y+z\geq\frac{1+xy}{1+y}+\frac{1+yz}{1+z}+\frac{1+zx}{1+x},$$

which is equivalent to

$$x^2+y^2+z^2+xy^2+yz^2+zx^2\geq 3+x+y+z.$$

This inequality is proved adding the inequalities

$$xy^2+yz^2+zx^2\geq 3,$$

$$x^2+y^2+z^2\geq\frac{(x+y+z)^2}{3}\geq x+y+z,$$

obvious by the AM–GM inequality.

160. By the Hölder's inequality,

$$\sum_{i=1}^{n}\frac{x_i^3}{ax_i+bx_{i+1}}\geq\frac{\left(\sum_{i=1}^{n}x_i\right)^3}{n\left(\sum_{i=1}^{n}ax_i+bx_{i+1}\right)}=\frac{1}{n(a+b)}.$$

161. Answer: $(A,B)=(\mathbb{Z},\mathbb{Z}),(2\mathbb{Z},2\mathbb{Z}+1)$.

If $0\in B$, then *iii)* implies $x+0\in A$ for all $x\in B$, so $B\subset A$. By *i)* it follows that $A=B=\mathbb{Z}$.

If $0\notin B$, then *i)* $0\in A$ and by *ii)*, $-1\in B$. Using *ii)* we get succesively $-2\in A$, $-3\in\in B$, $-4\in A$, ..., and by induction, $-2n\in A$, $-2n-1\in B$, for each positive integer n.

If $2\in B$, then $2+(-1)=1\in A$ and $1-1=0\in B$, contradiction. Hence $2\in A$ and $2-1=1\in B$.

We will prove that $2n\in A$ for all n. If not, let $n>1$ be minimal with $2n\in B$, then $2n-1\in A$ by *iii)* and $(2n-1)-1=2(n-1)\in B$, contradicting the minimality condition. It follows that A contains all even positive integers and not contains odd positive integers, while B contains all odd positive integers and not contains even positive integers. From $0\in A$, we deduce $0-1=-1\in B$. Hence $A=2\mathbb{Z}$, $B=2\mathbb{Z}+1$.

162. For any positive real numbers $x_1,\ldots,x_n$ the Cauchy–Schwarz inequality get us

$$\left((\sqrt{x_1})^2+\ldots+(\sqrt{x_n})^2\right)\left(\left(\frac{a_1}{\sqrt{x_1}}\right)^2+\ldots+\left(\frac{a_n}{\sqrt{x_n}}\right)^2\right)\geq(a_1+\ldots+a_n)^2,$$

hence

$$\frac{a_1^2}{x_1}+\ldots+\frac{a_n^2}{x_n}\geq\frac{(a_1+\ldots+a_n)^2}{x_1+\ldots+x_n}.$$

Applying this inequality to the proposed inequality, we have

$$\frac{a_1}{a_2^2+1}+\frac{a_2}{a_3^2+1}+\ldots+\frac{a_n}{a_1^2+1}=\frac{a_1^3}{a_1^2a_2^2+a_1^2}+\frac{a_2^3}{a_2^2a_3^2+a_2^2}+\ldots+\frac{a_n^3}{a_n^2a_1^2+a_n^2}$$
$$\geq\frac{\left(a_1\sqrt{a_1}+a_2\sqrt{a_2}+\ldots+a_n\sqrt{a_n}\right)^2}{a_1^2a_2^2+a_2^2a_3^2+\ldots+a_n^2a_1^2+1}.$$

Now it suffices to prove the inequality

$$a_1^2a_2^2+a_2^2a_3^2+\ldots+a_n^2a_1^2\leq\frac{1}{4}$$

for $n\geq 4$ and $a_1^2+\ldots+a_n^2=1$. More generally, if $n\geq 4$ and $x_1,x_2,\ldots,x_n$ are positive real numbers with $x_1+x_2+\ldots+x_n=1$, then

$$x_1x_2+x_2x_3+\ldots+x_nx_1\leq\frac{1}{4}.$$

If n is even, then

$$x_1x_2+x_2x_3+\ldots+x_nx_1\leq(x_1+x_3+\ldots x_{n-1})(x_2+x_4+\ldots+x_n)\leq\frac{1}{4},$$

by the AM–GM inequality. If n is odd, $n\geq 5$, then we can consider $x_1\geq x_2$ and we replace the n numbers x_1, …, x_n by the $n-1$ numbers x_1, x_2+x_3, x_4, …, x_n. From

$$x_1x_2+x_2x_3+x_3x_4\leq x_1(x_2+x_3)+(x_2+x_3)x_4$$

we see that LHS increases, the sum remains 1 and we have an even set of numbers, so the above argument applies again.

163. Define

$$f(a,b,c)=a+b+c+abc, g(a,b,c)=a^2+b^2+c^2+abc,$$
$$S=\{(a,b,c)|g(a,b,c)\leq 4, a,b,c\geq 0\}.$$

Then S is a closed subset of $[0,2]^3$ and f is a continuous function, so the desired maximum M exists and $M \geq f(1,1,1) = 4$. We will prove that $M = 4$.

Let $(x, y, z) \in S$ such that $f(x, y, z) = M$. If $x = 2$, then $y = z = 0$ and $f(x, y, z) = 2$, contradiction. If $x = 0$, then $y^2 + z^2 \leq 4$ and the AM–QM inequality implies

$$f(x,y,z)=y+z\leq 2\sqrt{\frac{y^2+z^2}{2}}\leq 2\sqrt{2}<M,$$

contradiction. Due to the symmetry of f, we conclude that $x, y, z \in (0, 2)$. Now we will show that

$$x\neq y \Rightarrow \left(\frac{x+y}{2},\frac{x+y}{2},z\right)\in S\ (1), f\left(\frac{x+y}{2},\frac{x+y}{2},z\right)>f(x,y,z) \qquad (2).$$

Indeed,

$$g\left(\frac{x+y}{2},\frac{x+y}{2},z\right)=x^2+y^2+z^2+xyz-\frac{(x-y)^2(2-z)}{4}\leq g(x,y,z),$$

which proves (1). By the AM–GM inequality,

$$f\left(\frac{x+y}{2},\frac{x+y}{2},z\right)-f(x,y,z)=\left[\left(\frac{x+y}{2}\right)^2-xy\right]z>0,$$

which proves (2). But M is the maximal value of f, hence $x = y$, and similarly, $x = z$. Let $t = x = y = z$. Then $3t + t^3 - 4 = (t - 1)(t^2 + t + 4) \leq 0$, hence $t \leq 1$. Therefore,

$$M=f(t,t,t)=3t+t^3\leq 4.$$

Finally, $M = 4$.

164. We will prove first the following assertion:

Let ε, s_1, s_2, …, s_n be real numbers such that

$$\varepsilon\geq 0, |s_1|\leq\varepsilon, |s_{i+1}-s_i|\leq\varepsilon, 1\leq i\leq n-1.$$

Then there exists $m \leq n$ for which $|2s_m - s_n| \leq \varepsilon$.

Proof. If $\varepsilon = 0$, then $s_1 = s_2 = \ldots = s_n = 0$ and the conclusion follows immediately. If $\varepsilon > 0$, then let

$$a = \min_{1 \le i \le n} s_i, b = \max_{1 \le i \le n} s_i.$$

The numbers $s_1, s_2, \dots, s_n$ divide the interval $[a,b]$ into intervals of length at most ε. Each number from $[a,b]$ is at a distance at most $\frac{\varepsilon}{2}$ to one of the numbers $s_1, \dots, s_n$. In particular, if $\frac{s_n}{2} \in [a,b]$, then the conclusion follows.

Assume that $\frac{s_n}{2} \notin [a,b]$. By definition, $s_n \in [a,b]$. If $0 \in [a,b]$, then $\frac{s_n}{2} \in [a,b]$, contradiction. Therefore, all numbers $s_1, s_2, \dots, s_n$ have the same sign and we can suppose that they are positive. Then $0 < a \le s_n \le b, 0 < a \le s_1 \le \varepsilon$. Then

$$\frac{s_n}{2} \notin [a,b] \Rightarrow 0 < \frac{a}{2} \le \frac{s_n}{2} < a \Rightarrow \left| a - \frac{s_n}{2} \right| < \frac{a}{2} \le \frac{\varepsilon}{2}.$$

As a is one of the numbers $s_1, s_2, \dots, s_n$, the conclusion follows.

Back to the problem, it suffices to take

$$\varepsilon = \max_{1 \le i \le n} |a_i|, s_i = a_1 + a_2 + \dots + a_i, i = 1, 2, \dots, n.$$

Then $|s_1| = |a_1| \le \varepsilon$, $|s_{i+1} - s_i| = |a_{i+1}| \le \varepsilon$, $i = 1, 2, \dots, n - 1$. The conditions of the above assertion are satisfied, hence there exists $m \le n$ such that $|2s_m - s_n| \le \varepsilon$, or

$$\left| \sum_{i=1}^{m} a_i - \sum_{i=m+1}^{n} a_i \right| \le \max_{1 \le i \le n} |a_i| = |a_k|.$$

3.32 2003

165. The constant solutions are $p = 0$ and $p = 1$.
Suppose that deg $(p) = n \ge 1$ and set $p(x - 1) = q(x)$. Then q satisfies the equation

$$q(x)q(x+2) = q(x^2).$$

We will prove that the solution of this equation is $q(x) = (x - 1)^n$, $n \in \mathbb{N}$, whence it follows that the solution of the given equation is $p(x) = q(x+1) = x^n$, $n \in \mathbb{N}$.

Let a be the coefficient of the leading term in q. Comparing leading terms in the functional equation, we obtain $a^2 = a$, whence $a = 1$. Now, we can write

$$q(x) = (x-1)^n + r(x),$$

where $r = 0$ or r is a polynomial of degree at most $n - 1$. Plugging this in the functional equation, we obtain (upon multiplying and cancelling)

$$r(x^2) = (x-1)^n r(x+2) + (x+1)^n r(x) + r(x)r(x+2).$$

If $q \neq 0$ and deg $(q) = k < n$, then deg $(r(x^2)) = 2k$, but the polynomial from RHS has degree $n + k > 2k$, contradiction.

166. The given equality implies $(a^2 - 4)(b^2 - 4) = (ab - 2c)^2 \geq 0$, so $|b| \geq 2$. the equation $a = x + \frac{1}{x}$ is equivalent to $x^2 - ax + 1 = 0$, which has discriminant $a^2 - 4 \geq \geq 0$, so that we can find x, and similarly, we can find y. Hence

$$(ab-2c)^2 = (a^2-4)(b^2-4) = \left[\left(x-\frac{1}{x}\right)\left(y-\frac{1}{y}\right)\right]^2,$$

so $ab - 2c = \pm\left(x - \frac{1}{x}\right)\left(y - \frac{1}{y}\right)$. Since we may interchange x and $\frac{1}{x}$, we may assume that the sign is minus. Then

$$c = \frac{1}{2}\left[\left(x+\frac{1}{x}\right)\left(y+\frac{1}{y}\right) + \left(x-\frac{1}{x}\right)\left(y-\frac{1}{y}\right)\right] = xy + \frac{1}{xy},$$

as desired.

167. Answer: $n = 1, 2, 3, 4, 6$.

We have $P_1 = P_2 = 1$, $P_3 = x + 1$, $P_4 = x^2 + 1$, $P_6 = x^4 + 1$. We will prove that these are the only solutions. For this, it suffices to show that for each $n \geq 3$, the polynomial $P_n(x)$ has a divisor of the form $x^r + 1$, $r \geq 1$, which is a proper divisor for $n = 5$ or $n \geq 7$. If $n \geq 3$ is a prime number, then

$$P_n(x) = (1+x)\left(1 + x^2 + x^4 + \ldots + x^{n-3}\right).$$

Also, $P_4(x) = x^2 + 1$.

Now we will use induction on n. Assume that the assertion is true for all m such that $3 \leq m < n$. If $n \geq 6$ is composite, then $n = mp$, with p prime and $m \geq 3$.

Case 1. $p|m$. Then

$$A_n = \left(\bigcup_{i=0}^{p-1}(A_m + im)\right) \Rightarrow P_n(x) = P_m(x)\sum_{i=0}^{p-1} x^{im}$$

and P_m has a divisor of the form $x^r + 1$, by the induction hypothesis.

Case 2. $p \nmid m$. Then

$$A_n = \left(\bigcup_{i=0}^{p-1} (A_m + im)\right) - pA_m \Rightarrow P_n(x) = P_m(x)\sum_{i=0}^{p-1} x^{im} - x^{p-1}P_m(x^p).$$

By the induction hypothesis, $P_m(x)$ has a divisor of the form $1 + x^r$, hence

$$1 + x^{pr} | P_m(x^p).$$

a) If $p \geq 3$, then $1 + x^r$ divides $1 + x^{pr}$, hence $1 + x^r$ divides $P_n(x)$.

b) If $p = 2$, then we can suppose that m is prime, otherwise it has an odd prime divisor and we are in the case a). Then

$$P_n(x) = \left(1 + x^{m+1}\right)\left(1 + x^2 + x^4 + \ldots + x^{m-3}\right).$$

168. Answer: $f(x) = x$ for all x.

From the given equation, we deduce that the function f is surjective. Also, $(f(x))^2 = (f(-x))^2$. Then we can choose a such that $f(a) = f(-a) = 0$. The substitutions $(x, y) = (0, \pm a)$ in the functional equation yield

$$0 = f(\pm a) = (f(0))^2 \pm 2a,$$

implying $a = 0$. Now,

$$y = \frac{(f(x))^2}{2} \Rightarrow f\left(x^2 + y + f(y)\right) \Rightarrow y + f(y) = -x^2.$$

It follows that the function $y + f(y)$ takes all negative values. Then

$$f\left(x^2\right) = (f(x))^2 \geq 0, f(y + f(y)) = 2y.$$

For $(z, t) = (x^2, y + f(y))$ we get $f(z + t) = f(z) + f(t)$, $z \geq 0 \geq t$. For $z = -t$ we get $f(-t) = -f(t)$ and then $f(z + t) = f(z) + f(t)$ for all z, t. Since $x \geq 0$ implies $f(x) \geq 0$, we deduce that f is increasing. If $f(y) > y$, then $f(f(y)) \geq f(y)$ and

$$2y = f(y + f(y)) = f(y) + f(f(y)) > 2f(y),$$

contradiction. We deduce $f(y) \leq y$ and analogously, $f(y) \geq y$, hence $f(x) = x$, $\forall\, x$.

169. The given equality is $abc = a + b + c$. Since a, b, c are positive, we can find $\alpha, \beta, \gamma \in \left(0, \frac{\pi}{2}\right)$ such that $a = \tan\alpha, b = \tan\beta, c = \tan\gamma$. Using the equality

$$\tan(x+y) = \frac{\tan x + \tan y}{1 - \tan x \tan y},$$

we get $\tan(\alpha + \beta + \gamma) = 0$, hence $\alpha + \beta + \gamma = \pi$. The required inequality becomes

$$\sin \alpha + \sin \beta + \sin \gamma \leq \frac{3\sqrt{3}}{2}.$$

Applying the Jensen's inequality to the concave sine function on $(0, \pi)$, we get

$$\sin \alpha + \sin \beta + \sin \gamma \leq 3 \sin \frac{\alpha + \beta + \gamma}{3} = 3 \sin \frac{\pi}{3} = \frac{3\sqrt{3}}{2}.$$

Equality holds only for $\alpha = \beta = \gamma = \frac{\pi}{3}$, that is $a = b = c = \sqrt{3}$.

170. If f is a solution of the functional equation, then f is surjective, because for $y = -f(x)$ we have

$$f(f(-f(x)) - x) = f(0) - 2x$$

and the RHS can take any real value. Then there exists x_0 such that $f(x_0) = 0$. Setting $x = x_0$ in the functional equation, we get $f(y) = 2x_0 + f(f(y) - x_0)$, so $f(z) = z - x_0$ for $z = f(y) - x_0$. The variable z takes any real value, since f is surjective, so

$$f(z) = z + c$$

for all real z and a fixed constant c.

Clearly, these functions satisfy the given functional equation.

171. We will show by induction on m that $x_i = i$, $1 \leq i \leq m$. The base case $x_1 = 1$ is given. Assume that the assertion is true for $m - 1$ and set $x_m = N + 1$. If $x_m > m$, then modulo N the sequence is $1, 2, \ldots, m - 1, 1, 2, \ldots, m - 1, \ldots$, so no number is divisible by N, contradiction. Hence $x_m = m$.

The polynomial $p(x) - x - 1$ has infinitely many roots; therefore, it is the zero polynomial.

172. Let $x_1, \ldots, x_{2n}$ be the roots of the polynomial. By the Viète's relations,

$$\sum_{i=1}^{2n} x_i = -2n + 10, \prod_{i=1}^{2n} x_i = 1, \sum_{i=1}^{2n} \frac{1}{x_i} = -2n + 10.$$

The AM–GM inequality implies $-2n + 10 \geq 2n$, hence $n \leq 2$.

If $n = 1$, then $P = x^2 - 8x + 1$, which has positive real roots.

If $n=2$, then $P=x^4-6x^3+(a+2)x^2-6x+1$. Divide the equation $P(x)=0$ by x^2 and set $y=x+\frac{1}{x}$. We have $y^2-6y+a=0$. Observe that y is positive if and only if x is positive. Also, $y=x+\frac{1}{x}\geq 2$, hence $\Delta=36-4a\geq 0$ and

$$\frac{6-\sqrt{\Delta}}{2}\geq 2,$$

therefore $8\leq a\leq 9$. Finally, $P=x^4-6x^3+bx^2-6x+1$, with $10\leq b\leq 11$.

173. Define $b_n=\frac{1}{a_n}$, then

$$b_1=1, b_{n+1}=b_n^2-b_n+1\ n\geq 1.$$

Multiplying $b_{n+1}-1=b_n(b_n-1)$, we get $b_{n+1}=b_1\ldots b_n+1$ or

$$\frac{1}{b_{n+1}}+\frac{1}{b_1\ldots b_n b_{n+1}}=\frac{1}{b_1\ldots b_n}.$$

Summing it, we obtain

$$\frac{1}{b_1}+\ldots+\frac{1}{b_n}+\frac{1}{b_1\ldots b_n}=1,$$

whence the conclusion.

174. Suppose that there are nonconstant polynomials $p,q\in\mathbb{Z}[X]$ such that $g=pq$, and $\deg(p),\deg(q)<2n$ where $n=\deg(f)$. If $\alpha\in\mathbb{C}$ is a root of f, then $p(\sqrt{\alpha})=0$ and there are polynomials $t,u\in\mathbb{Z}[X]$ such that

$$\deg(t),\deg(u)<2n\leq\frac{1}{2}\deg(p)<n, t(\alpha)+\sqrt{\alpha}u(\alpha)=0.$$

As $u\neq 0$ and $\gcd(u,f)=1$, there are $s,r\in\mathbb{Q}[X]$ such that $su+rf=1$. Then $s(\alpha)u(\alpha)=1$, hence $\sqrt{\alpha}=-t(\alpha)s(\alpha)$, so $\alpha=t^2(\alpha)s^2(\alpha)$. If $m=t^2s^2-x$, then $m(\alpha)=0$, hence $f|m$.

If $\alpha_1,\ldots,\alpha_n$ are the roots of f, then $m(\alpha_i)=0$, $1\leq i\leq n$ and $\alpha_i=t^2(\alpha_i)s^2(\alpha_i)$, so

$$\alpha_1\ldots\alpha_n=(ts)^2(\alpha_1)\ldots(ts)^2(\alpha_n)$$

is the square of a rational number. But $f(0)=(-1)^n\alpha_1\ldots\alpha_n$ is an integer and $|f(0)|$ is not the square of an integer, contradiction.

175. Answer: $(a,b,m,n)=(0,0,m,n)$, $(0,-1,kn,n)$, $(0,1,(2k+1)n,n)$,

$$(-1,0,kn-k+1,n),(1,0,(2k+1)(n-1)+1,n),(1,1,3k+2,2),(-1,1,6k+2,2).$$

Clearly, we have a solution with $a = b = 0$. If $a = 0$, $b \neq 0$, then $X^n + b|X^m + b$, hence $|b| = 1$. If $b = -1$, then from $X^n - 1|X^m - 1$ and $\gcd(X^n - 1, X^m - 1) == X^{\gcd(m,n)} - 1$, it follows that $n|m$, hence $(0, -1, kn, n)$ are solutions.

If $b = 1$, then f has the roots

$$z_k = \cos\left(\pi + \frac{2k\pi}{n}\right) + i\sin\left(\pi + \frac{2k\pi}{n}\right), k=0,1,\ldots,n-1,$$

176. Let x_0 be a root of the polynomial $p(x) - x$. Then x_0 is a root of the polynomial $p^n(x) - x$, for any positive integer n, hence $p(x) - x$ divides $p^n(x) - x$. Therefore,

$$p^{2003}(x) - 2p^{2002}(x) + p^{2001}(x) = \left(p^{2003}(x) - x\right) - 2\left(p^{2002}(x) - x\right) + \left(p^{2001}(x) - x\right)$$

is divisible by $p(x) - x$.

177. Denote by $f(a, b, c, d, e)$ the given expression. Then

$$f(a,b,c,d,e) > \frac{a}{a+b+c+d+e} + \ldots + \frac{e}{a+b+c+d+e} = 1.$$

For $t > 0$, we have

$$f(1,t,t^2,t^3,t^4) = \frac{4}{1+t} + 1 - \frac{1}{t^4+1} < \frac{4}{1+t} + 1.$$

Let $\varepsilon > 0$ and choose $t > \frac{4}{\varepsilon} - 1$, then

$$1 < f(1,t,t^2,t^3,t^4) < 1 + \varepsilon,$$

which shows that $C_1 = 1$.

From

$$f(a,b,c,d,e) = 5 - f(e,d,c,b,a) < 5 - 1 = 4$$

it follows that $C_2 = 4$ is a good candidate. We can show that $f(t^4, t^3, t^2, t, 1)$ is as closer as possible to 4, by the same method as above, therefore $C_2 = 4$.

3.33 2004

178. By the AM−GM inequality,

$$a^2 \tan x\sqrt[3]{\cos x} + b^2 \sin x \geq 2|ab| \cdot \frac{\sin x}{\sqrt[3]{\cos x}}.$$

The inequality

$$\frac{\sin x}{\sqrt[3]{\cos x}} \geq x$$

holds for $x \in \left(0, \frac{\pi}{2}\right)$. Indeed, by the Taylor's expansions,

$$\frac{\sin x}{x} \geq 1 - \frac{x^2}{6} \geq \sqrt[3]{1 - \frac{x^2}{2} + \frac{x^4}{24}} \geq \sqrt[3]{\cos x}.$$

Moreover, $|ab| \geq ab$ and $x \geq 0$, which ends the proof.

179. Answer: $k = 1, f(x, y) = \min\{x, y\}$; $k = 2, f(x, y) = xy$.

If $x \leq y$, then

$$f(x, y) = f\left(\frac{x}{y}, 1\right) y^k = xy^{k-1}.$$

If $x \geq y$, then $f(x, y) = f(y, x) = x^{k-1}y$.

Now, assume $x \geq y$ and choose z sufficiently small. Then

$$f(f(x, y), z) = f\left(x^{k-1}y, z\right) = \left(x^{k-1}y\right)^{k-1} z,$$
$$f(x, f(y, z)) = f\left(x, y^{k-1}z\right) = x^{k-1}y^{k-1}z,$$

whence $(k-1)^2 = k - 1$, so that $k \in \{1, 2\}$.

We see that $f(x, y) = \min\{x, y\}$ and $f(x, y) = xy$ satisfy the required conditions for $k = 1$ and $k = 2$, respectively.

180. Let $a = x^2, b = y^2, c = z^2$ and suppose that $x \geq y \geq z \geq 1$. Then $x^2 + y^2 + z^2 = 9$ and we have to prove that

$$(x + y + z)^2 \geq (xy)^2 + (yz)^2 + (zx)^2.$$

This inequality has the equivalent forms

$$\left(x^2+y^2+z^2\right)(x+y+z)^2 \geq 9\left[(xy)^2+(yz)^2+(zx)^2\right],$$

$$(x-y)^2\left(2x^2+9xy+3zx+2y^2+3yz-7z^2\right)+(z-x)(z-y)$$
$$\times\left(2z^2+6zx+6zy-x^2-y^2\right) \geq 0,$$

which is true, since

$$2z^2+6zx+6zy-x^2-y^2 \geq 0 \Leftrightarrow 3z^2+6zx+6zy-9 \geq 0.$$

The last inequality is true, since $x \geq y \geq z \geq 1$.

181. Answer: $(x_1, x_2, x_3, \ldots, x_{n-1}, x_n) = (n, 1, 0, \ldots, 0, 1)$.

Denote $x_1' = x_1 - n, x_2' = x_2 - 1, x_n' = x_n - 1$. By these substitutions, the given system becomes

$$\begin{cases} x_1' + x_2' + x_3 + \ldots + x_{n-1} + x_n' = 0 \\ x_1' + 2x_2' + 3x_3 + \ldots + (n-1)x_{n-1} + nx_n' = 0 \\ x_1' + 2^2x_2' + 3^2x_3 + \ldots + (n-1)^2x_{n-1} + n^2x_n' = 0 \\ x_1' + 2^3x_2' + 3^3x_3 + \ldots + (n-1)^3x_{n-1} + n^3x_n' = 0 \end{cases}.$$

We will eliminate the new variables: subtract the first equation from the three others,

$$\begin{cases} x_2' + 2x_3 + \ldots + (n-2)x_{n-1} + (n-1)x_n' = 0 \\ 3x_2' + 8x_3 + \ldots + n(n-2)x_{n-1} + (n^2-1)x_n' = 0 \\ 7x_2' + 26x_3 + \ldots + \left[(n-1)^3 - 1\right]x_{n-1} + (n^3-1)x_n' = 0 \end{cases}.$$

Multiply the first equation by 3 and subtract it from the second, then multiply the first equation by 7 and subtract it from the third:

$$\begin{cases} 2x_3 + \ldots + (n^2-3n+2)x_n' = 0 \\ 12x_3 + \ldots + (n^3-7n+6)x_n' = 0 \end{cases}.$$

Now, multiply the first equation by $n + 3$ and subtract from it the second equation. The coefficient of x_k, $3 \leq k \leq n - 1$, is

$$(k-1)(k-2)(n+3) - (k-1)(k-2)(k-3) = (k-1)(k-2)(n-k) > 0,$$

therefore $x_3 = \ldots = x_{n-1} = 0$. The system reduces to

$$\begin{cases} x_1' + x_2' + x_n' = 0 \\ x_1' + 2x_2' + nx_n' = 0 \\ x_1' + 2^2x_2' + n^2x_n' = 0 \\ x_1' + 2^3x_2' + n^3x_n' = 0 \end{cases}.$$

This homogenous system has only the solution $x_1' = x_2' = x_n' = 0$, since all minors of third order are nonzero. The conclusion follows.

182. By the AM−GM inequality,

$$\frac{a}{b} + \frac{a}{b} + \frac{b}{c} \geq 3\sqrt[3]{\frac{a^2}{bc}} = 3a.$$

Write two similar inequalities and add them to obtain the desired result, with equality only for $a = b = c$.

183. Answer: for all positive integers m.

Take $a = \cos\frac{\pi}{2(2m+1)} \neq 0$ and use the formula $2\cos x \cos y = \cos(x-y) + \cos(x+y)$. Then

$$\begin{aligned} f(m)\cdot a &= \left(\sum_{k=1}^{m}(-1)^k \cos\frac{k\pi}{2m+1}\right)\cdot a = \sum_{k=1}^{m}(-1)^k \cos\frac{k\pi}{2m+1}\cos\frac{\pi}{2(2m+1)} \\ &= \frac{1}{2}\sum_{k=1}^{m}(-1)^k\left(\cos\left(\frac{k\pi}{2m+1} - \frac{\pi}{2(2m+1)}\right) + \cos\left(\frac{k\pi}{2m+1} - \frac{\pi}{2(2m+1)}\right)\right) \\ &= \frac{1}{2}\sum_{k=1}^{m}(-1)^k\left(\cos\frac{(2k-1)\pi}{2(2m+1)} + \cos\frac{(2k+1)\pi}{2(2m+1)}\right) \\ &= \frac{1}{2}\left(-\cos\frac{\pi}{2(2m+1)} + (-1)^m\cos\frac{(2m+1)\pi}{2(2m+1)}\right) = \frac{1}{2}(-a + (-1)^m\cdot 0) = -\frac{1}{2}a, \end{aligned}$$

hence $f(m) = -\frac{1}{2}$ for all m.

184. Let

$$X = \frac{a_1^2}{a_1+b_1} + \ldots + \frac{a_n^2}{a_n+b_n},\ Y = \frac{b_1^2}{a_1+b_1} + \ldots + \frac{b_n^2}{a_n+b_n}.$$

Then

$$X - Y = \sum_{i=1}^{n}\frac{a_i^2 - b_i^2}{a_i+b_i} = \sum_{i=1}^{n}(a_i - b_i) = 0.$$

From

$$X+Y=\sum_{i=1}^{n}\frac{a_i^2+b_i^2}{a_i+b_i}\geq\frac{1}{2}\sum_{i=1}^{n}(a_i+b_i)=1,$$

it follows that

$$X=Y\geq\frac{1}{2}.$$

185. Multiplying $(a^2+b^2)(a^2+c^2)\geq(a^2+bc)^2$ by two similar inequalities, we get

$$\left(a^2+b^2\right)\left(b^2+c^2\right)\left(c^2+a^2\right)\geq\left(a^2+bc\right)\left(b^2+ca\right)\left(c^2+ab\right).$$

By the AM−GM inequality,

$$\frac{a^2+b^2}{c^2+ab}+\frac{b^2+c^2}{a^2+bc}+\frac{c^2+a^2}{b^2+ca}\geq 3\sqrt[3]{\frac{\left(a^2+b^2\right)\left(b^2+c^2\right)\left(c^2+a^2\right)}{\left(a^2+bc\right)\left(b^2+ca\right)\left(c^2+ab\right)}}\geq 3.$$

186. Answer: $f(x)=\pm\sqrt{c-1}\sin x+\cos x, x\in\mathbb{R}$.

The function $f(x)=0, \forall x\in\mathbb{R}$ is not a solution. Then $y=0$ implies $f(0)=0$. Now,

$$x=t, y=\pi\Rightarrow f(\pi+t)=f(\pi)f(t), t\in\mathbb{R},$$

$$x=\frac{\pi}{2}+t, y=\frac{\pi}{2}\Rightarrow f(\pi+t)=f\left(\frac{\pi}{2}\right)f\left(\frac{\pi}{2}+t\right)-c\cos t, t\in\mathbb{R},$$

$$x=\frac{\pi}{2}, y=t\Rightarrow f\left(\frac{\pi}{2}+t\right)=f\left(\frac{\pi}{2}\right)f(t)-c\sin t, t\in\mathbb{R}.$$

We have a system of equations, with solution

$$f(\pi)=\left(f\left(\frac{\pi}{2}\right)\right)^2-c, f(t)=f\left(\frac{\pi}{2}\right)\sin t+\cos t, t\in\mathbb{R}.$$

For $t=\pi$, we get $\left(f\left(\frac{\pi}{2}\right)\right)^2-c=-1$, whence the result.

187. Write the inequality as

$$\ln\frac{\prod x_j}{\prod(1-x_j)}\leq n\frac{\sum x_j}{\sum(1-x_j)}\Leftrightarrow\sum\ln\frac{x_j}{1-x_j}\leq n\ln\frac{\sum x_j}{\sum(1-x_j)}.$$

The function $f(x)=\ln\frac{x}{1-x}$ is concave on $\left(0,\frac{1}{2}\right)$, hence by the Jensen's inequality,

$$\sum \ln \frac{x_j}{1-x_j} = \sum f(x_j) \leq nf\left(\frac{\sum x_j}{n}\right) = n \ln \frac{\sum x_j}{\sum (1-x_j)}.$$

188. We will prove a stronger inequality

$$k\left|\sum_{cyc} \frac{a^3-b^3}{a+b}\right| \leq \sum_{cyc} (a-b)^2,$$

with $k = 2\sqrt{9+6\sqrt{3}}$. Indeed,

$$\left|\sum_{cyc} \frac{a^3-b^3}{a+b}\right| = \frac{|(a-b)(b-c)(c-a)|(ab+bc+ca)}{(a+b)(b+c)(c+a)}.$$

Let $c = \min\{a,b,c\}$, then

$$\frac{ab+bc+ca}{(a+b)(b+c)(c+a)} \leq \frac{1}{a+b-2c},$$

hence we need to prove that

$$\sum_{cyc} (a-b)^2 - \frac{k|(a-b)(b-c)(c-a)|}{a+b-2c} \geq 0.$$

Let $x = a - c$, $y = b - c$. If $x = y = 0$, or $xy = 0$ and $x + y > 0$, then the inequality is true, otherwise

$$\frac{x^3+y^3}{xy|x-y|} \geq \frac{k}{2} \Leftrightarrow \frac{k}{2} = \min_{t>1} \frac{t^3+1}{t(t-1)} = \sqrt{9+6\sqrt{3}},$$

which is true.

189. *a*) We will show by induction that

$$f\left(\left(\frac{1}{2}\right)^n\right) = 0$$

for any positive integer n. For $a = b \in [0,1]$, we have $f(a) \leq 2f(a)$, hence $f(a) \geq 0$. For $(a,b) = (0,1)$, we get

$$f\left(\frac{0+1}{2}\right) \leq f(0) + f(1) \Rightarrow f\left(\frac{1}{2}\right) \leq 0 \Rightarrow f\left(\frac{1}{2}\right) = 0.$$

Assume that for some positive integer n,

$$f\left(\left(\frac{1}{2}\right)^n\right)=0.$$

Then

$$f\left(\frac{0+\left(\frac{1}{2}\right)^n}{2}\right)\leq f(0)+f\left(\left(\frac{1}{2}\right)^n\right)=0\Rightarrow f\left(\left(\frac{1}{2}\right)^{n+1}\right)=0.$$

b) Answer: yes.
Define the function *f* as

$$f(x)=\begin{cases} 0, x\in[0,1]\cap\mathbb{Q} \\ c, x\in[0,1]\cap(\mathbb{R}-\mathbb{Q}) \end{cases},$$

where $c>0$ is a constant. If $x_1, x_2\in\mathbb{Q}$, then

$$f\left(\frac{x_1+x_2}{2}\right)=0\leq 0+0=f(x_1)+f(x_2).$$

If $x_1\in\mathbb{Q}$, $x_2\notin\mathbb{Q}$, then

$$f\left(\frac{x_1+x_2}{2}\right)=c\leq 0+c=f(x_1)+f(x_2).$$

If x_1, $x_2\notin\mathbb{Q}$, then

$$f\left(\frac{x_1+x_2}{2}\right)\leq c<c+c=f(x_1)+f(x_2).$$

190. If $n=2$ and $z^2=z_1z_2$, then $\pm z$ are the square roots of z_1z_2 and

$$|z_1-z||z_2+z|+|z_1+z||z_2-z|=2|z||z_1-z_2|,$$

hence
$z_1, z, z_2, -z$ are the vertices of a cyclic quadrilateral, which shows that z and $-z$ lie on the two arcs determined by z_1 and z_2 on a circle $\mathcal{C}$. Since at least one of the two arcs is included in $\mathcal{D}$, it follows that z or $-z$ belongs to $\mathcal{D}$.

If $n\geq 3$, then we may suppose that $r_k=|z_k|\neq 0$ for $k=1,2,\ldots,n$. Let $r=\sqrt[n]{r_1\ldots r_n}$. Then there exists w_1 on the segment $[z_1,z_2]$ such that $|w_1|^2=|z_1||z_2|$, there exists w_2 on the segment $[w_1,z_3]$ such that $|w_2|^3=|w_1|^2|z_3|$ and so on, …, which shows that there exists at least one complex number of modulus r in $\mathcal{D}$, hence the intersection of the circle $\mathcal{C}(O,r)$ with the interior of $\mathcal{D}$ is nonempty.

If $\mathcal{C}(O,r)\subset\mathcal{D}$, then every root of order n of $z_1\ldots z_n$ belongs to $\mathcal{D}$.

If $C(O,r) \not\subset \mathcal{D}$, then for some sufficiently small interval (a,b) containing r, the intersections of $\mathcal{D}$ with the circles $C(O,a), C(O,b)$ are arcs $\mathcal{A}_1, \mathcal{A}_2$ of these circles. Let $\mathcal{A}$ be the region of $\mathcal{D}$ between $\mathcal{A}_1$ and $\mathcal{A}_2$, that is, $\mathcal{A} = \{z \in \mathcal{D} | a < |z| < b\}$. Then there exist $\alpha, \beta \in \mathbb{R}$ such that every $z \in \mathcal{A}$ has an unique argument between α and β. We can replace in a finite number of steps $z_1, z_2, \ldots, z_n$ by $w_j \in \mathcal{A}, j = 1, 2, \ldots, n$ and $z_1 z_2 \ldots z_n = w_1 w_2 \ldots w_n$, so we can take

$$z = r(\cos\varphi + i\sin\varphi), \varphi = \frac{\sum_{j=1}^{n} \arg(w_j)}{n}.$$

This follows from the following lemma.

Lemma. A transformation of a family $r_1, r_2, \ldots, r_n$ of positive numbers consists of replacing $r_i = \max\{r_p | p = 1, 2, \ldots, n\}$ and $r_j = \min\{r_p | p = 1, 2, \ldots, n\}$ with $\sqrt{r_i r_j}$. If

$$a < r = \sqrt[n]{r_1 r_2 \ldots r_n} < b,$$

then after a finite number of such transformations, we get a family s_1, s_2, $\ldots$, s_n with $s_p \in (a,b)$, $p = 1, 2, \ldots, n$.

Proof. Let $\alpha = \ln a$, $\beta = \ln b$, $x = \ln r$ and $x_p = \ln r - p$, $p = 1, 2, \ldots, n$. Denote by d_s the value of the sum $\sum |x_p - x|$ after s transformations. Suppose that

$$x_1 \leq \ldots \leq x_k \leq x \leq x_{k+1} \leq \ldots \leq x_n.$$

Then $(x - x_1) + \ldots + (x - x_k) = (x_{k+1} - x) + \ldots + (x_n - x)$ and

$$d_s = 2\sum_{p \leq k} (x - x_p) = 2\sum_{p \geq k+1} (x_p - x).$$

When replacing (x_1, x_n) by $\left(\frac{x_1 + x_n}{2}, \frac{x_1 + x_n}{2}\right)$, d_s is modified by

$$2\left|\frac{x_1 + x_n}{2} - x\right| - (x - x_1) - (x_n - x) = -2\min\{x - x_1, x_n - x\}.$$

But $d_s \le 2n(x - x_1)$ and $d_s \le 2n(x_n - x)$, implies $2\min\{x - x_1, x_n - x\} \ge \frac{d_s}{n}$, whence

$$d_{s+1} \le \frac{n-1}{n} d_s.$$

This shows that $d_s \to 0$ and the conclusion follows.

191. Let $S = \{i_1, i_1 + 1, \dots, j_1, i_2, i_2 + 1, \dots, j_2, \dots, i_p, \dots, j_p\}$ be the ordering of S, where $j_k < i_{k+1} - 1$ for $k = 1, 2, \dots, p - 1$. Put $s_0 = 0$, $s_p = a_1 + a_2 + \dots + a_p$. Then

$$\sum_{i \in S} a_i = s_{j_p} - s_{i_p - 1} + s_{j_{p-1}} - s_{i_{p-1} - 1} + \dots + s_{j_1} - s_{i_1 - 1},$$

$$\sum_{1 \le i \le j \le n} (a_i + \dots + a_j)^2 = \sum_{0 \le i \le j \le n} (s_j - s_i)^2.$$

It suffices to prove an inequality of the form

$$\left(x_1 - x_2 + \dots + (-1)^{l+1} x_l\right)^2 \le \sum_{1 \le i < j < l} (x_j - x_i)^2 + \sum_{i=1}^{l} x_i^2. \qquad (1)$$

This inequality reduces to

$$4 \sum_{\substack{1 \le i < j \le l \\ j - i \text{ even}}} x_i x_j \le (l - 1) \sum_{i=1}^{l} x_i^2$$

and can be obtained by adding inequalities of the form $4x_i x_j \le 2\left(x_i^2 + x_j^2\right), i < j, j - i$ even.

192. Let $P(x) = P_n(x)$, $R(x) = R_n(x)$, $S(x) = S_n(x)$. We will prove by induction on n that $R_n(x)$ has only real roots; more precisely, if $x_1 > x_2 > \dots > x_n$ are the roots of R_n and $y_1 > y_2 > \dots > y_{n-1}$ are the roots of R_{n-1}, then

$$x_1 > y_1 > x_2 > y_2 > \dots > x_{n-1} > y_{n-1} > x_n.$$

The base case is obvious. Assume that the assertion is true for $n - 1$. From

$$R_n + iS_n = (x - i + n)(R_{n-1} + iS_{n-1}),$$

we deduce the recurrence relations

$$R_n = (x+n)R_{n-1} + S_{n-1}, S_n = (x+n)S_{n-1} - R_{n-1},$$

whence it follows

$$R_n = (2x+2n-1)R_{n-1} - \left[(x+n-1)^2 + 1\right]R_{n-2}.$$

If $z_1 > \ldots > z_{n-2}$ are the roots of $R_{n-2,}$ then by the induction hypothesis,

$$z_{i-1} > y_i > z_i, 2 \leq i \leq n-2.$$

Since R_{n-2} has alternately positive and negative values on the intervals (z_1, ∞), (z_2, z_1), …, (z_{n-2}, z_{n-1}), it follows that sgn $R_{n-2}(y_i) = (-1)^{i-1}$. From

$$R_n(y_i) = -\left[(x+n-1)^2 + 1\right]R_{n-2}(y_i),$$

we deduce sgn $R_n(y_i) = (-1)^i$, $1 \leq i \leq n-1$, therefore R_n has a root in each interval

$$(y_1, \infty), (y_2, y_1), \ldots, (-\infty, y_{n-1}).$$

193. The function $f(x) = \frac{x}{1-x^2}$ is convex on (0, 1), since

$$f''(x) = \frac{4x}{(1-x^2)^3} > 0.$$

By the Jensen's inequality,

$$\sum \frac{a}{1-a^2} \geq 3 \cdot \frac{\frac{p}{3}}{1-\left(\frac{p}{3}\right)^2} \geq \frac{3\sqrt{3}}{2} \Leftrightarrow \frac{1}{9}p^2 + \frac{2}{3\sqrt{s}}p \geq 1,$$

where $p = a + b + c$. The last inequality is true, since $p^2 \geq 3q = 3$, where $q = ab + + bc + ca = 1$

194. From $abc = 1$ and $(a^2 - b^2)(a^3 - b^3) = a^5 + b^5 - a^2b^2(a + b) \geq 0$, it follows

$$\frac{ab}{a^5 + ab + b^5} \leq \frac{ab}{a^2b^2(a+b) + ab} = \frac{c}{a+b+c}.$$

Adding with two similar inequalities, we get the result.

195. After squaring, we obtain $(a^2-45)^2+a-21=0$, hence a is a root of the polynomial $P(x)=x^4-90x^2+x+2004$. Analogously, b is a root of P. We find also that c,d are the roots of the polynomial $x^4-90x^2-x+2004$, so that $-c, -d$ are roots of P. We cannot have $a=b$, otherwise $-\sqrt{21-a}=\sqrt{21-b}$, contradiction. Analogously, the equality $c=d$ leads to a contradiction. It follows that $a, b, -c, -d$ are distinct, since a, b are positive numbers, and $-c, -d$ are negative.

By the Viète's relation, $abcd = ab(-c)(-d) = 2004$.

3.34 2005

196. Answer: $f(x) = x$ for all x; $f(x) = -x$ for all x.

Let $f(0) = k$, then $f(f(x)) = k^2 + x$, hence f is bijective. Let $f(a) = 0$, then $f(f(a)) = a$ implies $k = 0$. Now,

$$f(xf(x)+f(y))=f(f(x))f(xf(x)+f(y))\Leftrightarrow(f(x))^2+y=x^2+y\Leftrightarrow f(x)\in\{-x,x\}.$$

Choose $f(x) = x \neq 0$, $f(y) = -y \neq 0$, but at least one must be zero, contradiction.

Hence $f(x) = x$ for all x or $f(x) = -x$ for all x.

197. If P is constant, then $P = 0$. If deg $(P) = n \geq 1$, then $n^5 = 28 + n^2$, whence $n = 2$. Now we will prove that $x|P(x)$:

$$x=0\Rightarrow P^5(0)=0, x=P^3(0)\Rightarrow P^8(0)=0, x=P^5(0)\Rightarrow P^{10}(0)=0,$$
$$P^{10}(0)=P^2(P^8(0))=P^2(0)=0, P^5(0)=P(P^2(P^2(0)))=P(0)=0.$$

From this, it follows $x|P(x)|P^2(x)$, hence $P(x) = cx^2$.

Putting it in the hypothesis, we get $c = \pm 1$. The solutions are $P(x) = \pm x^2$.

198. Answer: $f(x) = 0, \forall x \in \mathbb{Z}$; $f(x) = x, \forall x \in \mathbb{Z}$; $f(x) = -x, \forall x \in \mathbb{Z}$.

If f is strictly monotonic, then $f(0) = 0$ implies $f(1) \in \{-1, 1\}$. Observe that if f is a solution, then $-f$ is also a solution. Suppose that $f(1) = 1$. Then $f(2) = 2$ and $f(a) = = a$ for some $a > 1$ implies $f(a^{2005}) = a^{2005}$, hence we have a strictly increasing sequence (a_n) such that $f(a_n) = a_n$ for all n. We deduce $f(x) = x$ for all positive integers x. From $f(-x) = -f(x)$, it follows that $f(x) = x$ for all integers x.

199. Answer: $(0, 0)$, $(0, 1)$, $(k, 0)$ and (k, n), with $k \geq n - 1 \geq 1$.

If $nk = 0$, then it is easy to find the above solutions. Suppose that $n, k \geq 1$. Then $k \geq n - 1$. Indeed, for $x = y$, we get $1 + x^{n-k} \geq (1 + x)^{n-k}$, which is true only if $n - k \leq 1$.

If $k = n - 1$, then

$$1+\frac{y^n}{x^{n-1}}\geq\frac{(1+y)^n}{(1+x)^{n-1}},$$

by the Radon's inequality.

If $k \geq n$, then the inequality is trivial for $x \geq y$; for $x < y$, we have

$$\frac{y^n}{x^k}\geq\frac{(1+y)^n}{(1+x)^k}.$$

200. By the Cauchy–Schwarz inequality,

$$a\sqrt{b}+b\sqrt{c}+c\sqrt{a}\leq\sqrt{(a+b+c)(ab+bc+ca)}\leq\sqrt{\frac{(a+b+c)^3}{3}}=\frac{1}{\sqrt{3}}.$$

201. We will use induction on $N = a_1 + a_2 + \ldots + a_m$. For $N = 1$ we have $m = 1$, $a_1 = 1$ and $b_1 = 1$ is the required number. Assume that the assertion is true for every $M < N = a_1 + a_2 + \ldots + a_m$. If all $a_1, a_2, \ldots, a_m$ are even, then

$$\frac{a_1}{2}+\frac{a_2}{2}+\ldots+\frac{a_m}{2}=\frac{N}{2}$$

and by the induction hypothesis, there exist $b_1, b_2, \ldots, b_n$ which satisfy the condition. Then the required numbers for $a_1, a_2, \ldots, a_m$ are $2b_1, 2b_2, \ldots, 2b_n$.

If at least one of $a_1, a_2, \ldots, a_m$ is odd, then we can suppose wlog that a_m is the smallest odd number among them. Define

$$a_i'=\begin{cases}\dfrac{a_i}{2}, 2|a_i,\\ \dfrac{a_i-a_m}{2}, 2\nmid a_i\end{cases}.$$

Then $a_1'+a_2'+\ldots+a_m'<N$ and the induction hypothesis implies the existence of numbers $b_1', b_2', \ldots, b_n'$ which satisfy the conditions. We will prove that the numbers $2b_1', 2b_2', \ldots, 2b_n'$, a_m are the required numbers for $a_1, a_2, \ldots, a_m$.

If two disjoint subsets of $\{2b_1', 2b_2', \ldots, 2b_n', a_m\}$ have equal sums, then a_m, as the only odd number of the set, does not belong to these subsets. Dividing by 2, we get two disjoint subsets of $\{b_1', b_2', \ldots, b_n'\}$ with equal sums, contradiction. Moreover, every a_i, $1 \leq i \leq m$, can be represented as a sum of some of $\{2b_1', 2b_2', \ldots, 2b_n', a_m\}$, which ends the proof by induction.

202. Denote the assertion by $P(x, y)$. Comparing

$$P(1,1), P(1,2), P(2,1), P(1,3), P(1,4),$$

we deduce $f(1) = 1$. Then $P(1, x)$ implies $f(x + 1) = f(x) + 1$ and $P(1, x^2)$ implies $f(x^2) = (f(x))^2$. It is known that this implies $f(x) = x$ for all x.

203. Let $S_a = \sum a_i$, $S_b = \sum b_i$. The inequality is homogenous, so we can suppose that $S_a + S_b = 1$. If $LHS < RHS$, then

$$\sum_{i=1}^{n} \frac{a_i^2}{a_i + b_i} < S_a^2, \sum_{i=1}^{n} \frac{b_i^2}{a_i + b_i} < S_b^2.$$

Multiplying and applying the Cauchy–Schwarz inequality, we get

$$\left(\sum_{i=1}^{n} \frac{a_i b_i}{a_i + b_i}\right)^2 \leq \left(\sum_{i=1}^{n} \frac{a_i^2}{a_i + b_i}\right)\left(\sum_{i=1}^{n} \frac{b_i^2}{a_i + b_i}\right) < (S_a S_b)^2,$$

contradiction.

204. Answer: $(a, b) = (-6, 12)$.

Let x_1, x_2, x_3 be the roots of the polynomial $x^3 + ax^2 + bx - 8$. Then

$$x_1 + x_2 + x_3 = -a, x_1 x_2 x_3 = 8,$$

by the Viète's relations. The AM–GM inequality gives

$$-a = x_1 + x_2 + x_3 \geq 3\sqrt[3]{x_1 x_2 x_3} = 6,$$

so $a \leq -6$. The polynomial $6x^2 - 24x - 4a$ has real roots if and only if

$$\Delta = 24^2 + 24 \cdot 4a \geq 0,$$

so $a \geq -6$, hence $a = -6$. This implies equality in the AM–GM inequality, hence

$$x_1 = x_2 = x_3 = 2,$$

implying $b = x_1 x_2 + x_2 x_3 + x_3 x_1 = 12$.

205. If $P(P(k)) = 0$, then $P(k) \in \{0, \alpha_1, \ldots, \alpha_n\}$. Let $P(k) = \alpha_i$ for some i. It follows that

$$a_n k(k - \alpha_2)\ldots(k - \alpha_n) = \alpha_i, \quad a_n, \alpha_i, k \neq 0.$$

We have $k|\alpha_i$, so $k|k - \alpha_i$. But $k - \alpha_i|\alpha_i$, hence $k - \alpha_i|k$. We deduce $|k - \alpha_i| = |k|$, so that $\alpha_i = 2k$. Now,

$$a_n(k-\alpha_2)\ldots(k-\alpha_i)\ldots(k-\alpha_n)=2.$$

All $k - \alpha_i$ are distinct and there are at least four, which belong to the set $\{-2,-1,1,2\}$. But -2 and 2 cannot occur together, contradiction.

In conclusion, k must be a root of P.

206. We will prove the inequality

$$\left(\frac{xy}{z}+\frac{yz}{x}+\frac{zx}{y}\right)^2 \geq 3\left(x^2+y^2+z^2\right),$$

whence it follows that the minimum is $\sqrt{3\cdot 25}=5\sqrt{3}$, realized when $x=y=z=\frac{5\sqrt{3}}{3}$. After squaring and cancelling terms, we have to prove that

$$\sum_{cyc}\frac{x^2y^2}{z^2} \geq \sum x^2.$$

This simplifies to

$$\sum_{cyc}x^4y^4 \geq \sum_{cyc}x^4y^2z^2,$$

which is Muirhead's inequality (or $x^4y^4 + y^4z^4 \geq 2x^2y^4z^2$, etc).

207. The general term of the expansion $(a + b)^n$ is

$$u_k=\binom{n}{k}a^kb^{n-k}, 0\leq k\leq n.$$

For $0 \leq k \leq n - 1$, we have

$$\frac{u_{k+1}}{u_k}=\frac{(n-k)a}{(k+1)b},$$

hence

$$\frac{u_{k+1}}{u_k} > 1 \Leftrightarrow (n-k)a > (k+1)b \Leftrightarrow k < \frac{na-b}{a+b}.$$

Case 1. $\frac{na-b}{a+b} > n-1$. Then the largest term is the last one, a^n.

Case 2. $\frac{na-b}{a+b} \leq 0$. Then the largest term is the first one, b^n.

Case

3. $0 < \frac{na-b}{a+b} \leq n-1$. If $\frac{na-b}{a+b} \notin N$, then the sequence strictly increases until $k = \left\lfloor \frac{na-b}{a+b} \right\rfloor + 1$, then it strictly decreases. The largest term is u_k and it is unique.

If $\frac{na-b}{a+b} = k \in \mathbb{N}$, then the largest terms are $u_k = u_{k+1}$.

208. Answer: no.

From the given condition we deduce $a_n > 0$ and $a_{n+1} = a_n + a_{n-1}$ or $a_n - a_{n-1}$, so that $a_n < a_{n-1}$ implies $a_{n+1} > \max\{a_n, a_{n-1}\}$. By removing all terms a_n for which $a_n < a_{n-1}$ we obtain a strictly increasing sequence $(b_m)_m$. The statement of the problem will follow if we prove that $b_{m+1} - b_m \geq b_m - b_{m-1}$ for $m \geq 2$.

Let $b_{m+1} = a_{n+2}$ for some n, then $a_{n+2} > a_{n+1}$.

Case 1. $a_{n+1} > a_n$. Then $b_m = a_{n+1}$, $b_{m-1} \geq a_{n-1}$, hence

$$b_{m+1} - b_m = a_{n+2} - a_{n+1} = a_n = a_{n+1} - a_{n-1} = b_m - a_{n-1} \geq b_m - b_{m-1}.$$

Case 2. $a_{n+1} < a_n$. Then $b_m = a_n$, $b_{m-1} \geq a_{n-1}$, hence

$$b_{m+1} - b_m = a_{n+2} - a_n = a_{n+1} = a_n - a_{n-1} = b_m - a_{n-1} \geq b_m - b_{m-1}.$$

209. We have

$$\frac{1+a}{1-a} = \frac{2a+b+c}{b+c} = \frac{2a}{b+c} + 1$$

and the inequality becomes

$$\sum_{cyc}\left(\frac{2a}{b} - \frac{2a}{b+c}\right) \geq 3 \Leftrightarrow \sum_{cyc}\frac{ac}{b(b+c)} \geq \frac{3}{2}. \qquad (1)$$

By the Cauchy–Schwarz inequality,

$$\sum_{cyc}\frac{ac}{b(b+c)} = \sum_{cyc}\frac{a^2c^2}{abc(a+b+c)} \geq \frac{(ab+bc+ca)^2}{2abc(a+b+c)} = \frac{(ab+bc+ca)^2}{2abc} \geq \frac{3}{2},$$

which follows from $(ab+bc+ca)^2 \geq 3abc(a+b+c)$.

210. By the power mean inequality, the LHS is less than

$$E=\frac{3}{\sqrt[3]{3}}\sqrt[3]{\frac{1}{a}+\frac{1}{b}+\frac{1}{c}+6(a+b+c)}.$$

From $ab + bc + ca = 1$, we obtain $3abc(a + b + c) \leq (ab + bc + ca)^2 = 1$. Since

$$\frac{1}{a}+\frac{1}{b}+\frac{1}{c}=\frac{ab+bc+ca}{abc}=\frac{1}{abc},$$

it follows that

$$E\leq\frac{3}{\sqrt[3]{3}}\sqrt[3]{\frac{3}{abc}}\leq\frac{1}{abc},$$

where the last inequality follows from the AM–GM inequality.

Equality holds only when $a=b=c=\frac{\sqrt{3}}{3}$.

211. Let

$$P(x,y):f(xy)f\left(\frac{f(y)}{x}\right)=1.$$

Then

$$P\left(\frac{f(x)}{x},x\right)\bigwedge P\left(\frac{1}{f(x)},f(x)\right)\Rightarrow f(1)=1, P(f(x),x)\Longrightarrow f(xf(x))=1.$$

Case 1. f increasing. We have $xf(x) \to \infty$ for $x \to \infty$, so $f(x) = 1$ for all x.

Case 2. f decreasing. Let $K=\{x\in\mathbb{R}^+|f(x)=1\}$. Then $k\in K$ implies $\frac{1}{k}\in K$. If K is bounded by some M, then $f(x)=1$ for $\frac{1}{M}<x<M$. Given $x\in\mathbb{R}^+$, choose y such that

$$\frac{1}{M}<xy<M,$$

so $\frac{1}{M}\leq\frac{f(x)}{y}\leq M$, whence $f(x)=\frac{1}{x}$.

If K is unbounded, then $f(x) = 1$ for all x.

212. Let $n! = 2^t(2k + 1)$, with k, t integers. It is easy to show that there is a positive integer u such that $2k + 1|2^u - 1 = a$ (take $u = \varphi(2k + 1)$). Define

$$P(x) = 2^t\left(1 + \sum_{i=1}^{n} \frac{x(x-1)\ldots(x-i+1)}{i!}a^i\right).$$

Since $i\,!\;|2^t a^i$, the coefficients of P are integers. For $1 \le m \le n$, we have

$$P(m) = 2^t\left(1 + \sum_{i=1}^{m}\binom{m}{i}a^i\right) = 2^t(1+a)^m = 2^{t+mu}.$$

213. Consider that the numbers are nonnegative. The equality holds if $a_i = b_i$ for some i and $a_j = b_j$ for $j \ne i$. Using

$$\sum a_i^2 + 2\sum a_i a_{i+1} < \left(\sum a_i\right)^2 = 1,$$

we have

$$2\sum a_i(a_{i+1} + b_i) = 2\sum a_i b_i + 2\sum a_i a_{i+1} \le \sum a_i^2 + \sum b_i^2 + 2\sum a_i a_{i+1} < 2.$$

214. By the Heron's formula,

$$S^2 = s(s-a)(s-b)(s-c) = s^4 - (a+b+c)s^3 + (ab+bc+ca)s^2 - abcs$$

is rational, since $2s = a + b + c$, $ab + bc + ca$, abc are rational. Let

$$P(x) = x^3 + mx^2 + nx + p$$

be the polynomial with roots a, b, c and h the altitude corresponding to side a. Then $2S = ah$ and

$$P(a) = P\left(\frac{2S}{h}\right) = \frac{8S^3 + 4mS^2h + 2nSh^2 + ph^3}{h^3} = 0,$$

whence $2S(4S^2 + nh^2) = -ph^3 - 4mS^2h$. It follows that h is a root of the polynomial of sixth degree with rational coefficients

$$Q(x) = \left(px^3 + 4ms^2x\right)^2 - 4S^2\left(4S^2 + nx^2\right)^2.$$

215. Define $g(x) = be^{2ix} - ice^{5ix} + de^{8ix}$. Then

$$g(x)+g\left(x+\frac{2\pi}{3}\right)+g\left(x+\frac{4\pi}{3}\right)=g(x)\left(1+e^{\frac{2\pi i}{3}}+e^{\frac{4\pi i}{3}}\right)=0.$$

From $f(x) = a + Re\ (g(x))$, we deduce

$$f(x)+f\left(x+\frac{2\pi}{3}\right)+f\left(x+\frac{4\pi}{3}\right)=3a. \ (1)$$

Case 1. $a < 0$. Then we can take $s = t$.
Case 2. $a > 0$. Then for $x = t$ in (1), we get

$$f\left(t+\frac{2\pi}{3}\right)+f\left(t+\frac{4\pi}{3}\right)=-a,$$

so we can choose $s=t+\frac{2\pi}{3}$ or $s=t+\frac{4\pi}{3}$.

Case 3. $a = 0$. Then in (1) at least one term is negative for some x, otherwise $f = 0$ for all x, which implies $a = b = c = d = 0$, contradiction.

216. We have

$$\frac{\sum_{j=1}^{i} a_j}{n}+\frac{\sum_{j=i+1}^{n} a_j}{n}=m\le a_i+\frac{\sqrt{(n-i)\left(\sum_{j=i+1}^{n} a_j^2\right)}}{n}\le a_i+\sqrt{\frac{n-i}{n}}.$$

217. We have

$$-1<\frac{a^2-b^2}{a^2+b^2}<1,$$

since $-2a^2 < 0 < 2b^2$. It is easy to prove the identity

$$(x-y)(x+z)(z+x)+(y-z)(z+x)(x+y)+(z-x)(x+y)(y+z)=-(x-y)\times(y-z)(z-x).$$

Choosing $x = a^2$, $y = b^2$, $z = c^2$, we deduce

$$\left|\frac{a^2-b^2}{a^2+b^2}+\frac{b^2-c^2}{b^2+c^2}+\frac{c^2-a^2}{c^2+a^2}\right|=\left|\frac{a^2-b^2}{a^2+b^2}\cdot\frac{b^2-c^2}{b^2+c^2}\cdot\frac{c^2-a^2}{c^2+a^2}\right|<1.$$

For

$$\frac{a^2-b^2}{a^2+b^2}=k\in(-1,1),$$

we have

$$\lim_{c\to 0}\left(\frac{a^2-b^2}{a^2+b^2}+\frac{b^2-c^2}{b^2+c^2}+\frac{c^2-a^2}{c^2+a^2}\right)=k-1+1,$$

whence it follows that the set of values of the expression is $(-1,1)$, since the expression is a continuous function in each variable.

218. Let A be the inverse of the given expression. Then

$$A=\left(1+\frac{x}{2}+\frac{x}{2}\right)\left(x+\frac{y}{2}+\frac{y}{2}\right)\left(y+\frac{z}{2}+\frac{z}{2}\right)(z+8+8)$$
$$\geq 81\sqrt[3]{\left(\frac{x}{2}\right)^2}\cdot\sqrt[3]{x\left(\frac{y}{2}\right)^2}\cdot\sqrt[3]{y\left(\frac{z}{2}\right)^2}\cdot\sqrt[3]{64z}=81xyz,$$

by the AM – GM inequality. The given expression is at most $\frac{1}{81}$, with equality for $(x,y,z)=(2,4,8)$.

219. Let p be the given polynomial and suppose that $p=rs$ for nonconstant polynomials r, s with integer coefficients. For $k=\pm 1,\pm 2,\ldots,\pm n$, we have

$$p(k)=r(k)s(k)=1.$$

Since the factors are integers, it follows that $r(k)=s(k)=\pm 1$, for $k=\pm 1,\pm 2,\ldots,\pm n$. But $\deg(r-s)\leq 2n-1$ and the polynomial $r-s$ has $2n$ roots, hence $r-s=0$. Then $p=r^2$, in particular $p(0)$ is a square.

We will show that this is impossible: if n is odd, then $p(0)=-(n!)^2+1<0$, for $n\leq 3$; if n is even, then $(n!)^2<p(0)<(n!+1)^2$.

3.35 2006

220. For $x=0$, we get $8P(0)=0$, hence $P(0)=0$, therefore $P(x)=xQ(x)$. Then

$$(x+1)^3(x-1)Q(x-1)-(x-1)^3(x+1)Q(x+1)=4(x^2-1)xQ(x)$$

or

$$(x+1)^2Q(x-1)-(x-1)^2Q(x+1)=4xQ(x).$$

This rewrites as

$$(x^2+1)(Q(x-1)-Q(x+1))=2x(2Q(x)-Q(x-1)-Q(x+1)).$$

If $\deg(Q) = n \geq 2$, then deg (LHS) $= n + 1$, deg (RHS) $= n - 1$, contradiction. If $Q(x) = ax + c$, then replacing in the equation, we obtain $-2a(x^2 + 1) = 0$, hence $a = 0$, $Q(x) = c$ and $P(x) = cx$, $c \in \mathbb{R}$.

221. We have $7n^2 - \lfloor n\sqrt{7}\rfloor^2 \neq 0, 1, 2,$ therefore $7n^2 - \lfloor n\sqrt{7}\rfloor^2 \geq 3$, so that

$$\left\{n\sqrt{7}\right\}=n\sqrt{7}-\left\lfloor n\sqrt{7}\right\rfloor=\frac{7n^2-\lfloor n\sqrt{7}\rfloor^2}{n\sqrt{7}+\lfloor n\sqrt{7}\rfloor}>\frac{7n^2-\lfloor n\sqrt{7}\rfloor^2}{2n\sqrt{7}}>\frac{3\sqrt{7}}{14n}.$$

222. Using $\cos(a + x) = \cos a \cos x - \sin a \sin x$, we get $f(x) = A \sin x + B \cos x$, where

$$A=-\sin a_1-\frac{\sin a_2}{2}-\ldots-\frac{\sin a_n}{2^{n-1}},B=\cos a_1+\frac{\cos a_2}{2}+\ldots+\frac{\cos a_n}{2^{n-1}}.$$

Numbers A, B cannot both be equal zero, otherwise f is identical 0, which contradicts

$$\begin{aligned}f(-a_1)&=\cos(a_1-a_1)+\frac{\cos(a_2-a_1)}{2}+\ldots+\frac{\cos(a_n-a_1)}{2^{n-1}}\\&\geq 1-\frac{1}{2}-\ldots-\frac{1}{2^{n-1}}=\frac{1}{2^{n-1}}>0.\end{aligned}$$

There exist C, φ such that $A = C\cos\varphi$, $B = C\sin\varphi$, hence $f(x) = C\sin(x+\varphi)$. Zeroes of f are of the form $-\varphi + k\pi$, $k \in \mathbb{Z}$, whence the result.

223. Suppose that a is rational. Then $a = \frac{p}{q}$, with $p < q$ positive integers. Denote

$$u=\sqrt[q]{x}, v=\sqrt[q]{y}.$$

Now we have to prove that

$$\frac{u^q-v^q}{1-(uv)^q}\leq\frac{u^p-v^p}{1-(uv)^p}$$

for $v \leq u$ and $p < q$. It suffices to prove the inequality

$$\frac{u^{p+1}-v^{p+1}}{1-(uv)^{p+1}}\leq\frac{u^p-v^p}{1-(uv)^p}$$

and then use induction. We have the equivalences

$$\frac{u^{p+1}-v^{p+1}}{1-(uv)^{p+1}} \le \frac{u^p-v^p}{1-(uv)^p} \Leftrightarrow \frac{1-u^{2p+1}}{u^p(1-u)} \le \frac{1-v^{2p+1}}{v^p(1-v)} \Leftrightarrow \sum_{i=1}^{p}\left(u^i+u^{-i}\right) \le \sum_{i=1}^{p}\left(v^i+v^{-i}\right).$$

The last inequality follows from $t + t^{-1} \le z + z^{-1}$, true for $0 < z \le t \le 1$.

If a is irrational, then choose a sequence (a_n) of rational numbers convergent to a and take $n \to \infty$ to obtain the desired inequality.

224. Write the given identity as

$$\frac{P(x)}{Q(x)} - \frac{1}{2}\left(\frac{1}{x} + \frac{1}{x+1}\right) = \frac{P(x+1)}{Q(x+1)} - \frac{1}{2}\left(\frac{1}{x+1} + \frac{1}{x+2}\right).$$

By induction, it follows that

$$\frac{P(x)}{Q(x)} - \frac{1}{2}\left(\frac{1}{x} + \frac{1}{x+1}\right) = \frac{P(x+n)}{Q(x+n)} - \frac{1}{2}\left(\frac{1}{x+n} + \frac{1}{x+n+1}\right)$$

for all positive integers n. Fix x and let $n \to \infty$. Then we deduce

$$\frac{P(x)}{Q(x)} - \frac{1}{2}\left(\frac{1}{x} + \frac{1}{x+1}\right) = c,$$

with c constant. From this, we conclude that there is a polynomial $R(x)$ such that

$$P(x) = \left(\frac{1}{2} + x + cx(x+1)\right)R(x), Q(x) = x(x+1)R(x).$$

225. Arrange $x_1, x_2, \ldots, x_n$ in nondecreasing order. By the Chebyshev's inequality,

$$\frac{1}{n^2}\left(\sum_{i=1}^{n}\sqrt{x_i}\right)\left(\sum_{i=1}^{n}\frac{1}{\sqrt{1+x_i}}\right) \le \frac{1}{n}\sum_{i=1}^{n}\sqrt{\frac{x_i}{1+x_i}}.$$

The function $f(x) = \sqrt{\frac{x}{1+x}}$ is concave for $x \ge 0$. By the Jensen's inequality,

$$\frac{1}{n}\sum_{i=1}^{n}\sqrt{\frac{x_i}{1+x_i}} = \frac{1}{n}\sum_{i=1}^{n} f(x_i) \le f\left(\frac{1}{n}\sum_{i=1}^{n} x_i\right) = f\left(\frac{1}{n}\right) = \frac{1}{\sqrt{n+1}},$$

hence

$$\left(\sum_{i=1}^{n} x_i\right)\left(\sum_{i=1}^{n} \frac{1}{\sqrt{1+x_i}}\right) \leq \frac{n^2}{\sqrt{n+1}}.$$

226. Denote $s = b + c$, $p = bc$. The given condition becomes

$$a^2 + sa + p = \frac{1}{s},$$

whence it follows

$$a = \frac{\sqrt{s^2 - 4\left(p - \frac{1}{s}\right)} - s}{2}.$$

Now, we have to prove that

$$as + p = \frac{\sqrt{s^2 - 4\left(p - \frac{1}{s}\right)} - s}{2} \cdot s + p \leq \frac{3}{4}.$$

It is easy to see, after simple calculations, that this inequality is equivalent to

$$\left(s^2 - 4p\right) + 2(s-1)^2 + \left(2p - \frac{1}{2}\right)^2 \geq 0,$$

which is true, with equality only when $s = 1, p = \frac{1}{4}$, that is, $a = b = c = \frac{1}{2}$.

227. From $a + 1 \geq 2\sqrt{a}$ and other two similar inequalities, we deduce

$$(a+1)(b+1)(c+1) \geq 8.$$

Now, we have

$$\begin{aligned} &\frac{a}{(a+1)(b+1)} + \frac{b}{(b+1)(c+1)} + \frac{c}{(c+1)(a+1)} \\ &= 1 - \frac{2}{(a+1)(b+1)(c+1)} \geq 1 - \frac{2}{8} = \frac{3}{4}. \end{aligned}$$

228. Consider the function $\varphi(x) = x + yf(x)$, $x, y > 0$. The function is injective:

$$\varphi(x_1) = \varphi(x_2) \Rightarrow f(x_1)f(y) = f(\varphi(x_1)) = f(\varphi(x_2)) = f(x_2)f(y) \Rightarrow f(x_1) = f(x_2),$$

so $x_1 = x_2$. If $x_1 > x_2$ and $f(x_1) < f(x_2)$, then $\varphi(x_1) = \varphi(x_2)$ for

$$y = \frac{x_1 - x_2}{f(x_2) - f(x_1)} > 0,$$

contradiction. It follows that f is nondecreasing. Now,

$$f(x)f(y) = 2f(x + yf(x)) \geq 2f(x) \Rightarrow f(y) \geq 2, y > 0 \Rightarrow f(x + yf(x))$$
$$= f(xy) = f(y + xf(y)) \geq f(2x)$$

for arbitrarily small $y > 0$, hence f is constant on $(x, \;\; 2x]$ for all $x > 0$. We deduce that f is constant on $\mathbb{R}^+$, therefore $f(x) = 2$ for all x.

229. Assume that $\max\{pq, pr, ps, qr, qs, rs\} = pq$ and $pq < rs + 2$. Since

$$pq + pr + ps + qr + qs + rs = 60,$$

it follows that $pq \geq 10$. Then $rs + 2 > 10$, hence $rs > 8$. The AM−GM inequality get us

$$8^2 < (rs)^2 \leq pqrs \leq \left(\frac{p^2 + q^2 + r^2 + s^2}{4}\right)^2 = \left(\frac{21}{4}\right)^2,$$

contradiction.

230. Answer: $f(x) = 2x - 1$; $f(x) = -x - 1$; $f(x) = x^2 - 1$.

For $y = 0$, we have $(f(0) + 1)(f(x) - 1) = 0$. Since $f(x) = 1$ for all x is impossible, we deduce $f(0) = -1$. If $(x, y) = (1, -1)$, we obtain $f(1) = 1$ or $f(-1) = 0$. In the first case, setting $x = 1$ in the functional equation yieds $f(-y + 1) = 2y + 1$, hence a solution is $f(x) = 2x - 1$.

If $f(1) = a \neq 1$ and $f(-1) = 0$, then plug $(x, y) = (z, 1)$ and $(x, y) = (-z, -1)$ in the functional equation to get

$$f(z + 1) = (1 - a)f(z) + 2z + 1; f(-z - 1) = f(z) + 2z + 1,$$

whence it follows $f(z + 1) = (1 - a)f(-z - 1) + a(2z + 1)$, that is

$$f(x) = (1 - a)f(-x) + a(2x - 1).$$

Analogously,

$$f(-x) = (1 - a)f(x) + a(-2x - 1).$$

From these equalities and the previous equation, we obtain

$$(a^2 - 2a)f(x) = -2a^2x - (a^2 - 2a).$$

We cannot have $a = 2$. If $a \neq 0$, 2, then

$$f(x) = -\frac{2ax}{a-2} - 1.$$

This function satisfies the condition only if $a = -2$, whence $f(x) = -x - 1$.

If $a = 0$, then $f(x) = f(-x)$. Setting $y = z$ and $y = -z$ in the functional equation and subtracting, we obtain $f(2z) = 4z^2 - 1$, so $f(x) = x^2 - 1$ is a solution.

231. Define $s_k := b_1 + b_2 + \ldots + b_k$. Then $s_1 \le s_2 \le \ldots \le s_n =: s_0$ and the inequality becomes

$$\sum_{k=1}^{n} \frac{s_k}{s_{k-1}} \le \sum_{k=1}^{n} \frac{s_{k-1}}{s_k}.$$

This can be proved by induction, the main step being

$$\frac{s_{n+1}}{s_n} + \frac{s_1}{s_{n+1}} - \frac{s_1}{s_n} \le \frac{s_n}{s_{n+1}} + \frac{s_{n+1}}{s_1} - \frac{s_n}{s_1},$$

which is equivalent to $(s_1 - s_n)(s_n - s_{n+1})(s_1 - s_{n+1}) \le 0$.

232. *a*) By the power mean inequality,

$$S_k = x^k + y^k + z^k \ge \frac{1}{3^{k-1}},$$

with equality only when $x = y = z = \frac{1}{3}$. It follows that

$$Q_{\min} = \frac{1}{3^2} \cdot \frac{1}{3^{2003}} = \frac{1}{3^{2005}}.$$

b) By the Newton's inequalities, $S_n^2 < S_{n-1}S_{n+1}$, hence

$$\frac{S_{n+1}}{S_n} > \frac{S_m}{S_{m-1}}, n \ge m \ge 1.$$

In particular, for $(m, n) = (3, 2004)$, we have $S_2S_{2005} > S_3S_{2004}$.

233. If P has a positive real root, then any multiple of P, including a multiple with real positive coefficients has the same root, contradiction.

Conversely, we may consider that the coefficients of P are real, otherwise we can work with its multiple $P\overline{P}$, obtaining the desired result. We will prove that P has a multiple of the form $(1+x)^n P$ with real positive coefficients. Since P has no positive real roots and $P(0) \neq 0$, it follows that P has only factors of the form $x+a$ and x^2-bx+c, with a,b,c reals, $a>0$ and $b^2-4c<0$. It suffices to prove the assertion for each, of these factors, since the product of polynomials with real positive coefficients has real positive coefficients. This is obvious for the factors of the form $x+a$, because we can choose an arbitrary positive integer n. For the factors of the second form, we have

$$(1+x)^n(x^2-bx+c) = \sum_{i=0}^{n+2}\left[\binom{n}{i-2} - b\binom{n}{i-1} + c\binom{n}{n}\right]x^i = \sum_{i=0}^{n+2} a_i x^i,$$

with

$$a_i = \frac{n!\left[(b+c+1)i^2 - ((b+2c)n+2b+3c+1)i + c(n^2+3n+2)\right]}{i!(n+2-i)!}.$$

As a quadratic polynomial in i the parenthesis from the numerator has discriminant

$$\Delta = (b^2-4c)n^2 - 2un + v, u = 2b^2 + b(c+1) - 4c, v = (2b+c+1)^2 - 4c.$$

Since $b^2 - 4c < 0$, it follows that$\Delta < 0$ for n sufficiently large and so, $a_i > 0$ for all i, as desired, since the coefficient of i^2 is

$$b+c+1 > b + \frac{b^2}{4} + 1 = \left(\frac{b}{2}+1\right)^2 > 0.$$

234. By the Bergström's inequality,

$$\frac{a_1^2}{a_1+a_2} + \frac{a_2^2}{a_2+a_3} + \ldots + \frac{a_n^2}{a_n+a_1} \geq \frac{(a_1+a_2+\ldots+a_n)^2}{2(a_1+a_2+\ldots+a_n)} = \frac{1}{2}.$$

235. Let $f_{k,\,l} = X^p + pX^k + pX^l + 1$, $1 \leq l < k \leq p-1$. If $k - l \equiv 1 (\mathrm{mod}\ 2)$, then

$$f_{k,l}(-1) = 0,$$

therefore $f_{k,\,l}$ is reducible. If $k - l \equiv 0 (\mathrm{mod}\ 2)$, then

$$f_{k,l}(X-1)=X^p+pXg(X)\pm 2p, g\in\mathbb{Z}[X].$$

By the Eisenstein's criterion, $f_{k,\ l}(X-1)$ is irreducible in $\mathbb{Z}[X]$, hence $f_{k,\ l}(X)$ is irreducible in $\mathbb{Z}[X]$.

We deduce that the number of polynomials $f_{k,\ l}$ irreducible in $\mathbb{Z}[X]$ is equal to the number of pairs (k,l), $k>l$, $k,l\in\{1,2,\ldots,p-1\}$ of the same parity, which is

$$2\binom{\frac{p-1}{2}}{2}=\frac{(p-1)(p-3)}{4}.$$

236. *a*) If $a_1=0$, then $k=1$ is a solution. If $a_1<0$, then we can change all signs, so we may suppose that $a_1>0$. Define $s_0=0, s_k=\sum\limits_{i=1}^{k} ia_i, 1\le k\le n$. Then

$$a_k=\frac{s_k-s_{k-1}}{k}\Rightarrow 0=\sum_{k=1}^{n}a_k=\frac{s_n}{n}+\sum_{k=1}^{n}\frac{s_k}{k(k+1)}.$$

From $s_1=a_1>0$, it follows that there exists the smallest k for which $s_k<0$, and so

$$k\ge|ka_k|=s_{k-1}-s_k=|s_{k-1}|+|s_k|.$$

Now,

$$|s_{k-1}|>\frac{2k-1}{4}, |s_k|>\frac{2k+1}{4}\Rightarrow|s_{k-1}|+|s_k|>k,$$

contradiction.

b) Case 1. n odd, $n>1$. For

$$\{a_i|1\le i\le n\}=\left\{\frac{3}{4},\frac{1}{4},-1,1,-1,1,\ldots\right\},$$

we have

$$\{s_i|1\le i\le n\}=\left\{\frac{3}{4},\frac{5}{4},-\frac{7}{4},\frac{9}{4},-\frac{11}{4},\frac{13}{4},-\frac{15}{4},\ldots\right\}.$$

Case 2. n even, $n>2$. For

$$\{a_i|1\le i\le n\}=\left\{1,\frac{1}{8},-1,-\frac{1}{8},1,-1,1,-1,\ldots\right\}$$

we have

$$\{s_i|1\le i\le n\}=\left\{1,\frac{5}{4},-\frac{7}{4},-\frac{9}{4},\frac{11}{4},-\frac{13}{4},\frac{15}{4},-\frac{17}{4},\ldots\right\}.$$

237. We have

$$\sum_{1\le i,j\le n}\left(|x_i|+|x_j|-|x_i+x_j|\right)=2\sum_{x_ix_j<0}\min\{|x_i|,|x_j|\}=4\sum_{x_i>0>x_j}\min\{x_i,-x_j\}$$

$$=\le 4\sum_{x_i>0>x_j}\sqrt{-x_ix_j}=4\left(\sum_{x_i>0}\sqrt{x_i}\right)\left(\sum_{x_j<0}\sqrt{-x_j}\right)\overbrace{\le}^{AM-GM}\left(\sum_{x_i>0}\sqrt{x_i}+\sum_{x_j<0}\sqrt{-x_j}\right)^2$$

$$==\left(\sum_{1\le k\le n}\sqrt{|x_k|}\right)^2\overbrace{\le}^{C-S}n\sum_{1\le k\le n}|x_k|,$$

hence

$$\sum_{1\le i,j\le n}|x_i+x_j|+n\sum_{1\le k\le n}|x_k|\ge\sum_{1\le i,j\le n}\left(|x_i|+|x_j|\right)=2n\sum_{1\le k\le n}|x_k|,$$

which is equivalent to the given inequality.

238. From the given conditions, we deduce

$$0=a_0\le a_1\le\ldots\le a_n=1,0\le a_1=a_1-a_0\le a_2-a_1\le\ldots\le a_n-a_{n-1}=1-a_{n-1}.$$

If the inequality is proved for $q=p+1$, then we can extent it for every $q>p$.

Define

$$S(m,r)=\sum_{k=1}^{m}a_k^r.$$

The Abel's summation formula implies

$$S(n,p+1)=\sum_{k=1}^{n}a_ka_k^p=a_nS(n,p)-\sum_{k=1}^{n-1}(a_{k+1}-a_k)S(k,p).$$

Then

$$a_n = 1 \Rightarrow S(n-1, p+1) = S(n-1, p) - \sum_{k=1}^{n-1} (a_{k+1} - a_k) S(k, p),$$

$$a_j - a_{j-1} \le a_{k+1} - a_k, 1 \le j \le k \Rightarrow (a_{k+1} - a_k) S(k, p) \ge \sum_{j=1}^{k} (a_j - a_{j-1}) a_j^p.$$

By the generalized AM–GM inequality,

$$p a_j^{p+1} + a_{j-1}^{p+1} \ge (p+1) a_{j-1} a_j^p \Rightarrow (a_j - a_{j-1}) a_j^p \ge \frac{a_j^{p+1} - a_{j-1}^{p+1}}{p+1},$$

hence

$$S(n-1, p+1) \le S(n-1, p) - \frac{1}{p+1} \sum_{k=1}^{n-1} \sum_{j=1}^{k} \left(a_j^{p+1} - a_{j-1}^{p+1} \right) = S(n-1, p)$$

$$- \frac{1}{p+1} S(n-1, p+1) \Rightarrow (p+2) S(n-1, p+1) \le (p+1) S(n-1, p).$$

Equality cases are

$$p = q; a_1 = \ldots = a_{n-1} = 0; p = 0, q = 1, a_k = \frac{k}{n}, 0 \le k \le n.$$

239. Let

$$H_n = 1 + \frac{1}{2} + \ldots + \frac{1}{n}, n \ge 1.$$

We will prove that

$$n \le \sqrt{n} x_n < n + \frac{H_n}{8}, n \ge 1.$$

Squaring the recurrence relation yields

$$x_{n+1}^2 = x_n^2 + 1 + \frac{1}{4x_n^2}.$$

As $x_1^2 = 1$, we will prove by induction that $x_n^2 \ge n : x_{n+1}^2 \ge n + 1 + \frac{1}{4x_n^2} \ge n + 1$, so

$$\sqrt{n}x_n \geq n \text{ for all } n.$$

Now, iterating the squared relation,

$$x_n^2 = x_{n-1}^2 + 1 + \frac{1}{4x_{n-1}^2} = \ldots = x_1^2 + (n-1)$$
$$+ \sum_{k=1}^{n-1} \frac{1}{4x_k^2} \leq n + \frac{1}{4}\sum_{k=1}^{n-1} \frac{1}{k} < n + \frac{H_n}{4} < \left(\sqrt{n} + \frac{H_n}{8\sqrt{n}}\right)^2,$$

so $\sqrt{n}x_n < n + \frac{H_n}{8}$.

By a well-known inequality, $H_n \leq 1 + \ln n$, so it suffices to show that $\ln 625 < 7$. But

$$e^7 > 2.6 \cdot \left(\frac{5}{2}\right)^6 > 5^4 \cdot \frac{65}{64} > 5^4,$$

hence $\ln 625 < 7$ and $\lfloor 25x_{625} \rfloor = 625$.

240. We have the implications

$$\sum_{i=1}^{n} |x_i| = 1 \Rightarrow \sum_{i=1}^{n} x_i^2 + 2\sum_{1 \leq i < j \leq n} |x_i x_j| = 1,$$

$$\sum_{i=1}^{n} x_i = 0 \Rightarrow \sum_{i=1}^{n} x_i^2 + 2\sum_{1 \leq i < j \leq n} x_i x_j = 0.$$

Adding these equalities, we get

$$1 = 2\sum_{i=1}^{n} x_i^2 + 2\sum_{1 \leq i < j \leq n} \left(x_i x_j + |x_i x_j|\right) \geq 2\sum_{i=1}^{n} x_i^2.$$

By the Cauchy–Schwarz inequality, we have

$$LHS \leq \sqrt{\left(x_1^2 + x_2^2 + \ldots + x_n^2\right)\left(\frac{1}{1^2} + \frac{1}{2^2} + \ldots + \frac{1}{n^2}\right)} < \frac{\pi}{2\sqrt{3}} < \frac{1}{2}\left(1 - \frac{1}{n}\right).$$

241. The Viète's relations are $a + b + c = 2$, $ab + bc + ca = -1$, $abc = -1$. From the hypothesis, we deduce $c < 0 < b < a$. Denote

$$u = a^2b + b^2c + c^2a, v = ab^2 + bc^2 + ca^2.$$

Then $u - v = (a - c)(c - b)(b - a) > 0$, whence $u > v$, and

$$u+v=(a+b+c)(ab+bc+ca)-3abc=1,$$
$$uv=(ab)^3+(bc)^3+(ca)^3+3(abc)^2+abc\left(a^3+b^3+c^3\right),$$
$$(ab)^3+(bc)^3+(ca)^3=(ab+bc+ca)^3-3abc(u+v)-6(abc)^2=-4,$$
$$a^3+b^3+c^3=(a+b+c)^3-3(u+v)-6abc=8-3+6=11,$$
$$uv=-4+3-11=-12.$$

Numbers u, v are solutions of the quadratic equation $z^2-z-12=0$, hence $u=4$, since $u>v$.

242. We have

$$\left(\sum x\right)\left(\sum \frac{1}{x}\right)-6\sum\frac{x}{y+z}=\frac{1}{4xyz}\sum\frac{(y-z)^2}{(x+y)(x+z)}$$
$$\times\left[(x+y+z)(3x-y-z)^2+(-x+y+z)(x-y+z)(x+y-z)\right]\geq 0,$$

with equality for $x=2$, $y=z=1$ and permutations.

3.36 2007

243. Denote $a=\frac{x^2}{x-1}, b=\sqrt{x-1}, c=\frac{\sqrt{x-1}}{x^2}$, then $abc=1$ and

$$a+b+c=\frac{1}{a}+\frac{1}{b}+\frac{1}{c},$$

so $(a-1)(b-1)(c-1)=0$. The only solution of the equations $a=1$, $b=1$, $c=1$ is $x=2$, since the equations $a=1$ and $c=1$ have no real roots.

244. Answer: $(x,a)\in\{(2,1),(2,2)\}$.

Define

$$f:\{-2\}\cup[2,\infty)\to\mathbb{R},\ f(x)=\sqrt{x^2-4}+\sqrt{x+2}-x-\frac{1}{4}.$$

The equation becomes

$$\left(\sqrt{x-a}-\frac{1}{2}\right)^2+f(x)=0.$$

We see that $x=-2$ is not a solution, since $f(-2)=\frac{7}{4}>0$. If $x\geq 2$, then $f'(x)>0$, hence f is strictly increasing on $[2,\infty)$. From $f(3)>0$, it follows that $f(x) > 0$ on $[3, \infty\ )$, so the equation has no solutions in this interval.

Therefore $x = 2$ is the only solution, so

$$\left(\sqrt{2-a}-\frac{1}{2}\right)^2-\frac{1}{4}=0,$$

whence $a \in \{1, 2\}$.

245. Let $P(x) = (x - x_1)(x - x_2)(x - x_3)$. Using the Viète's formulae,

$$x_1 + x_2 + x_3 = 2, x_1x_2 + x_1x_3 + x_2x_3 = b, x_1x_2x_3 = -c,$$

we have to prove that

$$8(x_1x_2 + x_1x_3 + x_2x_3) - 9x_1x_2x_3 \leq 8,$$

which is equivalent to the homogenous inequality

$$4(x_1x_2 + x_1x_3 + x_2x_3)(x_1 + x_2 + x_3) - 9x_1x_2x_3 \leq (x_1 + x_2 + x_3)^3.$$

This inequality reduces to

$$x_1^3 + x_2^3 + x_3^3 + 3x_1x_2x_3 \geq x_1^2(x_2 + x_3) + x_2^2(x_1 + x_3) + x_3^2(x_1 + x_2),$$

which is true, by the Schur's inequality. Equality holds for $x_1 = x_2 = x_3 = \frac{2}{3}$.

246. Taking $x = y = 1$, we get

$$a \leq f(1) - (f(1))^2 = \frac{1}{4} - \left(f(1) - \frac{1}{2}\right)^2 \leq \frac{1}{4}.$$

For $a < \frac{1}{4}$, we can take $f : (0, 1] \rightarrow \left[\frac{1}{2}, \min\left\{\frac{3}{4} - a, 1\right\}\right]$ arbitrary. Then

$$a + f(x + y - xy) + f(x)f(y) - (f(x) + f(y)) = a + f(x + y - xy) + (1 - f(x))$$
$$\times (1 - f(y)) - 1 \leq a + \left(\frac{3}{4} - a\right) + \frac{1}{2} \cdot \frac{1}{2} - 1 = 0, x, y \in (0, 1].$$

For $a = \frac{1}{4}$, we have $\left(f(1) - \frac{1}{2}\right)^2 \leq 0$, therefore $f(1) = \frac{1}{2}$. Taking $y = 1$, we get

$$f(x) \geq \frac{1}{2}, \forall x \in (0, 1].$$

Taking $x = y$, we have

$$f(2x - x^2) \leq 2f(x) - (f(x))^2 - \frac{1}{4} = \frac{3}{4} - (1 - f(x))^2 \leq \frac{3}{4}, \forall x \in (0, 1].$$

The quadratic function $2x - x^2$ takes all values from $(0, 1]$. Let

$$t = \sup\{f(x) | x \in (0, 1]\}.$$

Then

$$\frac{1}{2} \leq t \leq \frac{3}{4}, t = \sup\{f(2x - x^2) | x \in (0, 1]\} = \sup\left\{\frac{3}{4} - (1 - f(x))^2 | x \in (0, 1]\right\} = \frac{3}{4} - (1 - t)^2,$$

whence $t = \frac{1}{2}$, which implies that f is constant, contradiction.

247. Answer: $C(n) = \frac{1}{n-1}$.

Changing $x - e, x + e$ by x, x, doesn't change LHS, but increases RHS. The result is obvious if $x_i \leq \frac{1}{2}$, $1 \leq i \leq n$. From the given condition, it follows that at most one of the numbers $x_1, x_2, \ldots, x_n$ is greater than $\frac{1}{2}$. If $x_1 = x_2 = \ldots = x_{n-1} \leq \frac{1}{2} \leq x_n$, then

$$(1 - x_1)(1 - x_n) \geq \frac{1}{4} \Rightarrow \sum_{i=1}^{n} x_i \leq \frac{n}{2}.$$

Since

$$\left(1 - \frac{\sum_{i=1}^{n} x_i}{n}\right)^2 \geq \frac{1}{4},$$

the maximum is attained when $x_1 = x_2 = \ldots = x_n = \frac{1}{2}$.

248. Assume such function exists. In that case, we claim there would exist real z such that $f(f(z)) > 0$. Otherwise, $f(f(x)) \leq 0$ for all x, so for all $y \leq 0$, we have

$$f(x + y) \geq f(x) + yf(f(x)) \geq f(x).$$

Then f is decreasing, so $f(0) > 0 \geq f(f(x))$ for all x, which implies $f(x) > 0$. This contradicts $f(f(x)) \leq 0$.

Now, we see that $x \to \infty$ implies $f(z + x) \geq f(z) + xf(f(z))$, so $f(x) \to \infty$ and $f(f(x)) \to \infty$. Then there are $x, y > 0$ such that

$$f(x) \geq 0, f(f(x)) > 1, f(x+y) > 0, f(f(x+y+1)) > 0, y \geq \frac{x+1}{f(f(x))-1} \tag{1}$$

Define $A = x + y + 1$, $B = f(x + y) - (x + y + 1)$. Then $f(f(A)) > 0$ and by (1),

$$f(x+y) \geq f(x) + yf(f(x)) \geq x+y+1,$$

so $B \geq 0$. Next,

$$f(f(x+y)) = f(A+B) \geq f(A) + Bf(f(A)) \geq f(A) = f(x+y+1) = f(x+y) + f(f(x+y)) > f(f(x+y))$$

contradiction.

249. Answer: $f(x) = \frac{1}{x^2}$.

Let p, q be positive integers and denote by $P(x, y)$ the given equation. Then

$$P(x,1) \Rightarrow f(x+1) = \frac{f(x)}{f(x) + 2xf(x) + f(1)}.$$

Since

$$P(1,1) \Rightarrow f(2) = \frac{1}{4},$$

we can apply this twice to get $f(1) = 1$. We have by induction

$$Q(x,n) : \frac{1}{f(x+n)} = n^2 + 2nx + \frac{1}{f(x)}, n \in \mathbb{N}^*.$$

Now,

$$Q(1,n) \Rightarrow f(n) = \frac{1}{n^2},$$

$$Q\left(\frac{1}{n}, n\right) \Rightarrow \frac{1}{f\left(n + \frac{1}{n}\right)} = n^2 + 2 + \frac{1}{f\left(\frac{1}{n}\right)}.$$

If $x = \frac{p}{q}$, then

$$Q\left(\frac{1}{q}, p\right) \Rightarrow \frac{1}{f\left(p + \frac{1}{q}\right)} = p^2 + \frac{1}{q^2} + 2x,$$

$$P\left(p, \frac{1}{q}\right) \Rightarrow f(x) = \frac{\frac{q^2}{p^2} + q^4}{p^2q^2 + 1} \Rightarrow f(x) = \frac{1}{x^2}.$$

250. Follows from

$$\frac{a^2}{b} + \frac{b^2}{c} + \frac{c^2}{a} - 3\left(a^2 + b^2 + c^2\right) = \sum \frac{(a+c)(a-b)^2}{b} \geq 0.$$

251. The inequality

$$6x^3 - x^2 \geq \frac{5}{8}x - \frac{1}{8}$$

implies

$$6\sum a^3 - \sum a^2 \geq \frac{5}{8}\sum a - \frac{1}{8} \cdot 4 = \frac{1}{8}.$$

252. Suppose that $c \leq b \leq a$, then $a + c \leq a + b \leq c + 1$, so $a \leq 1$, hence

$$0 \leq a, b, c \leq 1.$$

It follows that $a \geq b \geq bc$. Write the proposed inequality as

$$a^2 - 2abc + b^2c^2 \leq 1 - b^2 - c^2 + b^2c^2 \Leftrightarrow (a - bc)^2 \leq \left(1 - b^2\right)\left(1 - c^2\right)(*).$$

Since $a \leq 1 + b - c$, $a \leq 1 + c - b$, we have

$$0 \leq a - bc \leq 1 + b - c - bc = (1 + b)(1 - c),$$
$$0 \leq a - bc \leq 1 + c - b = (1 - b)(1 + c).$$

Multiplying these two inequalities, we get $(*)$.

253. The only constant polynomials that are solutions are $f(x) = c$, $c \in \{-1, 0, 1\}$.

Let f be a nonconstant solution. If f is a solution, then $f(x + a)$ is a solution for any integer a. We can assume that $\deg f = k \geq 1$ and the coefficient of x^{k-1} is 0. Then for sufficiently large $|x|$,

$$(x - 1)^k < f(x) < (x + 1)^k.$$

It follows that if f is a kth power, then $f(x) = x^k$. Since $k \geq 1$, $f(\mathbb{Z})$ is infinite. For $x \in f(\mathbb{Z})$, we know that $x^k \in f(\mathbb{Z})$, so $f(x) = x^k$ for infinitely many x. We deduce that $f(x) = x^k$ for all x, hence nonconstant solutions are

$$f(x) = (x + a)^k, a \in \mathbb{Z}, k \in \mathbb{N}^*.$$

254. It suffices to prove that

$$\sum_{i=1}^{n} \frac{a_i}{2n + S} \le \frac{1}{3},$$

which is equivalent to $3(a_1 + a_2 + \ldots + a_n) \le 2n + a_1^3 + a_2^3 + \ldots + a_n^3$. This follows from the AM – GM inequality $3a_i \le 1 + 1 + a_i^3, 1 \le i \le n$, by summation.

255. Define $p = a + b + c$, $q = ab + bc + ca$, $r = abc$. Then by the condition we have

$$pr \ge q \ge \sqrt{3pr} \Rightarrow pr \ge 3 \ (*).$$

The AM–GM inequality implies $p \ge 3$ $(**)$.

Case 1. $r \le 1$. Then the inequality is equivalent to

$$p^2 r \ge 3r + 2p.$$

Using $(*)$, it suffices to prove that $3p \ge 3r + 2p$, or $p \ge 3r$. But $p \ge 3\sqrt[3]{r} \ge 3r$ holds.

Case 2. $r > 1$. It suffices to prove that

$$p \ge \frac{3}{p} + 2 \Leftrightarrow (p - 3)(p + 1) \ge 9,$$

which is true by $(**)$.

256. Let $a, b \in \mathbb{Q}$, $a < b$, and let $a = x_0 < x_1 < \ldots < x_n = b$ be an equidistant division of the interval $[a, b]$. Then

$$|f(a) - f(b)| = \left| \sum_{k=0}^{n-1} (f(x_{k+1}) - f(x_k)) \right| \le \sum_{k=0}^{n-1} |f(x_{k+1}) - f(x_k)| \le \frac{(b - a)^2}{n}.$$

If $f(a) \ne f(n)$, then for $n > \frac{(b-a)^2}{|f(a) - f(b)|}$ we get a contradiction.

257. It suffices to find a set of values x_i for $n = 4$ such that

$$E = \sum_{i=1}^{4} \frac{1}{x_i^p} - \sum_{i=1}^{4} x_i^p < 0,$$

since for $n > 1$ we can extend this set of values by taking $n - 4$ values equal to 1.

Choose $x \in (0, 1)$ and take $x_1 = x_2 = x_3 = x$, then $x_4 = y = 4 - 3x$ and

$$E = \frac{3}{x^p} + \frac{1}{y^p} - 3x^p - y^p = \frac{3 - (xy)^p}{x^p} + \frac{1 - 3(xy)^p}{y^p}.$$

By the AM–GM inequality,

$$xy = \frac{3x(4-3x)}{3} \leq \frac{4}{3},$$

with equality for $3x = 4 - 3x$, that is, $x = \frac{2}{3}, y = 2$. We have

$$\left(\frac{4}{3}\right)^p \geq \left(\frac{4}{3}\right)^4 = \frac{256}{81} > 3,$$

hence $E < 0$ for the values $\frac{2}{3}, \frac{2}{3}, \frac{2}{3}, 2$ and 1 repeated $n - 4$ times.

258. Denote

$$A = \sum_{i=1}^{n} a_i, B = \sum_{i=1}^{n} b_i.$$

Then

$$\begin{aligned} 0 \leq \sum_{i=1}^{n} (1 - Aa_i - Bb_i)^2 &= \sum_{i=1}^{n} \left(1 + A^2 a_i^2 + B^2 b_i^2 - 2Aa_i - 2Bb_i + 2ABa_i b_i\right) \\ &= \sum_{i=1}^{n} 1 + A^2 \sum_{i=1}^{n} a_i^2 + B^2 \sum_{i=1}^{n} b_i^2 - 2A \sum_{i=1}^{n} a_i - 2B \sum_{i=1}^{n} b_i + 2AB \sum_{i=1}^{n} a_i b_i \\ &= n + A^2 + B^2 - 2A^2 - 2B^2 + 0 = n - \left(A^2 + B^2\right). \end{aligned}$$

259. Denote $S = a_1 + a_2 + \ldots + a_n$. For $i \neq j$,

$$\frac{4a_i a_j}{a_i + a_j} = a_i + a_j - \frac{\left(a_i - a_j\right)^2}{a_i + a_j} \leq a_i + a_j - \frac{\left(a_i - a_j\right)^2}{S} = \frac{\sum_{i \neq k} a_i a_k + \sum_{j \neq k} a_j a_k + 2a_i a_j}{S}.$$

Summing these inequalities for all pairs (i, j), $i \neq j$, we get

$$\sum_{i<j}\frac{a_ia_j}{a_i+a_j}=\frac{1}{2}\sum_i\sum_{j\neq i}\frac{a_ia_j}{a_i+a_j}\leq\frac{1}{8S}\sum_i\sum_{j\neq i}\left(\sum_{k\neq i}a_ia_k+\sum_{k\neq j}a_ja_k+2a_ia_j\right)$$
$$=\frac{1}{8S}\left(\sum_k\sum_{i\neq k}\sum_{j\neq i}a_ia_k+\sum_k\sum_{j\neq k}\sum_{j\neq i}a_ja_k+2\sum_i\sum_{j\neq i}a_ia_j\right)$$
$$=\frac{1}{8S}\left((n-1)\sum_k\sum_{i\neq k}a_ia_k+(n-1)\sum_k\sum_{j\neq k}a_ja_k+2\sum_i\sum_{j\neq i}a_ia_j\right)$$
$$=\sum_i\sum_{j\neq i}a_ia_j=\frac{n}{2S}\sum_{i<j}a_ia_j.$$

260. Answer: $(1,1)$, $(1,2)$, $(1,3)$, $(2,1)$, $(2,2)$, $(2,3)$, $(3,1)$, $(3,2)$.

A pair (r,s) is good if and only if the pair (s,r) is good. To see this, it suffices to replace the polynomial P by $7-P$.

If (r,s) is a good pair, then a_1, …, a_r are roots of $P(x)-2$, therefore

$$P(x)-2=(x-a_1)\ldots(x-a_r)Q(x),$$

where Q is a polynomial with integer coefficients. Then

$$3=P(b_1)-2=(b_1-a_1)\ldots(b_1-a_r)Q(b_1)$$

Each factor from the right side is a nonzero integer and the factors b_1-a_i are distinct. Since 3 is a prime number, it follows that $r\leq 3$. In case $r=3$, we must have

$$a_1=b_1-1, a_2=b_1+1, a_3=b_1\pm 3.$$

A similar argument applies to $b_i (i\geq 2)$, so $b_i=b_1$ or $b_i=b_1\pm 2$. Since b_i are distinct, it follows $s\leq 2$. The polynomial $P(x)=-x^2(x-2)(x+2)+2$, with

$$P(-2)=P(0)=P(2)=2, P(-1)=P(1)=5,$$

shows that the pair $(3,2)$ is good. Then all pairs (u,v), with $1\leq u\leq 3$, $1\leq v\leq 2$ are also good.

3.37 2008

261. Answer: $f(x)=\frac{1}{x}$ for all x.

Let $P(x,y)$ be the assertion. If $f(u)=f(v)$, then from $P(u,z)$ and $P(v,z)$ we get

$$u^2(v+z)=v^2(u+z),$$

whence $u=v$. Now, $P(1,1)$ implies $f(1)=1$; $P(1,x)$ implies $f(x)=\frac{1}{x}$.

262. Answer: $f(x)=0$ for all x; $f(x)=x^2$ for all x.

Let $P(x,y)$ be the assertion. Then $P\left(x,\frac{x^2-f(x)}{2}\right)$ implies $f(x)\in\{0,x^2\}$. If $f(u)=0$ and $f(v)=v^2$, then $P(u,v)$ implies $f(u^2-v)=v^2\neq 0$, hence $u^2=2v$, which is not always true. We conclude that $f(x)=0, \forall x$, or $f(x)=x^2, \forall x$.

263. Answer: $a\le 1$.

For $a\le 1$, we can find a function $f(x)=kx^2$ such that $3k^2-2k^3=a$, for some $k>0$. Indeed, consider the function $g\colon(0,\infty)\to\mathbb{R}$, $g(x)=3x^2-2x^3$. From

$$g'(x)=6x(1-x^2),$$

we see that g has a maximum at $x=1$. Since

$$\lim_{x\to 0}g(x)=0, g(1)=1, \lim_{x\to\infty}g(x)=-\infty,$$

it follows that the image of g is $(-\infty,1]$.

Suppose $a>1$ and define the sequence (y_n) of nonnegative real numbers by

$$y_0=0, y_{n+1}^2=\frac{2y_n^3+a}{3}, n\ge 0.$$

Then $y_n>1$ for all sufficiently large n. Indeed, suppose that $y_n\le 1$ for all sufficiently large n and define $b_n=y_n^3-a$. We have $b_{n+1}\le\frac{2}{3}b_n$, which implies

$$\lim_{n\to\infty}b_n=0,$$

that is,

$$\lim_{n\to\infty}y_n=\sqrt[3]{a}>1,$$

contradiction. By induction on the given equation, we get $f(x)>y_nx^2$, hence (y_n) is bounded and

$$y_{n+1}^2-y_n^2=\frac{2y_n^3-3y_n^2}{3}>0,$$

hence there exists $\lim_{n\to\infty}y_n=l$. This implies $3l^2-2l^3=a$, contradicting $a>1$.

264. Lemma. Let P be a polynomial function of degree n and suppose that there is a complex number z_0 such that $|P''(z_0)| \geq (n-1)|P'(z_0)|$. Then P' has a root in the disk $|z - z_0| \leq 1$.

Proof. If all roots $w_1, \ldots, w_{n-1}$ of P' lie outside the disk, then

$$\left|\frac{P''(z_0)}{P'(z_0)}\right| \leq \sum_{i=1}^{n-1} \frac{1}{|z_0 - w_i|} < n-1,$$

contradiction.

Consider the polynomial functions $Q(z) = (z - z_2)(z - z_3)$, $P(z) = (z - z_1)$ $Q(z)$. Then

$$\frac{P''(z_1)}{P'(z_1)} = \frac{2Q'(z_1)}{Q(z_1)} = 2\left(\frac{1}{z_1 - z_2} + \frac{1}{z_1 - z_3}\right) = \frac{2(2z_1 - z_2 - z_3)}{(z_1 - z_2)(z_1 - z_3)}.$$

Let ABC be the triangle with vertices $A(z_1)$, $B(z_2)$, $C(z_3)$. Then

$$|z_1 - z_2| = c, |z_1 - z_3| = b, |2z_1 - z_2 - z_3| = 2m_a.$$

By the lemma, it suffices to show that $4m_a \geq 2bc$. Since $R \leq 1$, we can homogenize to $2Rm_a \geq bc$. We have

$$\frac{1}{2}am_a \geq \frac{abc}{4R} = K = \frac{1}{2}ah_a,$$

which is true, since $m_a \geq h_a$.

265. Answer: $a)\ \sqrt{2}; b)\ \frac{2\sqrt{6}}{3}$.

By the AM–GM inequality,

$$a)\ x^2 + y^2 + z^2 = x^2 + \frac{1}{2}y^2 + \frac{1}{2}y^2 + z^2 \geq \sqrt{2}(xy + yz),$$

with equality for $(x, y, z) = \left(\sqrt{2}, 2, \sqrt{2}\right)$.

$$b)\ x^2 + y^2 + 2z^2 = x^2 + \frac{2}{3}y^2 + \frac{1}{3}y^2 + 2z^2 \geq \frac{2\sqrt{6}}{3}(xy + yz),$$

with equality for $(x, y, z) = \left(2\sqrt{6}, 6, \sqrt{6}\right)$.

266. Suppose that all digits of $\sqrt[3]{3}$ between the 1000000th and the 3141592nd position after the decimal point are equal to 2. Let $s = \left\lfloor 10^{1000000}\sqrt[3]{3}\right\rfloor$. Then

$$\left|\sqrt[3]{3\cdot\left(9\cdot 10^{1000000}\right)^3}-(9s+2)\right|$$
$$=\frac{\left|3\cdot\left(9\cdot 10^{1000000}\right)^3-(9s+2)^3\right|}{\left(\sqrt[3]{3\cdot\left(9\cdot 10^{1000000}\right)^3}\right)^2+(9s+2)\sqrt[3]{3\cdot\left(9\cdot 10^{1000000}\right)^3}+(9s+2)^2}$$
$$>\frac{1}{3\cdot\left(10^{1000002}\right)^2}>\frac{1}{10^{2000005}}\qquad(1)$$

The estimate for the numerator holds because the numbers are different, since one is divisible by 3 and the other is not, the estimate for the denominator holds because

$$\sqrt[3]{3\cdot\left(9\cdot 10^{1000000}\right)^3}=\sqrt[3]{3}\cdot 9\cdot 10^{1000000}<10^{1000002}$$

and

$$9s+2<10s<10\cdot 10^{1000000}\cdot\sqrt[3]{3}<10^{1000002}.$$

On the other hand,

$$\left|\sqrt[3]{3}-\left(s+\frac{2}{9}\right)\cdot 10^{-1000000}\right|<10^{-3141591},$$

which is equivalent to

$$\left|\sqrt[3]{3\cdot\left(9\cdot 10^{1000000}\right)^3}-(9s+2)\right|<9\cdot 10^{-2141591}<10^{-22141590}.\qquad(2)$$

From (1) and (2), we have a contradiction.

267. Answer: 1, 2, …, 2008.

Let f be a function which satisfies the given condition. Then

$$f(m)=f(n+(m-n))\geq f(n)+f(f(m-n))-1\geq f(n)$$

for all $m>n$, hence f is nondecreasing. The function $f(n)=1$ for all n is clearly a solution. Let f be another solution and let a the smallest positive integer such that $f(a)>1$. Then $b\geq a$ implies $f(b)\geq f(a)>1$. If n is a positive integer such that $f(n)>n$, then

$$f(f(n))=f((f(n)-n)+n)\geq f(f(n)-n)+f(f(n))-1,$$

so $f(f(n)-n) \leq 1$, hence $f(n)-n<a$. Then there exists $c=\max\{f(n)-n\}$ and let k such that $f(k)-k=c \geq 1$. We have

$$2k+c \geq f(2k)=f(k+k) \geq f(k)+f(f(k))-1 \geq f(k)+f(k)-1=2k+2c-1,$$

hence $c \leq 1$ and $f(n) \leq n+1$ for all n. We deduce $f(2007) \leq 2008$.

A family of examples for which $f_j(2007)=j$, $1 \leq j \leq 2007$ that satisfy the given condition is

$$f_j(n)=\max\{1,n+j-2007\}, 1 \leq j \leq 2007, \quad f_{2008}(n)=\begin{cases} n, & 2007 \nmid n \\ n+1, & 2007 \mid n \end{cases}.$$

268. Answer: $f(x)=cx$, $c \geq 0$.

Let $P(x,y)$ be the given inequality and denote $b=f(a)$. Then

$$P(a-y,y):\quad 2by+(a-2y)(f(a-y)+f(y)) \geq 0,$$
$$P(y,a-y):\quad 2b(a-y)-(a-2y)(f(a-y)+f(y)) \geq 0.$$

Divide both inequalities by $a-2y$ and let $y \to \infty$. Then

$$\lim_{y \to \infty}(f(a-y)+f(y))=b=f(a).$$

Since f is a polynomial function, we see that $f(a-y)+f(y)=f(a)$. As a is arbitrary, this is a Cauchy equation, hence f is linear.

269. Let

$$S=\sum_{k=1}^{100} a_k^2 a_{k+1},$$

with indices considered modulo 100. By the Cauchy–Schwarz inequality, and then by the AM–GM inequality, we have

$$(3S)^2=\left(\sum_{k=1}^{100} a_{k+1}\left(a_k^2+2a_{k+1}a_{k+2}\right)\right)^2 \leq \left(\sum_{k=1}^{100} a_{k+1}^2\right)\left(\sum_{k=1}^{100}\left(a_k^2+2a_{k+1}a_{k+2}\right)^2\right)$$
$$=\sum_{k=1}^{100}\left(a_k^2+2a_{k+1}a_{k+2}\right)^2=\sum_{k=1}^{100}\left(a_k^4+4a_k^2a_{k+1}a_{k+2}+4a_{k+1}^2a_{k+2}^2\right)$$
$$\leq \sum_{k=1}^{100}\left(a_k^4+2a_k^2\left(a_{k+1}^2+a_{k+2}^2\right)+4a_{k+1}^2a_{k+2}^2\right)=\sum_{k=1}^{100}\left(a_k^4+6a_k^2a_{k+1}^2+2a_k^2a_{k+2}^2\right).$$

Easy estimates show that

$$\sum_{k=1}^{100}\left(a_k^4+2a_k^2a_{k+1}^2+2a_k^2a_{k+2}^2\right)\le\left(\sum_{k=1}^{100}a_k^2\right)^2,\ \sum_{k=1}^{100}a_k^2a_{k+1}^2\le\left(\sum_{i=1}^{50}a_{2i-1}^2\right)\left(\sum_{j=1}^{50}a_{2j}^2\right),$$

so that

$$(3S)^2\le\left(\sum_{k=1}^{100}a_k^2\right)^2+4\left(\sum_{i=1}^{50}a_{2i-1}^2\right)\left(\sum_{j=1}^{50}a_{2j}^2\right)\le 1+\left(\sum_{i=1}^{50}a_{2i-1}^2+\sum_{j=1}^{50}a_{2j}^2\right)^2=2,$$

and finally

$$S\le\frac{\sqrt{2}}{3}<\frac{12}{25}.$$

270. Notice that

$$x^2-x+1=\left(x^2-1\right)+(2-x)\Big|\left(1-x^2\right)^{2n+1}+(x-2)^{2n+1},$$

therefore $x^2-x+1|(x-1)^n+a$. But

$$(x-1)^3=\left(x^2-x+1\right)(x-2)+1.$$

Case 1. $n=3c$. Then $(x-1)^{3c}\equiv 1(\text{mod}\, x^2-x+1)$, which means $a=-1$.

Case 2. $n=3c+1$. Then $(x-1)^{3c+1}\equiv x-1\ (\text{mod}\ x^2-x+1)$. No solution.

Case 3. $n=3c+2$. Then $(x-1)^{3c+2}\equiv(x-1)^2\equiv x\ (\text{mod}\, x^2-x+1)$. No solution.

271. We apply the inequality $2xy\le x^2+y^2$ to each term of the LHS and then use the identity $(\sin\alpha)^2+(\cos\alpha)^2=1$.
The maximum value of the sum is 1004 and is achieved when $\alpha_1=\ldots=\alpha_{2008}=\frac{\pi}{4}$.

272. Answer: *a*) $f(x)=x^3+4x^2-4$; *b*) $m=-1$.

a) We have

$$x=y=0\Rightarrow f(0)=-4; x=0\Rightarrow c=0;$$

$$(x,y)=(-1,1)\Rightarrow f(0)=f(-1)+f(1)-b+4, f(-1)=b-9;$$

$$(x,y)=(-1,-1)\Rightarrow f(-2)=2f(-1)-2a+b+4=-2a+3b-14;$$

$$(x,y)=(-2,-2)\Rightarrow f(-4)=2f(-2)-16a+4b+4,$$

$$-4=-20a+10b-24,\ -2=2a-b;\qquad(1)$$

$$(x, y) = (1, 1) \Rightarrow f(2) = 2f(1) + 2a + b + 4, 14 = 2a + b; \qquad (2)$$

$$(1) \wedge (2) \Rightarrow a = 3, b = 8 \Rightarrow f(x + y) = f(x) + f(y) + 3xy(x + y) + 8xy + 4;$$

$$y = 1 \Rightarrow f(x + 1) - f(x) = 3x^2 + 11x + 5,$$

$$\sum_{t=1}^{x-1} (f(t + 1) - f(t)) = \sum_{t=1}^{x-1} \left(3t^2 + 11t + 5\right) \Rightarrow f(x) = x^3 + 4x^2 - 4.$$

$$b)\, f(x) = x^3 + 4x^2 - 4 \geq mx^2 + (5m + 1)x + 4m \Rightarrow (x + 1)(x + 4)(x - 1) \geq m(x + 1)(x + 4), \forall x \geq 0 \Rightarrow x - 1 \geq m \Rightarrow x \geq m + 1,$$

hence $m \geq -1$.

273. Answer: $x = 1$, $y = \pm 4$.

Clearly, $x \neq 0$. If $x = 1$, then from the first equation we get $y = \pm 4$. Suppose that $x \neq 1$. We will eliminate the term y^2:

$$y^2 = \frac{49 - x^3}{3x} = -x^2 - 8xy + 8y + 17x,$$

whence we obtain

$$y = -\frac{2x^3 - 51x^2 + 49}{24x(x - 1)} = -\frac{2x^2 - 49x - 49}{24x}.$$

Replacing this value in the first equation, we have

$$x^3 + \frac{(2x^2 - 49x - 49)^2}{24^2 x} = 49.$$

After simple calculations, this equation becomes

$$49(x - 1)^2 \left(4x^2 + 4x + 49\right) = 0.$$

No other solutions.

274. Assume $f(x + y) \geq yf(f(x))$ for all $x, y > 0$. Let $a > 1$, then $t = f(f(a)) > 0$. For $b \geq a(1 + t^{-1} + t^{-2}) > a$, we have

$$f(b) = f(a + (b - a)) \geq (b - a)f(f(a)) = (b - a)t \geq a\left(1 + t^{-1}\right) > a.$$

Then

$$f(f(b)) = f(a + (f(b) - a)) \geq (f(b) - a)t \geq a.$$

If we take

$$x \geq \frac{ab+2}{a-1} > b,$$

then

$$f(x) = f(b + (x-b)) \geq (x-b)f(f(b)) \geq (x-b)a \geq x+2.$$

Hence $f(x) > x + 1$ (1). However,

$$f(f(x)) = f(x + (f(x) - x)) \geq (f(x) - x)f(f(x)).$$

Cancelling $f(f(x))$ on *both* sides, we get $f(x) \leq x + 1$, which contradicts (1).

275. Choose for (x, y) pairs of the form $(x, -1)$ and $(x - 1, 1)$ to obtain the system

$$\begin{cases} (x-1)f(x) - (x+1)f(x-1) = (x-1)f(-1) \\ (x-2)f(x) - xf(x-1) = -xf(1) \end{cases},$$

with solution $2f(x) = x(x + 1)f(1) + x(x - 1)f(-1)$, that is,

$$f(x) = ax^2 + bx, a, b \in \mathbb{R}.$$

Conversely, any function of this type satisfies the given functional equation.

276. Let $a \geq b \geq c \geq d$ be the roots of the polynomial. By the Viète's relations,

$$a^2 + b^2 + c^2 + d^2 = (a+b+c+d)^2 - 2(ab+ac+ad+bc+bd+cd) = 4.$$

We need to prove that $1 \leq a \leq 2$. If $a > 2$, then $a^2 + b^2 + c^2 + d^2 \geq a^2 > 4$, contradiction. If $a < 1, d \leq -1$, then $a + b + c + d < 1 + 1 + 1 - 1 = 2$, contradiction. If $a < 1$, $d > -1$, then $a, b, c, d \in (-1, 1)$, hence $a^2 + b^2 + c^2 + d^2 < 4$, contradiction.

Note. Both limits can be attained: $P(x) = x^4 - 2x^3 + 2x - 1$, respectively, $P(x) = x^4 - - 2x^3$.

277. Let x, y, z be the roots of the equation. Then

$$x + y + z = a, xy + yz + zx = b, xyz = c.$$

We will prove that

$$\frac{1+a+b+c}{3+2a+b} - \frac{c}{b} \geq \frac{1}{3}.$$

Indeed, this is equivalent to $ab + 2b^2 \geq 9c + 6ac$, or

$$(x+y+z)(xy+yz+zx)+2(xy+yz+zx)^2 \geq 9xyz+6(x+y+z)xyz,$$

which is true, with equality only for $x=y=z$.

278. Rewrite the inequality as

$$\prod_{cyc} \frac{a+b}{1+a} \geq \frac{16abcd}{\left(1+\sqrt[4]{abcd}\right)^4}.$$

We will use the inequality

$$\frac{x+y}{(1+x)(1+y)} \geq \frac{2\sqrt{xy}}{\left(1+\sqrt{xy}\right)^2}.$$

By the Cauchy–Schwarz and the above inequality,

$$\frac{a+b}{(1+a)(1+b)} \cdot \frac{c+d}{(1+c)(1+d)}(b+c)(a+d) \geq \frac{4\sqrt{abcd}\left(\sqrt{ab}+\sqrt{cd}\right)^2}{\left(1+\sqrt{ab}\right)^2\left(1+\sqrt{cd}\right)^2}$$

$$\geq \frac{16abcd}{\left(1+\sqrt[4]{abcd}\right)^4}$$

3.38 2009

279. Answer: $f(x) = ax + b$.

Write f as a formal power series

$$f(x) = \sum_{n=0}^{\infty} a_n x^n.$$

Then we have

$$\left(\frac{n}{2} + \frac{1}{2^n} - 1\right) a_n = 0, \forall n \geq 0,$$

hence $a_n = 0$ for all $n \geq 2$.

280. Answer: $\frac{4}{9}$.

If i, j do not divide each other and $a_i a_j \neq 0$, then replacing (a_i, a_j) by $(0, a_i + a_j)$ will not decrease the sum, but will increase the number of zero terms.

Apply this operation until it's no longer possible. Let $I=\{i|a_i\neq 0\}=\{i_1,\ldots, i_k\}$. If $i_1<\ldots<i_k$, then $i_1|i_2|\ldots|i_k$.

The given sum is at most $\frac{k-1}{2k}$. We want to maximize k, which is $k=9$. Equality holds when

$$a_1=a_2=a_4=\ldots=a_{256}=\frac{1}{9}.$$

281. Answer: $20\sqrt{2}$.

Clearly, $a_2>0$. From $a_2(a_3+a_4+\ldots+a_{100})\geq a_3^2+a_4^2+\ldots+a_{100}^2\geq 100$, we have

$$a_3+a_4+\ldots+a_{100}\geq\frac{100}{a_2},$$

$$a_1+a_2+\ldots+a_{100}\geq a_1+a_2+\frac{100}{a_2}\geq 2a_2+\frac{100}{a_2}\geq 2\sqrt{2a_2\cdot\frac{100}{a_2}}=20\sqrt{2}.$$

This value can be attained, for example, when

$$a_1=\ldots=a_4=5\sqrt{2}, a_5=\ldots=a_{100}=0.$$

282. From

$$\begin{aligned}&\frac{P(a+b)+P(c)+P(0)}{a+b+c}-\frac{P(a)+P(b)+P(c)}{a+b+c}\\&=\frac{P(a+b)+P(0)-P(a)-P(b)}{a+b+c},\end{aligned}$$

taking c sufficiently large, it follows $P(a+b)+P(0)-P(a)-P(b)=0$, hence $P(x)=kx+l$. Replacing in the given condition, we get $l=0$, hence $P(x)=kx$, $k\in\mathbb{Z}$.

283. We will use the inequality

$$\frac{1}{x}+\frac{1}{y}\geq\frac{4}{x+y}, x, y>0.$$

The given inequality is equivalent to

$$E:=\frac{a+c}{b+c}+\frac{b+d}{c+d}+\frac{c+a}{d+a}+\frac{d+b}{a+b}\geq 4.$$

Then

$$E=(a+c)\left(\frac{1}{b+c}+\frac{1}{d+a}\right)+(b+d)\left(\frac{1}{a+b}+\frac{1}{c+d}\right)\geq\frac{4(a+b+c+d)}{a+b+c+d}=4.$$

284. Answer: $\pm\left(\frac{1}{3},\frac{1}{3},\frac{1}{3}\right)$, $\left(0,0,\pm\frac{\sqrt{3}}{3}\right)$ and permutations.

By the AM–GM and triangle inequalities,

$$x^2y^2+y^2z^2+z^2x^2=|xyz(x+y+z)|^3\geq|xyz|(|x|+|y|+|z|)\geq|xyz|$$
$$\times|x+y+z|\Rightarrow|xyz|\left((x+y+z)^2-1\right)\geq 0.$$

If $xyz(x+y+z)=0$, then two of x, y, z are zero, whence the solutions $\left(0,0,\pm\frac{\sqrt{3}}{3}\right)$. Otherwise, $(x+y+z)^2\geq 1$, so

$$xy+yz+zx\geq\frac{1}{3}=x^2+y^2+z^2\geq xy+yz+zx.$$

The equality case implies $x=y=z$, whence the solutions $\pm\left(\frac{1}{3},\frac{1}{3},\frac{1}{3}\right)$.

285. Define

$$E:=3\left(\frac{a}{b}+\frac{b}{c}+\frac{c}{d}+\frac{d}{a}\right)+\left(\frac{b}{a}+\frac{c}{b}+\frac{d}{c}+\frac{a}{d}\right).$$

By the AM–GM inequality, we have

$$E=\left(\frac{a}{b}+\frac{a}{b}+\frac{b}{c}+\frac{a}{d}\right)+\left(\frac{b}{c}+\frac{b}{c}+\frac{c}{d}+\frac{b}{a}\right)+\left(\frac{c}{d}+\frac{c}{d}+\frac{d}{a}+\frac{c}{b}\right)$$
$$+\left(\frac{d}{a}+\frac{d}{a}+\frac{a}{b}+\frac{d}{c}\right)\geq 4\sqrt[4]{\frac{a^3}{bcd}}+4\sqrt[4]{\frac{b^3}{cda}}+4\sqrt[4]{\frac{c^3}{dab}}+4\sqrt[4]{\frac{d^3}{abc}}$$
$$=4(a+b+c+d)>3\left(\frac{a}{b}+\frac{b}{c}+\frac{c}{d}+\frac{d}{a}\right)+(a+b+c+d),$$

whence we get the required inequality.

286. Answer: $f(x)=x$ for all x.

Let f be a function which satisfies the given condition. We say that the triple (x,y,z) is *good* if it satisfies $x^3+f(y)x+f(z)=0$. The problem will be solved in 7 steps.

1. There exists $x_0\in\mathbb{R}$ such that $f(x_0)=0$.

Proof. Let x_0 be a real solution of the equation $x^3+f(0)x+f(0)=0$. Then $(x_0,0,0)$ is good, so $(f(x_0))^3=0$, whence $f(x_0)=0$.

2. $f(0)=0$.

Proof. We see that $(0,y,x_0)$ is good if $f(x_0)=0$, so $(f(0))^3+yf(0)+x_0=0$ for all $y\in\mathbb{R}$, hence $f(0)=0$.

3. f is injective.

Proof. If $f(a) = f(b)$ and (x, y, a) is good, then (x, y, b) is also good, so

$$(f(x))^3 + f(x)y + a = (f(x))^3 + f(x)y + b = 0,$$

whence $a = b$.

4. f is surjective.

Proof. Consider the polynomial $P(X) = (X - t)(X^2 + tX + 1 + t^2)$, $t \in \mathbb{R}$. For $y_0 = 1$, $z_0 = -t(1 + t^2)$, choose $x_0 \in \mathbb{R}$ such that (x_0, y_0, z_0) is good. Then

$$P(x_0) = (f(x_0))^3 + y_0 f(x_0) + z_0 = 0.$$

Since P has only one real root, we conclude that $f(x_0) = t$.

5. $f(y) \geq 0$ for all $y \geq 0$.

Proof. Choose $y_0 > 0$ such that $f(y_0) < 0$. Let x_0 be a solution of $x^2 = -f(y_0)$; we see that $(x_0, y_0, 0)$ is good, so $(f(x_0))^3 + y_0 f(x_0) = 0$, contradiction, since $f(x_0) \neq 0$ (f is injective) and $(f(x_0))^2 + y_0 > 0$.

6. f is additive.

Proof. Choose a, b, c distinct real numbers such that $a + b + c = 0$. Since f is surjective, we can choose y_0, z_0 such that a, b, c are the roots of the polynomial

$$X^3 + f(y_0)X + f(z_0).$$

Then $f(a), f(b), f(c)$ are distinct, since f is injective, and they are the roots of the polynomial $X^3 + y_0X + z_0$, hence $f(a) + f(b) + f(c) = 0$. Then

$$f(-(a+b)) = -(f(a) + f(b)), a, b \in \mathbb{R}.$$

For $b = 0$, we get $f(-x) = -f(x)$ for all x, hence $f(a + b) = f(a) + f(b)$ for all distinct a, b. If $a = b$, then choose m sufficiently large, so

$$f(2a) = f(a+m) + f(a-m) = (f(a) + f(m)) + (f(a) - f(m)) = 2f(a),$$

therefore f is additive.

7. $f(x) = x$ for all x.

Proof. From 5. and 6. we deduce that f is additive and bounded on any bounded interval, so f is linear. Since $f(0) = 0$, we have $f(x) = cx, c \neq 0$. The triple $(1, y, z)$ is good, if $z = -\frac{1}{c} - y = c^3 - cy$. For $y \to \infty$, we conclude that $c = 1$.

287. Answer: $(x, y, z) = (10, 8, 6)$.

By the Bergström's inequality,

$$36=\frac{(x+2)^2}{y+z-2}+\frac{(y+4)^2}{z+x-4}+\frac{(z+6)^2}{x+y-6}$$
$$\geq\frac{(x+y+z+12)^2}{2(x+y+z)-12}\Leftrightarrow(x+y+z-24)^2\leq 0\Leftrightarrow x+y+z=24.$$

Equality holds when

$$\frac{x+2}{y+z-2}=\frac{x+2}{22-x}=\frac{y+4}{20-y}=\frac{z+6}{18-z}=\frac{x+2+y+4+z+6}{22-x+20-y+18-z}=1,$$

whence the conclusion.

288. Answer: $P(x,y)=Q(x+y,xy(x-y)^2)$, $Q(x,y)\in\mathbb{R}[x,y]$.

Both $P(x,y)=x+y$ and $P(x,y)=xy(x-y)^2$ satisfy the given equation.

Conversely, changing y to $-y$, we see that P is symmetric. If we subtract from P any polynomial $R(x+y)$, then the resulting P also satisfies the equation. Now, we can assume that $P(x,0)=0$. As P is symmetric, it follows that $xy|P$. For $y=x$, we have

$$P(x^2,x^2)=P(2x^2,0)=0,$$

so $x-y|P$. As P is symmetric, we deduce $(x-y)^2|P$. Cancel a factor $xy(x-y)^2$ from P, then P still satisfies the given equation. Repeating this procedure, we find the above solution.

289. Let $f(x)+2009=(x-a_1)\ldots(x-a_{50})$ and $f(x)=g(x)h(x)$. Then

$$g(x)h(x)=(x-a_1)\ldots(x-a_{50})-2009,$$

so $g(a_1),\ldots,g(a_{50})$ divide $2009=7^2\cdot 41$, which has $2\cdot 3\cdot 2=12$ divisors. By the Pigeonhole principle, there exist $\left\lceil\frac{50}{12}\right\rceil=5$ values a_i,a_j,a_ka_l,a_m such that

$$g(a_i)=g(a_j)=g(a_k)=g(a_l)=g(a_m),$$

thus $\deg g\geq 5$.

290. We have

$$a^2+b^2+c^2\geq ab+bc+ca\geq\sqrt{3(a+b+c)abc}\geq\sqrt{3}abc.$$

291. Answer: $f(n)=n^2+1$.

Let $P(m,n)$ be the assertion. Then $P(0,n)$ implies $f(n)(f(0)-1)=f(-1)-2$. Since the constant function doesn't satisfy the equation, we deduce $f(0)=1$, $f(-1)==2$. Comparing $P(1,n)$, $P(-1,n)$, we see that $f(1)=2, f$ is even and

$$f(n+1)+f(n-1)=2f(n)+2.$$

By induction, we get the result.

292. The Cauchy–Schwarz inequality gives

$$\left(x^2+y+1\right)\left(z^2+y+1\right) \geq (x+y+z)^2 \Leftrightarrow \frac{1}{x^2+y+1} \leq \frac{z^2+y+1}{(x+y+z)^2}.$$

Then

$$\frac{1}{x^2+y+1}+\frac{1}{y^2+z+1}+\frac{1}{z^2+x+1} \leq \frac{3+x+y+z+x^2+y^2+z^2}{(x+y+z)^2}.$$

Now, it suffices to prove that $(x+y+z)^2 \geq 3+x+y+z+x^2+y^2+z^2$, or, equivalently, $x+y+z \geq 3$, which is true.

Equality holds only when $x=y=z=0$.

293. Answer: *a)* n prime, $P(x) = -(n+1)x+n^2+n+1$; *b)* $n=1$, $P(x)=1$; *c)* $n=6$, $P(x)=-2x^3+23x^2-82x+97$.

If $n>1$, then let $1=d_1<\ldots<d_k=n$ be the positive divisors of n. Then

$$d_k d_{1-i+k}=n, d_{k-1}(d_2-1)=n-d_{k-1}|P(n)-P(d_{k-1})=1-d_2^2,$$

hence $d_{k-1}|d_2+1$. We deduce $k=2$ or $k=4$ and $d_3=d_2+1$. In the first case n is a prime number, in the second case n is the product of two consecutive primes, hence $n=6$.

294. We will prove a more general result: Let $n \geq 3$ be a positive integer, let $a_1, \ldots, a_n$ be nonnegative real numbers and

$$k \geq \frac{n-5+\sqrt{(n-1)(n+7)}}{4(n-1)}.$$

Then

$$\prod_{i=1}^{n}\left(k+\frac{a_i}{s-a_i}\right) \geq \left(k+\frac{1}{n-1}\right)^n,$$

where $s=a_1+\ldots+a_n$.

Define $f(x)=\ln\left(k+\frac{x}{1-x}\right)$. Then we have to show that

$$f(a_1)+f(a_2)+\ldots+f(a_n) \geq nf\left(\frac{1}{n}\right),$$

where $a_1+a_2+\ldots+a_n=1$. For $k \geq \frac{1}{2}$, this inequality follows from the Jensen's inequality, since f is convex. If $0<k<\frac{1}{2}$, then f is convex for

$\frac{1-2k}{2(1-k)} \leq x < 1$, hence f is convex for $x \geq \frac{1}{n}$. Now, we will apply a result of *Vasile Cârtoaje:*

Half-Convex Functions Theorem. Let $f: I \to \mathbb{R}$ be a function on an interval, which is convex for $x \leq s$ or $x \geq s$, $s \in I$. If $f(x) + (n-1)f(y) \geq nf(s)$, $\forall\, x, y \in I$ such that

$$x + (n-1)y = ns,$$

then $f(x_1) + \ldots + f(x_n) \geq ns$, $\forall\, x_1, \ldots, x_n \in I$ for which $x_1 + \ldots + x_n = ns$.

By the HCF Theorem, it suffices to prove the inequality for $a_1 + \ldots + a_n = 1$ and $a_2 = \ldots = a_n$, that is, to prove the given inequality for $a_2 = \ldots = a_n = 1$.

For a geometric proof, observe that there is a triangle with sidelengths

$$x = b + c, y = c + a, z = a + b.$$

Let R, r, s be the circumradius, the inradius, and the semiperimeter of this triangle, respectively. Using the identities

$$x + y + z = 2s, xy + yz + zx = s^2 + r^2 + 4Rr,$$
$$x^3 + y^3 + z^3 = 2s\left(s^2 - 3r^2 - 6Rr\right), xyz = 4sRr,$$

and the Gerretsen's inequality $s^2 - 16Rr + 5r^2 \geq 0$, the proposed inequality becomes $r(R - 2r)(4k^2 + 2k - 1) \geq 0$, whence

$$k \geq \frac{-1 + \sqrt{5}}{4}.$$

3.39 2010

295. If f is constant, then $f(x) = -1, 0, 1$ for all x.
If f is not constant, then

$$f(x+y) + 1 = \frac{(f(x)+1)(f(y)+1)}{f(x)f(y)+1}, f(x+y) - 1$$
$$= \frac{(f(x)-1)(f(y)-1)}{f(x)f(y)+1} \Rightarrow \frac{f(x)+1}{f(y)-1} = e^{g(x)} \Rightarrow g(x+y) = g(x) + g(y) \Rightarrow g(x)$$
$$= cx$$

for all x, where c is constant.

296. *a*) Points of the form (r, α), with $r \in \mathbb{Q}$, $\alpha \in \mathbb{R} - \mathbb{Q}$, cannot lie on a line passing through two rational points, hence *a*) is false.

b) Choose an arbitrary line through the point, not parallel to the coordinate axes. It suffices to show that there are at least two irrational points on the line.

The projections of the line onto the coordinate axes are bijections. The preimages of the rational numbers by these projections are countable sets on the line, hence cannot cover it completely. The parts not covered contain irrational points.

297. Answer: $f(x) = x + \frac{1}{2}$.

Let $P(x, y)$ be the given assertion. We have $f(f(-1)) = \left(f(x) + \frac{1}{2}\right) \times \left(f(-1) + \frac{1}{2}\right)$, hence $f(-1) = -\frac{1}{2}, f\left(-\frac{1}{2}\right) = 0$. Suppose that $f(a) = -\frac{1}{2}$ for some a. Then for $y = a$ we get $a = -1$. Suppose that $f(a) = 0$ for some a. We have

$$P(x, a) : f((a+1)x) = \frac{1}{2}\left(f(x) + \frac{1}{2}\right),$$

$$P(-1, a) : f(-a-1) = 0,$$

$$P(x, -a-1) : f(-ax) = \frac{1}{2}\left(f(x) + \frac{1}{2}\right) = f((a+1)x).$$

Then $f(-ax) = f((a+1)x)$. For $x = -\frac{1}{a+1}$ we get $\frac{a}{a+1} = -1$, so $a = -\frac{1}{2}$. It follows that $f(a) = 0$ if and only if $a = -\frac{1}{2}$. Now,

$$P(-1, y) : f(f(y) - y - 1) = 0 \Rightarrow f(y) = y + \frac{1}{2}.$$

298. Answer: $f(n) = n^2$; $f(n) = n^2 - n$.

The function is even, by *a*), so it suffices to determine it on positive integers. Putting $n = 1$ into *a*), we get $f(1)(f(-1) - 1) = 0$.

Case 1. $f(1) = 0$. Then putting $n = 1$ into *b*), yields $f(m + 1) = f(m) + 2m$. Summing from 1 to m, we have $f(m + 1) = f(1) + m^2 + m$, that is, $f(m) = m^2 - m$.

Case 2. $f(-1) = 1$. Then putting $n = -1$ into *b*), yields $f(m) = f(m - 1) + 2m - 1$. Summing from 0 to m, we have $f(m) = f(-1) + m^2 - 1 = m^2$.

299. Denote $b_n = \frac{1}{a_n^2}, n \geq 1$. Then

$$a_{n+1} = \frac{a_n}{1 + a_n^2} \Leftrightarrow b_{n+1} = b_n + 2 + \frac{1}{b_n}.$$

Now, we have to prove that

a') $b_n > 2n$, for $n > 2$.
b') There is n such that

$$b_n < \frac{100n}{49}.$$

a') Follows from

$$b_2 \geq 4, b_{n+1} = b_n + 2 + \frac{1}{b_n} > b_n + 2$$

by induction.
b') Denote

$$h_n = \sum_{k=1}^{n} \frac{1}{k}..$$

We have the implications

$$b_n \geq 2n, n \geq 2 \Longrightarrow b_{n+1} = b_n + 2 + \frac{1}{b_n} \leq b_n + 2 + \frac{1}{2n}, n \geq 2 \Longrightarrow b_{n+1} - b_2 = \sum_{k=2}^{n}$$
$$\times (b_{k+1} - b_k) \leq \sum_{k=2}^{n} \left(2 + \frac{2}{2n}\right) = 2(n-1) + \frac{h_n - 1}{2} \Longrightarrow b_{n+1} \leq b_n - \frac{5}{2} + \frac{h_n}{2}$$
$$+ b_2 < 2(n+1) + \frac{h_{n+1}}{2} + b_2, n \geq 2 \Longrightarrow b_n < 2n + \frac{h_n}{2} + b_2, n \geq 3.$$

Also $h_n < \sqrt{2n}$ for any positive integer n. Indeed, by the Cauchy–Schwarz inequality,

$$h_n^2 \leq n \sum_{k=1}^{n} \frac{1}{k^2},$$

and

$$\sum_{k=1}^{n} \frac{1}{k^2} < 1 + \sum_{k=2}^{n} \frac{1}{(k-1)k} = 1 + \sum_{k=2}^{n} \left(\frac{1}{k-1} - \frac{1}{k}\right) = 1 + 1 - \frac{1}{n} < 2.$$

Since $\dfrac{100n}{49} = 2n + \dfrac{2n}{49}$ and $h_n < \sqrt{2n}$, it suffices to prove that there is n such that

$$\frac{\sqrt{2n}}{2} + b_2 < \frac{2n}{49} \Longleftrightarrow 49b_2 < \sqrt{2n}\left(\sqrt{2n} - \frac{49}{2}\right).$$

This inequality holds for

$$n \geq \max\left\{\frac{49^2 b_2^2}{2}, \frac{51^2}{8}\right\}.$$

300. Answer: $\lambda \geq e$.

Suppose that $\lambda < e$ and rewrite the inequality as

$$\sum a_1 \ldots a_{i-1} a_{i+1} \ldots a_n - \lambda \leq (n-\lambda) a_1 \ldots a_n. \quad (1)$$

Set $a_1 = \ldots = a_{n-1} = \frac{n}{n-1}, a_n = 0$. Then we get

$$\left(\frac{n}{n-1}\right)^{n-1} - \lambda \leq 0 \Longrightarrow \lambda \geq \left(\frac{n}{n-1}\right)^{n-1}.$$

Taking $n \to \infty$, we have $\lambda \geq e$, contradiction.

Suppose that $\lambda \geq e$ and define

$$f(a_1, \ldots, a_n) = \sum_{i=1}^{n} \frac{1}{a_i} - \lambda \prod_{i=1}^{n} \frac{1}{a_i}.$$

Then $f(a_1, \ldots, a_n) \leq f(1, a_1 + a_2 - 1, \ldots, a_n)$, (2), so that

$$f(a_1, \ldots, a_n) \leq \ldots \leq f(1, \ldots, 1) = n - \lambda.$$

Proof of (2).

$$\sum_{i=1}^{n} \frac{1}{a_i} - \lambda \prod_{i=1}^{n} \frac{1}{a_i} \leq 1 + \frac{1}{a_1 + a_2 - 1} + \sum_{i=3}^{n} \frac{1}{a_i} - \frac{\lambda}{a_1 + a_2 - 1} \prod_{i=3}^{n} \frac{1}{a_i} \Leftrightarrow \frac{a_1 + a_2}{a_1 a_2}$$
$$- \frac{a_1 + a_2}{a_1 + a_2 - 1} \leq \lambda \left(\frac{1}{a_1 a_2} - \frac{1}{a_1 + a_2 - 1}\right) \prod_{i=3}^{n} \frac{1}{a_i} \Leftrightarrow (a_1 + a_2) a_3 \ldots a_n \leq \lambda.$$

By the AM–GM inequality,

$$(a_1 + a_2) a_3 \ldots a_n \leq \left(\frac{n}{n-1}\right)^{n-1} < e \leq \lambda.$$

301. Let $n > 0$ be the degree of the polynomial P, i.e.,

$$P(x, y) = a_n x^n + a_{n-1} x^{n-1} y + \ldots + a_1 x y^{n-1} + a_0 y^n.$$

If n is odd, then choosing $t = 0$, respectively, $t = \pi$, we get $a_0 = 0$, respectively, $a_0 = -1$. Hence n is even. Also, $P(0, 0) = 0$. Now choose x, y such that $x \neq 0$ or $y \neq 0$ and take $c = \sqrt{x^2 + y^2}$. Since

$$\left(\frac{x}{\sqrt{x^2+y^2}}\right)^2+\left(\frac{y}{\sqrt{x^2+y^2}}\right)^2=1,$$

there exists a real number t such that

$$\sin t=\frac{x}{\sqrt{x^2+y^2}},\ \cos t=\frac{y}{\sqrt{x^2+y^2}},$$

therefore $P(\sin t, \cos t) = 1$. By homogeneity,

$$P(x,y)=c^nP\left(\frac{x}{c},\frac{y}{c}\right)=\left(\sqrt{x^2+y^2}\right)^nP\left(\frac{x}{\sqrt{x^2+y^2}},\frac{y}{\sqrt{x^2+y^2}}\right)=\left(\sqrt{x^2+y^2}\right)^n$$

for all x, y. Since n is even, we have $n = 2k$ and $P(x, y) = (x^2 + y^2)^k$.

302. The AM–HM inequality gives

$$(x+y+z+w)\left(\frac{1}{x}+\frac{1}{y}+\frac{1}{z}+\frac{1}{w}\right)\geq 16,$$

so that

$$4\left(5-\frac{1}{xyzw}\right)\geq 16,$$

which implies $xyzw \geq 1$. On the other hand, using the first equality and the AM–GM inequality, we deduce

$$xyzw\leq\left(\frac{x+y+z+w}{4}\right)^4=1.$$

Then all inequalities are equalities and $(x, y, z, w) = (1, 1, 1, 1)$.

303. We will use the inequality $m_q + m_g \leq 2m_a$ for two variables, that is,

$$\sqrt{\frac{a^2+b^2}{2}}+\sqrt{ab}\leq a+b, a, b>0.$$

Denote $x=\sqrt{\frac{a}{b}}$. The inequality becomes

$$\sqrt{\frac{1}{2}\left(x^2+\frac{1}{x^2}\right)}+1\leq x+\frac{1}{x}.$$

Taking squares yields

$$\left(x+\frac{1}{x}-2\right)^2 \geq 0,$$

which is obvious, with equality only when $x=1$, that is, $a=b$.

Divide the above inequality by $\sqrt{a+b}$, multiply by $\sqrt{2}$ and write two analogous inequalities. Summing the three inequalities, we get

$$\sum_{cyc}\sqrt{\frac{a^2+b^2}{a+b}}+\sqrt{2}\sum_{cyc}\sqrt{\frac{ab}{a+b}} \leq \sqrt{2}\sum_{cyc}\sqrt{a+b}.$$

Now it suffices to prove the inequality $\sqrt{\frac{ab}{a+b}}+\sqrt{\frac{bc}{b+c}}+\sqrt{\frac{ca}{c+a}} \geq \frac{3}{\sqrt{2}}$. By the mean inequality $m_1 \geq m_{-2}$ for the variables $\sqrt{\frac{ab}{a+b}}, \sqrt{\frac{bc}{b+c}}, \sqrt{\frac{ca}{c+a}}$, we have

$$\sqrt{\frac{ab}{a+b}}+\sqrt{\frac{bc}{b+c}}+\sqrt{\frac{ca}{c+a}} \geq 3\sqrt{\frac{3}{\frac{a+b}{ab}+\frac{b+c}{bc}+\frac{c+a}{ca}}}=3\sqrt{\frac{3abc}{2(ab+bc+ca)}} \geq \frac{3}{\sqrt{2}}.$$

Equality holds only when $a=b=c$. This completes the proof of the proposed inequality.

304. We will prove both inequalities by induction. From

$$a_2^2-a_1^2=\frac{16}{9}-1=\frac{7}{9}>\frac{1}{2}, a_3^2-a_2^2=1+\frac{4}{3}-\frac{16}{9}=\frac{5}{9}>\frac{1}{2},$$

it follows that the inequality is true for $n=2, 3$. Suppose that it is true for $n-2$ and $n-1$. Then by the Cauchy–Schwarz inequality,

$$a_n^2=1+a_{n-1}a_{n-2}>1+\sqrt{\left(a_{n-2}^2+\frac{1}{2}\right)\left(a_{n-3}^2+\frac{1}{2}\right)} \geq 1+a_{n-2}a_{n-3}+\frac{1}{2}=a_{n-1}^2 +\frac{1}{2}.$$

For the second inequality, we have

$$1+\frac{1}{a_1}+\frac{1}{a_2}-2a_2=1+1+\frac{3}{4}-\frac{8}{3}=\frac{1}{12}>0,$$

which proves the initial case. Assume that the inequality holds for $n-1$ and add $\frac{1}{a_n}$ to both sides:

$$1+\frac{1}{a_1}+\ldots+\frac{1}{a_{n-1}}+\frac{1}{a_n}>2a_{n-1}+\frac{1}{a_n}.$$

Now, it suffices to show that $2a_{n-1}+\frac{1}{a_n}>2a_n$, which follows immediately from the first inequality:

$$2\left(a_{n-1}a_n-a_n^2\right)+1=2\left(a_{n+1}^2-a_n^2\right)-1>0.$$

305. If we assume $\mathbb{H}$ minimal, then $\mathbb{H}\subset\mathbb{Q}$, hence it is countable, so that there is a bijection $f\colon\mathbb{Z}\to\mathbb{H}$. Otherwise, $\mathbb{H}=(0,\infty)$ is a counterexample, since it is uncountable, that is, there is no bijection $f\colon\mathbb{Z}\to\mathbb{H}$.

306. Denote by $A(a,b,c)$ the assertion. Then

$$A(0,0,0)\Longrightarrow P(0,1)=0; A(0,0,c)\Longrightarrow P(0,y)=0, \forall y>1\Longrightarrow x|P(x,y);$$
$$A(a,b,0)\Longrightarrow P(x,1)=0, \forall x\in\mathbb{R}\Longrightarrow(y-1)|P(x,y).$$

Put $P(x,y)=x(y-1)Q(x,y)$ and define

$$B(a,b,c): cQ\left(ab,c^2+1\right)+aQ\left(bc,a^2+1\right)+bQ\left(ca,b^2+1\right)=0.$$

Then

$$B(0,0,c)\Longrightarrow Q(0,y)=0, \forall y>1\Longrightarrow x|Q(x,y).$$

Put $Q(x,y)=xR(x,y)$. Then

$$R\left(ab,c^2+1\right)+R\left(bc,a^2+1\right)+R\left(ca,b^2+1\right)=0.$$

We conclude that P is divisible by $x^{2n}(y-1)^n$ for all positive integers n, therefore

$$P(x,y)=0, \forall x,y.$$

307. We have

$$\sin(x+2010^\circ)=-\frac{\sqrt{3}}{2}\sin x-\frac{1}{2}\cos x.$$

Let

$$\max\left\{|\sin x|,\left|\frac{\sqrt{3}}{2}\sin x+\frac{1}{2}\cos x\right|\right\}>c.$$

If $|\sin x|\le c$, then $|\cos x|\ge\sqrt{1-c^2}$ and

$$\left|\frac{\sqrt{3}}{2}\sin x+\frac{1}{2}\cos x\right|>\left|\frac{\sqrt{3}}{2}c-\frac{1}{2}\sqrt{1-c^2}\right|.$$

For

$$c\leq\frac{\sqrt{6}-\sqrt{2}}{4}<\frac{1}{\sqrt{17}},$$

it follows that

$$\left|\frac{\sqrt{3}}{2}c-\frac{1}{2}\sqrt{1-c^2}\right|\geq c,$$

which ends the proof.

308. Assume that $a_1 < a_2 < \ldots < a_8$ and define

$$s_1=a_2-a_1, s_2=a_3-a_2, \ldots, s_7=a_8-a_7,$$
$$s_8=a_3-a_1, s_9=a_4-a_2, \ldots, s_{13}=a_8-a_6.$$

Then

$$\sum_{i=1}^{13}s_i=2a_8+a_7-a_2-2a_1\leq 2\cdot 17+16-2-2\cdot 1=46.$$

If for any $k>0$ there are at most two values i such that $s_i=k$, then

$$\sum_{i=1}^{13}s_i\geq 2(1+2+3+4+5+6)+7=49,$$

contradiction.

An example of a set which satisfies the required property is $\{1,2,3,5,8,12,17\}$.

309. Denote $s=a+b+c+d$. The equation becomes

$$\frac{s(a^2+c^2-2bd)+(a-c)^2(a+c)}{s^2-(a-c)^2}+\frac{s(b^2+d^2-2ac)+(b-d)^2(b+d)}{s^2-(b-d)^2}=0.$$

Since $s>0$, it is equivalent to

$$\left[(a-c)^2+(b-d)^2\right]s^3+\left[(a-c)^2(a+c)+(b-d)^2(b+d)\right]$$
$$-\left[(a-c)^2\left(b^2+d^2-2ac\right)+(b-d)^2\left(a^2+c^2-2bd\right)\right]s$$
$$-(a-c)^2(b-d)^2\underbrace{(a+b+c+d)}_{s}=0\Leftrightarrow\left[(a-c)^2+(b-d)^2\right]s^2$$
$$+\left[(a-c)^2(a+c)+(b-d)^2(b+d)\right]s$$
$$-(a-c)^2\left[b^2+d^2-2ac-\frac{(b-d)^2}{2}\right]-(b-d)^2\left[a^2+c^2-2bd-\frac{(a-c)^2}{2}\right]$$
$$=0\Leftrightarrow(a-c)^2\left[s^2+(a+c)s+2ac-\frac{(b+d)^2}{2}\right]$$
$$+(b-d)^2\left[s^2+(b+d)s+2bd-\frac{(a+c)^2}{2}\right]=0.$$

From $s > a + c$, $s > b + d$, it follows that equality holds only when $a = c$, $b = d$.

310. We will use the following result:

$$\frac{x^3+y^3}{2}\leq\left(\frac{x^2+y^2}{x+y}\right)^3, \forall x,y>0\Leftrightarrow(x-y)^4\left(x^2-xy+y^2\right)\geq 0, \forall x,y>0.$$

We have

$$\sum_{\text{cyc}}\sqrt[4]{\frac{\left(a^2+b^2\right)\left(a^2-ab+b^2\right)}{2}}=\sum_{\text{cyc}}\sqrt[4]{\frac{\left(a^2+b^2\right)\left(a^3+b^3\right)}{2(a+b)}}\leq\sum_{\text{cyc}}\frac{a^2+b^2}{a+b}.$$

Now we have to prove the inequality

$$\sum_{\text{cyc}}\frac{a^2+b^2}{a+b}\leq\frac{2}{3}\left(a^2+b^2+c^2\right)\left(\frac{1}{a+b}+\frac{1}{b+c}+\frac{1}{c+a}\right)\Leftrightarrow\sum_{\text{cyc}}\frac{2a^2-b^2-c^2}{b+c}\geq 0.$$

The last inequality follows from the Chebyshev's inequality.

3.40 2011

311. We have

$$(-2)^2a-2b+c=1, 2^2a+2b+c=9,$$

whence we deduce $b = 2$, $c = 5 - 4a$. The parabola does not intersect the x–axis, so

$$\Delta = b^2 - 4ac = 2^2 - 4a(5 - 4a) = 4\left(4a^2 - 5a + 1\right) = 4(a - 1)(4a - 1) < 0,$$

hence $a \in \left(\frac{1}{4}, 1\right)$. The x – coordinate of the vertex is $x_V = -\frac{b}{2a} = -\frac{1}{a} \in (-4, -1)$.

312. Answer: all $a \neq 1$.

If $a \neq 1$, then we can take $f(x) = \frac{2x^2 - 1}{a - 1}$.

If $a = 1$, then

$$f(\sin x) + f(\cos x) = \cos 2x,$$

$$f(\cos x) + f(\sin x) = f\left(\sin\left(\frac{\pi}{2} - x\right)\right) + f\left(\cos\left(\frac{\pi}{2} - x\right)\right) = \cos 2\left(\frac{\pi}{2} - x\right) = -\cos 2x,$$

for all x, contradiction.

313. By the AM–GM inequality,

$$3a\sqrt[3]{1 + b - c} \le a(1 + b - c + 1 + 1).$$

Adding with two similar inequalities, we get

$$3\left(a\sqrt[3]{1 + b - c} + b\sqrt[3]{1 + c - a} + c\sqrt[3]{1 + a - b}\right) \le 3(a + b + c) = 3.$$

314. Answer: $\left(\frac{3}{4}\right)^n$.

Choose $x_i = 2$ for all i, then $\lambda \le \left(\frac{3}{4}\right)^n$. Denote

$$S = \left(\frac{1}{2n}\sum_{i=1}^{2n}(x_i + 2)^n\right)^{\frac{1}{n}}, \quad S' = \left(\frac{1}{2n}\sum_{i=1}^{2n}(x_i + 1)^n\right)^{\frac{1}{n}}, \quad P = \left(\prod_{i=1}^{2n} x_i\right)^{\frac{1}{n}}.$$

Now, it remains to show that $S \ge P$ implies $S' \ge \frac{3}{4}P$.

Case 1. $P \ge 4$. Then by the Minkowski's inequality,

$$S' + 1 = \left(\frac{1}{2n}\sum_{i=1}^{2n}(x_i + 1)^n\right)^{\frac{1}{n}} + \left(\frac{1}{2n}\sum_{i=1}^{2n}1^n\right)^{\frac{1}{n}} \ge \left(\frac{1}{2n}\sum_{i=1}^{2n}(x_i + 2)^n\right)^{\frac{1}{n}} = S \ge P$$
$$\ge \frac{3}{4}P + 1.$$

Case 2. $0 \le P \le 4$. Then by the AM–GM and Hölder's inequalities,

$$S' \geq \left(\prod_{i=1}^{2n}(x_i+1)\right)^{\frac{1}{2n}} \geq \left(\prod_{i=1}^{2n}x_i\right)^{\frac{1}{2n}} + \left(\prod_{i=1}^{2n}1\right)^{\frac{1}{2n}} = \sqrt{P}+1.$$

From $0 \leq \sqrt{P} \leq 2,$ we deduce

$$1+\sqrt{P}-\frac{3}{4}P = \frac{1}{4}\left(2-\sqrt{P}\right)\left(2+3\sqrt{P}\right) \geq 0,$$

hence $S' \geq \frac{3}{4}P$.

315. Answer: yes.

Define an operation $\bigoplus$ on the set of nonnegative integers, which maps two nonnegative integers a, b to a nonnegative integer $a \bigoplus b$, such that for all $i = 0, 1, 2, \ldots,$ $(a \bigoplus b)_i \equiv a_i + b_i \pmod 2$, where n_i denotes the binary digit corresponding to 2^i in the binary representation of n. This operation satisfies condition (1) for all nonnegative integers, because addition modulo 2 satisfies it. The operation also satisfies condition (2), because for $x, y \in \{0, 1\}$,

$$x+x+y \equiv y \equiv y+x+ \pmod 2.$$

The set of nonnegative integers and the set of all integers are countable, hence there exists a one-to-one correspondence between these sets, i.e.,

$$f(x) = (-1)^x \left\lfloor \frac{x+1}{2} \right\rfloor.$$

Therefore, we can define the operation $*$by $f(x) * f(y) = f(x \bigoplus y)$. By the above considerations, both conditions (1), (2) hold.

316. Answer: $f(x) = cx,\ g(x) = \frac{x}{c},\ c \in \mathbb{Q}^*$.

Choose $y = f(g(x))$ in (1). Then $f(g(x) - g(f(g(x)))) = 0$. Let $f(t) = 0$ for some t. Choose $y = t$ in (2). Then $g(f(x)) = g(f(x)) - t$, which implies $t = 0$, so $f(0) = 0$.

Choose $x = y$ in (1). Then $f(g(x)) - x = 0$, hence $f(g(x)) = x$. From this, it is easy to prove that f, g are bijective.

Choose $x = f(u)$, $y = f(v)$ in (1). Then

$$f(g(f(u)) - g(f(v))) = f(u-v) = f(g(f(u))) - f(v) = f(u) - f(v).$$

The function f satisfies a linear functional equation, so $f(x) = cx$ for some $c \in \mathbb{Q}^*$. As g is the inverse of f, it follows $g(x) = \dfrac{x}{c}$.

317. Changing the variable, we have

$$x \to \frac{x-3}{x+1}:\ f(x) + f\left(\frac{3+x}{1-x}\right) = \frac{x-3}{x+1},$$

$$x \to \frac{3+x}{1-x}:\ f\left(\frac{x-3}{x+1}\right)+f(x)=\frac{3+x}{1-x}.$$

Adding these equalities, we get

$$f(x)=\frac{x^3+7x}{2(1-x^2)}, x\neq \pm 1.$$

318. Select randomly a subset with x elements of $\{1, 2, \ldots, 2m\}$. Then the expected value of x is m and the variance is $\sigma^2=\frac{m}{2}$. By the Chebyshev's inequality,

$$Pr\left(|x-m|\geq k\sqrt{\frac{m}{2}}\right)\leq\frac{1}{k^2}, k>0.$$

Take $k=(t+1)\sqrt{\frac{2}{m}}$, then

$$\frac{2\left(\binom{2m}{0}+\binom{2m}{1}+\ldots+\binom{2m}{m-t-1}\right)}{2^{2m}}\leq\frac{m}{2(t+1)^2}\leq\frac{1}{2k}.$$

319. Answer: $n\in\{-2, 26, 38, 130\}$.
If $P(x)$ has an integer root x_0, then

$$n=\frac{3x_0^3-2}{x_0+1}=3\left(x_0^2-x_0+1\right)-\frac{5}{x_0+1}$$

is an integer, so $x_0+1\in\{-5,-1,1,5\}$, therefore $x_0\in\{-6,-2,0,4\}$. We find for n the values 130, 26, -2, 38, respectively. If $P(x)$ does not have integer roots and is reducible in $\mathbb{Z}[x]$, then there is a decomposition

$$P(x)=(3x+a)\left(x^2+bx+c\right), a, b, c\in\mathbb{Z}, 3\nmid a.$$

Identifying the coefficients of x^2, we get $a=-3b$, contradiction.

320. Denote $x=a-b$, $y=b-c$, then $c-a=-(x+y)$. The denominator is

$$x^3+y^3-(x+y)^3=-3xy(x+y)=3(a-b)(b-c)(c-a).$$

Similarly, the numerator is $3(a^2-b^2)(b^2-c^2)(c^2-a^2)$, so that LHS is

$$(a+b)(b+c)(c+a).$$

The desired inequality follows immediately, applying the AM–GM inequality to each factor.

321. For a given positive integer m, by the Pigeonhole Principle, there exists a positive integer n such that $\{n\pi\} \leq \frac{1}{m}$. Then $\lfloor kn\pi \rfloor, 1 \leq k \leq m$, is an arithmetic progression of length m.

Suppose that $\{a + nd | n \in \mathbb{N}\} \subset S$ for some a, d. Denote $[m] = m - \lfloor \frac{m}{\pi} \rfloor \pi$. Then $[d] > 0$, since π is irrational. We have $m \in S$ only if $[m] > \pi - 1$. The sequence $\{[a + nd]\}_{n \geq 0}$ is dense in the interval $(0, \pi)$, because π is irrational. Thus, not every element of the sequence can be in S, contradiction.

322. Since $a + \frac{1}{a} = \lfloor a \rfloor + \lfloor \frac{1}{a} \rfloor + 1$, it follows that $a + \frac{1}{a}$ is an integer. Denote

$$s_n = a^n + \frac{1}{a^n}$$

for any positive integer n. Then $s_1 s_n = s_{n+1} + s_{n-1}, n \geq 2$. Since $s_2 = s_1^2 - 2$ is an integer, we deduce inductively that s_n is an integer for all n. Then

$$\{a^n\} + \left\{\frac{1}{a^n}\right\} = s_n - \lfloor a^n \rfloor - \left\lfloor \frac{1}{a^n} \right\rfloor$$

is an integer for all n. As $0 \leq \{x\} < 1$, we deduce that $\{a^n\} + \{\frac{1}{a^n}\} \in \{0, 1\}$.

If $\{a^n\} + \{\frac{1}{a^n}\} = 0$, then $a = \pm 1$. In this case, $\{a\} + \{\frac{1}{a}\} = 0 \neq 1$, contradiction. It remains

$$\{a^n\} + \left\{\frac{1}{a^n}\right\} = 1.$$

323. The polynomial $P(x) = x(x + 2)\ldots(x + 2n) + (x + 1)(x + 3)\ldots(x + 2n - 1)$ is monic, has degree $n + 1$ and integer coefficients. Every rational root of P must be an integer, since P is monic. We will prove that P has a real root in each interval

$$(0, -1), (-2, -3), (-4, -5), \ldots, (-2n + 2, -2n + 1), (-2n, -2n - 1).$$

We have

$$P(-2i) = (-2i + 1)(-2i + 3)\ldots(-1) \cdot 1 \cdot 3 \cdot \ldots (2(n - i) - 1) = (-1)^i M,$$

$$P(-2i - 1) = (-2i - 1)(-2i + 1)\ldots(-1) \cdot 1 \cdot 3 \cdot \ldots \cdot (2(n - i) - 1) = (-1)^{i+1} N,$$

for $0 \leq i \leq n - 1$, where M, N are positive integers. Moreover,

$$P(-2n) = (-2n+1)\ldots(-1) = (-1)^n \cdot 1 \cdot 3 \cdot \ldots \cdot (2n-1),$$
$$P(-2n-1) = (-2n-1)\ldots(-1) + (-2n)\ldots(-2) =$$
$$= (-1)^{n+1}(1 \cdot 3 \cdot \ldots \cdot (2n+1) - 2 \cdot 4 \cdot \ldots \cdot (2n)).$$

Now we will prove the inequality $1 \cdot 3 \cdot \ldots \cdot (2n + 1) > 2 \cdot 4 \cdot \ldots \cdot (2n)$ for all positive integers n. Indeed, let

$$a_n = \frac{1 \cdot 3 \cdot \ldots \cdot (2n+1)}{2 \cdot 4 \cdot \ldots \cdot (2n)}, n \geq 1.$$

Then $a_1 = \frac{3}{2} > 1$ and

$$\frac{a_{n+1}}{a_n} = \frac{2n+3}{2n+2} > 1,$$

for all $n \geq 1$, hence $a_n > 1$ for all $n \geq 1$. From this we deduce

$$P(-2i)P(-2i-1) < 0, 0 \leq i \leq n,$$

therefore P has $n + 1 = \deg(P)$ real roots, which are not integers. It follows that all roots are irrational.

324. Let

$$a = \frac{x_1 + x_2}{2}, b = \frac{x_3 + x_4 + \ldots + x_n}{n-2}.$$

Then $2a + (n - 2)b = 0$ and

$$n(n-1) = \left(x_1^2 + x_2^2\right) + \left(x_3^2 + x_4^2 + \ldots + x_n^2\right) \geq 2a^2 + (n-2)b^2 = \frac{2na^2}{n-2},$$

therefore $a \leq \sqrt{\frac{(n-1)(n-2)}{2}}$, so $S \leq 2\sqrt{(n-1)(n-2)}$, with equality when

$$x_1 = x_2 = \sqrt{\frac{(n-1)(n-2)}{2}}, x_3 = x_4 = \ldots = x_n = -2\sqrt{\frac{n-1}{n-2}}.$$

Fix $S = x_1 + x_2$. Then $x_1 + x_2 = -(x_3 + x_4 + \ldots + x_n) \geq -(n-2)x_2$, so $-\frac{S}{n-2} \leq x_2$. Since $S = x_1 + x_2 \geq 2x_2$, we have

$$-\frac{S}{n-2} \leq x_2 \leq \frac{S}{2}.$$

By the Karamata's inequality,

$$n(n-1)=x_1^2+x_2^2+\ldots+x_n^2\leq x_1^2+(n-2)x_2^2+(-x_1-(n-2)x_2)^2=(S-x_2)^2 \\ +(n-2)x_2^2+(S+(n-3)x_2)^2.$$

Since the RHS is convex in x_2, it attains the maximum at $x_2=-\frac{S}{n-2}$ or $x_2=\frac{S}{2}$. Therefore, the inequality must hold in at least one of these two points.

For $x_2=-\frac{S}{n-2}$, this becomes

$$n(n-1)\leq\left(\left(\frac{n-1}{n-2}\right)^2+\frac{1}{n-2}+\frac{1}{(n-2)^2}\right)S^2=\frac{n(n-1)S^2}{(n-2)^2}.$$

Then $S\geq n-2$.

For $x_2=\frac{S}{2}$, the inequality becomes

$$n(n-1)\leq\left(\frac{1}{4}+\frac{n-2}{4}+\frac{(n-1)^2}{4}\right)S^2=\frac{n(n-1)S^2}{4},$$

therefore $S\geq 2$.

If $n\geq 4$, then $S\geq 2$ and $S=2$ is attained for

$$x_1=x_2=\ldots=x_{n-1}=1, x_n=-(n-3).$$

If $n=3$, then $S\geq 1$ and $S=1$ is attained for $x_1=2$, $x_2=x_3=-1$.

3.41 2012

325. Adding

$$\frac{1}{x^2-4x+9}\leq\frac{x+2}{18}$$

with two similar inequalities, we get

$$\sum\frac{1}{x^2-4x+9}\leq\sum\frac{x+2}{18}=\frac{7}{18}.$$

326. *a)* Let $P(x,y)$ be the given functional equation. If $y\neq 1$, $(y-1)(f(y)-1)>0$, then $P\left(\frac{f(y)-1}{y-1},y\right)$ implies $y=1$, contradiction.

b) From *a)* it follows that $f(y)\leq 1$ for $y>1$ and $f(y)\geq 1$ for $y<1$. The following implications lead to contradictions:

$$f(y) < \frac{1}{y} \Rightarrow 1 - \frac{1}{y} + f(y) < 1 \Rightarrow yf(y) = f\left(1 - \frac{1}{y} + f(y)\right) \geq 1 \Rightarrow f(y) \geq \frac{1}{y},$$

$$f(y) > \frac{1}{y} \Rightarrow 1 - \frac{1}{y} + f(y) > 1 \Rightarrow yf(y) = f\left(1 - \frac{1}{y} + f(y)\right) \leq 1 \Rightarrow f(y) \leq \frac{1}{y}.$$

We deduce $f(y) = \frac{1}{y}$ for all $y > 0$.

327. We have

$$\sum_{\text{cyc}} \frac{a^3}{b^2 + c} = \sum_{\text{cyc}} \frac{a^4}{ab^2 + ac} \geq \frac{1}{\sum_{\text{cyc}} (ab^2 + ac)},$$

hence we need to prove the inequality

$$\sum_{\text{cyc}} (ab^2 + ac) \leq 1 + \frac{\sqrt{3}}{3}.$$

This follows from $ab + bc + ca \leq 1$ and

$$\left(\sum_{\text{cyc}} ab^2\right)^2 \leq (a^2 + b^2 + c^2)(a^2b^2 + b^2c^2 + c^2a^2) = a^2b^2 + b^2c^2 + c^2a^2$$

$$\leq \frac{(a^2 + b^2 + c^2)^2}{3} = \frac{1}{3}.$$

328. Using the AM–QM inequality, it suffices to prove that

$$\sum_{i=1}^{n} |z_i|^2 \leq n.$$

We have

$$\sum_{i=1}^{n} |z_i|^2 = \sum_{i=1}^{n} (xy_i + x_i y - x_i y_i)\left(\frac{\overline{x}}{y_i} + \frac{\overline{y}}{x_i} - \frac{1}{x_i y_i}\right) = n|x|^2 + \sum_{i-1}^{n} x\overline{y} \cdot \frac{y_i}{x_i} - n|x|^2$$

$$+ n|y|^2 + \sum_{i=1}^{n} \overline{x}y \cdot \frac{y_i}{x_i}$$

By the triangle's inequality,

$$\left|\sum_{i=1}^{n} x\bar{y}\cdot\frac{y_i}{x_i}+\bar{x}y\cdot\frac{x_i}{y_i}\right|\leq|x\bar{y}|\sum_{i=1}^{n}\left|\frac{y_i}{x_i}\right|+|\bar{x}y|\sum_{i=1}^{n}\left|\frac{x_i}{y_i}\right|=2n|xy|,$$

hence

$$\sum_{i=1}^{n}|z_i|^2=\sum_{i=1}^{n}\left(x\bar{y}\cdot\frac{y_i}{x_i}+\bar{x}y\cdot\frac{x_i}{y_i}\right)-n\left(|x|^2+|y|^2\right)+n\leq n-n\left(|x|^2+|y|^2\right)\leq n.$$

329. For any integer $m \geq 2$ the numbers $2m^2 + 1$, $m(2m^2 + 1)$, $-m(2m^2 + 1)$ satisfy the required conditions, since they are distinct and

$$\begin{aligned}&\left(2m^2+1\right)^2+\left(m\left(2m^2+1\right)\right)^2+\left(-m\left(2m^2+1\right)\right)^2\\&\quad=\left(1+m^2+m^2\right)\left(2m^2+1\right)^2=\left(2m^2+1\right)^3,\\&\left(2m^2+1\right)^3+\left(m\left(2m^2+1\right)\right)^3+\left(-m\left(2m^2+1\right)\right)^3=\left(2m^2+1\right)^3.\end{aligned}$$

When m increases, the numbers of these triples get arbitrary large in absolute value, hence for any such triple we can find a new triple with components larger in absolute value.

For any $k = 3q + r$, with q nonnegative integer and $0 \leq r \leq 2$, we can choose q triples with distinct components, as above. If $r = 1$, then add 0 and if $r = 2$, then add 0, 1 to obtain k numbers having the required property.

330. Answer: 2012.

We will prove the inequality

$$\left(x^2+y^2+z^2\right)\left(x^3+y^3+z^3\right)\leq(x+y+z)\left(x^4+y^4+z^4\right).$$

Multiplying, simplifying, and rearranging the terms, we get

$$xy(x+y)(x-y)^2+yz(y+z)(y-z)^2+zx(z+x)(z-x)^2\geq 0.$$

331. Answer: for all positive integers k.

For $k = 1$, the identic function is $1-$tastrophic.

Suppose $k \geq 2$ and define f as follows: $f(1) = 1$, $f(a^k) = (f(a))^k$, $a \geq 2$, and if n is the smallest positive integer for which $f(n)$ is not yet defined, then denote $n_1 = n, n_2, \ldots, n_k$ the k smallest positive integers in this order, which are not kth powers and for which f is not yet defined and define

$$f(n_1)=n_2, f(n_2)=n_3, \ldots, f(n_{k-1})=n_k, f(n_k)=n_1^k.$$

Clearly, this is a function $f\colon \mathbb{N}^* \to \mathbb{N}^*$ and it remains to verify that it is $k-$tastrophic. If $a \geq 2$ is an integer, then

$$f_2\left(a^k\right) = f\left(f\left(a^k\right)\right) = f\left(\left(f(a)\right)^k\right) = \left(f\left(f(a)\right)\right)^k = \left(f_2(a)\right)^k,$$

hence, inductively, $f_p(a^k) = (f_p(a))^k$ for all integers $p \geq 1$.

Now, we will prove by induction that f is $k-$tastrophic. Clearly, $f_k(1) = 1^k$. Suppose that there is $n \geq 2$ such that $f_k(m) = m^k$ for all $m \leq n-1$. If $n = a^k$ for an integer $a \geq 2$, then $a \leq n-1$, hence

$$f_k(n) = f_k\left(a^k\right) = \left(f_k(a)\right)^k = \left(a^k\right)^k = n^k.$$

If n is not a kth power, then there are positive integers $n_1, n_2, \ldots, n_k$ such that $n = n_i$ for some i, $1 \leq i \leq k$. Then

$$f_k(n) = f_k(n_i) = f_{k+i-1}(n_1) = f_{i-1}\left(f_k(n_1)\right) = f_{i-1}\left(f(n_k)\right) = f_{i-1}\left(n_1^k\right)$$
$$= \left(f_{i-1}(n_1)\right)^k = n_i^k = n^k,$$

which ends the proof.

332. Answer: $P(x) = x^n$; $P(x) = x^{n-2}(x+2)(x-1)$; $P(x) = x^{n-3}(x+1)^2(x-1)$.

We see that $P(x) = x^n$ is a solution. If $a_0 \neq 0$, then by the Viète's relations, we have $a_1 a_2 \ldots a_{n-1} = (-1)^n$, hence $a_1, a_2, \ldots, a_{n-1} \in \{-1, 1\}$.

Case 1. $n = 2$. Then $x^2 + a_1 x + a_0 = x^2 - (a_0 + a_1)x + a_0 a_1$. Since $a_0 \neq 0$, we have $a_1 = 1$, hence $a_0 = -2$ and $P(x) = (x+1)(x-2)$.

Case 2. $n = 2t$, $t > 1$. By the Viète's relations,

$$a_0 + a_1 + \ldots + a_{n-1} = -a_{n-1} \in \{-1, 1\}.$$

On the left, there are an even number of summands, contradiction.

Case 3. $n = 2t + 1$. Then

$$\left(\sum a_i\right)^2 - 2\sum_{i<j} a_i a_j = \sum a_i^2 = n,$$

hence $1 - 2a_{n-2} = n > 0$. We deduce $a_{n-2} = -1$, $n = 3$, so that

$$P(x) = (x-1)(x+1)^2.$$

Multiplying by x^k, we obtain the above solutions.

333. Using the tangent line method, we have

$$\sum \frac{a^2}{(3-a)^3} \geq \sum \left(\frac{7a}{16} - \frac{5}{16} \right) = \frac{3}{8}.$$

334. We can assume that $b \geq c$, so $a \geq b \geq c \geq d$. Consider the function

$$f(x) = (x+a)(x+d) + (x+b)(x+c).$$

Then $f(-d) = (b-d)(c-d) > 0$, $f(-b) = (a-b)(d-b) < 0$. By the Intermediate Value Theorem, f has at least one real root. But then both roots are real, since f is a quadratic function with real coefficients.

335. Answer: $f(x) = cx^2$ for all x, with $c \leq 0$.

It is easy to see that these are solutions. Conversely, let f be a solution and denote $g(x) = \frac{f(x)}{x}$. Then

$$2(xg(x) + yg(y)) \leq (x+y)g(x+y), g(x+y) \leq g(x) + g(y),$$

hence

$$2(xg(x) + yg(y)) \leq (x+y)g(x+y) \leq (x+y)(g(x) + g(y)).$$

It follows that $(x-y)(g(x) - g(y)) \leq 0$, so g is a nonincreasing function. For $x = y$ we obtain $g(2x) = 2g(x)$ and by induction, $g(2^n x) = 2^n g(x)$ for any positive integer n. We will prove by induction that $g(nx) = ng(x)$ for any positive integer n. The assertion holds for all powers of 2. Let $n = 2^q + r$, $0 \leq r < 2^q$. By the induction hypothesis, $g(kx) = kg(x)$ for all $k < n$. Then

$$\begin{aligned} 2^{q+1}g(x) &= g(2^{q+1}x) \leq g((2^q+r)x) + g((2^q-r)x) = g(nx) + (2^q-r)g(x) \\ &\leq (2^q+r)g(x) + (2^q-r)g(x) = 2^{q+1}g(x), \end{aligned}$$

whence $g(nx) + (2^q - r)g(x) = 2^{q+1}g(x)$, so $g(nx) = ng(x)$.

Now,

$$g\left(\frac{m}{n}\right) = mg\left(\frac{1}{n}\right) = \frac{m}{n}g(1)$$

for all positive integers m, n, so $g(x) = xg(1)$ for all positive rational numbers x. Since g is nonincreasing, we deduce $g(1) \leq 0$.

Let x be a positive real number and (y_k), (z_k) be sequences of positive rational numbers such that

$$y_k < x < z_k, \forall k \in \mathbb{N}^*, \quad \lim_{k\to\infty} y_k = \lim_{k\to\infty} z_k = x.$$

We have $g(z_k) \le g(x) \le g(y_k)$, since g is non-increasing. From

$$\lim_{k\to\infty} g(y_k) = \lim_{k\to\infty} g(z_k) = xg(1),$$

we deduce $g(x) = xg(1)$, so $f(x) = x^2 g(1)$.

336. The given polynomial can be written as a product of two polynomials with integer coefficients:

$$\begin{aligned}
X^8 + 98X^4 + 1 &= \left(X^4+1\right)^2 + 96X^4 = \left(X^4+1\right)^2 + 16\left(X^4+1\right)X^2 + 64x^4 \\
&- 16\left(X^4+1\right)X^2 + 32x^4 = \left(x^4 + 8X^2 + 1\right)^2 - 16\left(X^4 - 2X^2 + 1\right)X^2 \\
&= \left(X^4 + 8X^2 + 1\right)^2 - \left(4X^3 - 4X\right)^2 = \left(X^4 - 4X^3 + 8X^2 + 4X + 1\right) \\
&\times \left(X^4 + 4X^3 + 8X^2 - 4X + 1\right).
\end{aligned}$$

337. Answer: no.

Let x_1, x_2 be the roots of P. Then the roots of $P(Q)$ are the roots of $Q - x_1$ and $Q - x_2$. By the Vìete′s relations, the sum of the roots of $Q - x_1$ equals the sum of the roots of $Q - x_2$.Since $2+3+5+7=17$ is odd, we cannot divide the set of the roots into two subsets with equal sums, whence the conclusion.

338. From $0 \le x^3 = \{(x+1)^3\} < 1$, it follows that $0 \le x < 1$. If the fractional part of $(x+1)^3 = x^3 + 3x^2 + 3x + 1$ is x^3, then $3x^2 + 3x + 1$ is an integer, that is, $3x^2 + 3x$ is an integer. For $0 \le x < 1$, we have $0 \le 3x^2 + 3x < 6$. The solutions are

$$x = \frac{-3 + \sqrt{9 + 12k}}{6}, k \in \{0, 1, 2, 3, 4, 5\}.$$

339. Let $c = g(a)$, $d = g(b)$. Then c, d are fixed points for $g \circ f$. If $a = b$, then the required equality is true. If $a \ne b$, then

$$a - b | g(a) - g(b) = c - d | f(c) - f(d) = a - b,$$

therefore $|a - b| = |c - d|$, hence $a + d = b + c$ or $a + c = b + d$.

340. We have

$$\begin{aligned}
\text{RHS} &\le \frac{16abc(a+b+c)}{(a+b)(b+c)(c+a)} \le \frac{18abc(a+b+c)}{(a+b+c)(ab+bc+ca)} \\
&\le 2\sqrt{3(ab+bc+ca)} \le \text{LHS},
\end{aligned}$$

which is true, since $\sqrt{3(ab+bc+ca)} \le a+b+c$ and $ab+bc+ca \le 1$.

341. The matrices

$$A=\begin{pmatrix} 1 & c & b \\ c & 1 & a \\ b & a & 1 \end{pmatrix},\ B=\begin{pmatrix} 1 & z & y \\ z & 1 & x \\ y & x & 1 \end{pmatrix}$$

are positive semidefinite, since all principal minors are nonnegative. The Hadamard product is defined componentwise:

$$A \bigodot B=\begin{pmatrix} 1 & cz & by \\ cz & 1 & ax \\ by & ax & 1 \end{pmatrix}.$$

By the Schur product theorem, the matrix $A \bigodot B$ is also positive semidefinite, so

$$1+2(ax)(by)(cz)\geq (ax)^2+(by)^2+(cz)^2.$$

3.42 2013

342. The assertion is false. Choose $a>1, b=\frac{1}{a}, c=d=1$. Then $abcd=1$ and the inequality becomes

$$a^x+\left(\frac{1}{a}\right)^x+1+1\geq a+\frac{1}{a}+1+1.$$

We have

$$a^x+\frac{1}{a^x}-a-\frac{1}{a}=\frac{(a^x-a)(a^{x+1}-1)}{a^{x+1}}\geq 0, \forall x\geq 1.$$

343. Let $P(x,y)$ be the given equation. Then

$$P(x,-x)\Rightarrow f(x^3)+f(-x^3)=0\Rightarrow f(x)+f(-x)=0\Rightarrow f(0)=0,$$
$$P(x,0)\Rightarrow f(x^3)=xf(x^2).$$

Now $P(x,y)$ becomes $(x+y)f(xy)=xf(y^2)+yf(x^2)$.

$$P(x,-y)\Rightarrow (y-x)f(xy)=xf(y^2)-yf(x^2)\Rightarrow yf(xy)=xf(y^2),$$

$$P(x,1) \Rightarrow f(x) = xf(1) \Rightarrow f(x) = cx, c \in \mathbb{R}.$$

344. We will apply the Lagrange multipliers' method. Let

$$U = [0,1]^n, f(x_1, \ldots, x_n) = F_n, g(x_1, \ldots, x_n) = x_1 + \ldots + x_n, G = \{x \in U | g(x) = 1\}.$$

Now,

$$\nabla g(x) \neq 0, \forall x \in U, \nabla f(x) = \lambda \nabla g(x) \Rightarrow x_1 = \ldots = x_n.$$

The minimum is achieved for $x_i = \frac{1}{n}, 1 \leq \leq n,$ hence $F_n \geq -\frac{1}{n}$.

$$a) \min F_3 = -\frac{1}{3}; b) \min F_4 = -\frac{1}{4}; c) \min F_5 = -\frac{1}{5}.$$

345. Let $f(x) = x^2 - 2$ and assume that $\frac{m}{n}$ is a rational approximation of $\sqrt{2}$ (that is, if n is fixed, then m is optimal). By the Mean Value Theorem, there exists r between $\frac{m}{n}$ and $\sqrt{2}$ such that

$$\frac{f\left(\frac{m}{n}\right) - f\left(\sqrt{2}\right)}{\frac{m}{n} - \sqrt{2}} = f'(r),$$

which implies

$$\left|\sqrt{2} - \frac{m}{n}\right| = \frac{|m^2 - 2n^2|}{2|r|n^2}.$$

The minimal value of n is 1, which means that $|r| \leq \sqrt{2} - 1$, therefore

$$\left|\sqrt{2} - \frac{m}{n}\right| < \frac{|m^2 - 2n^2|}{n^2}$$

for all rational approximations of $\frac{m}{n}$.

Consider the sequence $\frac{m_i}{n_i}$. Then $|m_i^2 - 2n_i^2| \in \{1, 2, \ldots, \lfloor C \rfloor\}$. The smallest solution of the equation $|x^2 - 2y^2| = 1$ is $(x, y) = (1, 1)$. Since the fundamental unit of $\mathbb{Z}\left[\sqrt{2}\right]$ is $3 + 2\sqrt{2}$, the kth smallest solution y of the Pell's equation $x^2 - 2y^2 = 1$ is

$$\frac{\left(1+\sqrt{2}\right)\left(3+2\sqrt{2}\right)^{k-1} - \left(1-\sqrt{2}\right)\left(3-2\sqrt{2}\right)^{k-1}}{2\sqrt{2}} > 3^{k-1},$$

so we can take $q = \sqrt[\lfloor C \rfloor]{3}$.

346. Answer: 1.

We have

$$\sum_{i=1}^{n} c_i a_i^k = a_n^k \sum_{i=1}^{n} c_i + \sum_{i=1}^{n-1}\left(\sum_{k=1}^{i} c_i\right)\left(a_i^k - a_{i+1}^k\right) \le a_n^k n^k + \sum_{i=1}^{n-1} i^k\left(a_i^k - a_{i+1}^k\right) = \sum_{i=1}^{n} \times \left(i^k - (i-1)^k\right)a_i^k.$$

Now,

$$(a_1 + a_2)^k \ge a_1^k + (2^k - 1)a_2^k,$$
$$(a_1 + a_2 + a_3)^k \ge (a_1 + a_2)^k + (3^k - 2^k)a_3^k,$$
$$\vdots$$
$$1 = \left(\sum_{i=1}^{n} a_i\right)^k \ge \left(\sum_{i=1}^{n-1} a_i\right)^k + \left[n^k - (n-1)^k\right]a_n^k$$

Adding these inequalities, we get the result. Equality holds when

$$\sum_{i=1}^{s} c_i = s^k, a_1 = a_2 = \ldots = a_s = \frac{1}{s}, a_{s+1} = \ldots = a_n = 0, 1 \le s \le n.$$

347. Answer: $r = \frac{a}{b}, a, b \in \mathbb{Z}, b \neq 0, 3|a+b; r$ is irrational.

Case 1. r rational. If $r = \frac{a}{b}, a, b \in \mathbb{Z}, b \neq 0, 3|a+b,$ then take

$$S = \bigcup_{n \ge 0}\left[\frac{n}{b}, \frac{n+1}{b}\right).$$

Suppose that $3 \nmid a+b$. If $t \in S$, then $t + r, t + 1 \notin S$. Exactly one of $t + r + 1$, $t + 1$, $t + + r$ belongs to S, so $t + r + 1 \in S$. Conversely, $t \in S$ if and only if $t + r + 1 \in S$. Now, either $t + 2r$ or $t + r + 1$ belongs to S, but $t + r + 1 \in S$, so $t + 2r \notin S$. Either $t + 3r$ or $t + 2r + 1$ is in S, but $t + 2r + 1 \notin S$, so $t + 3r \in S$. Continuing, we conclude that $t \in S$ if and only if $t + cr + d \in S$, for $3|c - d$. Then $t \in S$ implies $t + dr - c \notin S$, since $3 \nmid a + b$, but $dr - c = 0$, so $t \notin S$, contradiction.

Case 2. r irrational. Let

$$Q = \{cr + d | c, d \in \mathbb{Z}\}, Q' = \{cr + d | c, d \in \mathbb{Z}, 3|c - d\},$$
$$R = \{x | x \in [0,1), \nexists y \in \mathrm{R}, q \in Q, x = y + q\}.$$

Then

$$S=\{x+q'|x\in R, q'\in Q'\}$$

satisfies the required conditions, since $cr + r = er + f$, $c, d, e, f \in \mathbb{Z}$, implies $c = e$, $d = f$.

348. Setting $y = x - 1$, we get $f(x) - f(x - 1) \geq f(1) \geq 0$, so

$$f(x)\geq f(x-1). \quad (1)$$

Setting $y = 0$, we get $f(x) - f(0) \geq xf(x)$ (2). Replacing (x, y) by $(0, x)$, we get

$$f(0)-f(x)\geq -xf(-x). \quad (3)$$

Adding (2) and (3), we get $0 \geq xf(x) - xf(-x)$. Then for $x > 0$, we get

$$f(-x)\geq f(x). \ (4)$$

Setting $(x, y) = (1, 0)$ in the enounce, we get $f(0) \leq 0$ (5). By (5), (1), (4), we get

$$0\geq f(0)\geq f(-1)\geq f(1)\geq 0,$$

so $f(0) = f(-1) = f(1) = 0$. Using (1) repeatedly, we get

$$f(x)\geq f(x-1)\geq f(x-2)\geq \ldots, (6)$$

that is, $f(x) \geq f(x - k)$ for all real x and positive integer k. Using (6), (1) and replacing (x, y) by $(x - 1, -1)$ in the enounce, we get

$$f(x)\geq f(x-1)=f(x-1)-f(-1)\geq xf(x).$$

Then $f(x)(x - 1) \leq 0$. It follows that $f(x) \leq 0$ for $x > 1$ and $f(x) \geq 0$ for $x < 1$. For $x > 1$, there is $y < 1$ such that $k = x - y$ is a positive integer. Then

$$0\geq f(x)\geq f(x-k)=f(y)\geq 0.$$

So, $f(x) = 0$ for $x > 1$. Similarly, for $x < 1$, there is $y > 1$ such that $k = y - x$ is a positive integer. As above, $f(x) = 0$ for all x, which is indeed a solution.

349. (*i*) By the AM–GM inequality,

$$\left(\sum_{k=1}^{n}\frac{x_k}{k}\right)\left(\sum_{k=1}^{n}kx_k\right)=\frac{1}{n}\left(\sum_{k=1}^{n}\frac{nx_k}{k}\right)\left(\sum_{k=1}^{n}kx_k\right)\leq\frac{1}{n}\cdot\frac{1}{4}\left(\sum_{k=1}^{n}\frac{nx_k}{k}+\sum_{k=1}^{n}kx_k\right)^2$$
$$=\frac{1}{4n}\left(\sum_{k=1}^{n}\left(\frac{n}{k}+k\right)x_k\right)^2\leq\frac{(n+1)^2}{4n}\left(\sum_{k=1}^{n}x_k\right)^2,$$

which is true, since the inequality $\frac{n}{k} + k \leq n + 1$ is equivalent to $(n-k)(k-1) \geq 0$.

By the Cauchy–Schwarz inequality,

$$\left(\sum_{k=1}^{n} \frac{x_k}{k}\right)\left(\sum_{k=1}^{n} kx_k\right) \geq \left(\sum_{k=1}^{n} \sqrt{\frac{x_k}{k}} \cdot \sqrt{kx_k}\right)^2 = \left(\sum_{k=1}^{n} x_k\right)^2.$$

(*ii*) The upper bound is achieved, for instance, when $x_1 = x_n = 1$, $x_2 = \ldots = x_{n-1} = 0$. The lower bound is attained if exactly one of the x_is is nonzero.

350. Induction on n. For $n = 1$ let $x_1, y_1 \in [0, 1]$ and $x_2 = 1$. We have to prove that $(x_1 - y_1) + (x_2 - x_1)y_1 \geq 0$:

$$(x_1 - y_1) + (x_2 - x_1)y_1 = (x_1 - y_1) + (1 - x_1)y_1 = x_1(1 - y_1) \geq 0.$$

Assume that for some $n \geq 1$ and for any real numbers $0 \leq a_1 \leq a_2 \leq \ldots \leq a_n \leq 1$ and $0 \leq b_1 \leq b_2 \leq \ldots \leq b_n \leq 1$ with $a_{n+1} = 1$ it holds

$$\sum_{i=1}^{n} (a_i - b_i) + n\sum_{i=1}^{n} (a_{i+1} - a_i)b_i \geq 0.$$

Consider real numbers $0 \leq x_1 \leq \ldots \leq x_{n+1} \leq 1$ and $0 \leq y_1 \leq \ldots \leq y_{n+1} \leq 1$ with $x_{n+2} = 1$. Then

$$S_{n+1} := \sum_{i=1}^{n+1} (x_i - y_i) + (n+1)\sum_{i=1}^{n+1} (x_{i+1} - x_i)y_i = \sum_{i=2}^{n+2} (x_i - y_i) + n\sum_{i=2}^{n+1} (x_{i+1} - x_i)y_i$$

$$+ (x_1 - y_1) + + \sum_{i=2}^{n+1} (x_{i+1} - x_i)y_i + (n+1)(x_2 - x_1)y_1 \geq (x_1 - y_1) + \sum_{i=2}^{n+1}$$

$$\times (x_{i+1} - x_i)y_i + (n+1)(x_2 - x_1)y_1,$$

by the induction hypothesis. Further,

$$S_{n+1} \geq (x_1 - y_1)\sum_{i=2}^{n} (x_{i+1} - x_i)y_1 + (n+1)(x_2 - x_1)y_1 = x_1$$

$$+ y_1\left[-1 + (n+1)(x_2 - x_1) + \sum_{i=2}^{n+1} (x_{i+1} - x_i)\right] = x_1 + y_1[nx_2 - (n+1)x_1]$$

$$= x_1(1 - y_1) + ny_1(x_2 - x_1) \geq 0,$$

which ends the proof.

351. Answer: $\frac{3}{\sqrt{3}+1}$.

By the Cauchy–Schwarz inequality,

$$\sum \frac{x}{1+\frac{yz}{x}} = \sum \frac{x^2}{x+yz} \geq \frac{(x+y+z)^2}{x+y+z+1}.$$

The function $f(x) = \frac{x^2}{x+1}$ is increasing and $x+y+z \geq \sqrt{3}$, so

$$\frac{(x+y+z)^2}{x+y+z+1} \geq \frac{3}{\sqrt{3}+1}.$$

Equality holds only when $x=y=z=\frac{1}{\sqrt{3}}$.

352. Expanding, we have $P = x^4 - 6x^3 + 33x^2 - 60x + 101$. Now, the required result follows from a more general assertion:

Let $f(x) = a_n x^n + a_{n-1}x^{n-1} + \ldots + a_1 x + a_0$, $a_n \neq 0$, be a polynomial with integer coefficients. If $|a_0|$ is a prime number and $|a_0| > |a_1| + \ldots + |a_n|$, then f is irreducible.

Indeed, if $\alpha \in \mathbb{C}$ is a root of f and $|\alpha| \leq 1$, then

$$|a_0| = |a_1\alpha + \ldots + a_n\alpha^n| \leq |a_1| + \ldots + |a_n|,$$

contradiction. So all roots of f satisfy $|\alpha| > 1$. Suppose that $f = gh$, with g, h nonconstant polynomials with integer coefficients. Then $a_0 = f(0)$-$= g(0)h(0)$, so $|g(0)| = 1$ or $|h(0)| = 1$, since $|a_0|$ is a prime number. If $|g(0)| = 1$, then let b be the leading coefficient of g and $\alpha_1, \ldots, \alpha_k$ be the roots of g. By the Viète's relations,

$$|\alpha_1 \ldots \alpha_k| = \frac{1}{|b|} \leq 1,$$

contradiction, since $\alpha_1, \ldots, \alpha_k$ are also roots of f. We conclude that f is irreducible.

353. Observe that $\bigoplus$ is the addition modulo 2 and replace it by" + ". Let $n = (a_1 a_2 \ldots a_k)$, then $\lfloor \frac{n}{2} \rfloor = (0a_1 \ldots a_{k-1})$ and so

$$f(n) = (a_1(a_2 + a_1)(a_3 + a_2)\ldots(a_{k-1} + a_k)).$$

- f is injective: if $x = (b_1 \ldots b_k)$ and $f(x) = f(n)$, then comparing the digits one by one, we see that x and n have the same digits;
- f is surjective: this is true, because the representation in base 2 exists and is unique.

So f is bijective.

354. Clearly, $x \neq 0, 1$. From the first equation, we get

$$y = \frac{1 - 2x^2}{x}.$$

Replacing this in the second equation, we obtain, after calculations,

$$\left(2x^2 - 4x + 5\right)\left(2x^2 + 2x - 1\right) = 0.$$

The first factor has no real roots, the roots of the second are

$$x_{1,2} = \frac{-1 \pm \sqrt{3}}{2}.$$

Now, it is easy to find the corresponding values of y.

355. We will prove a more general statement:

$$\sum_{n=0}^{m} \frac{(2m)!}{[n!(m-n)!]^2} = \binom{2m}{m}^2. \quad (1)$$

We have

$$\frac{(2m)!}{[n!(m-n)!]^2} = \frac{(m!)^2}{[n!(m-n)!]^2} \cdot \frac{(2m)!}{(m!)^2} = \binom{m}{n}^2 \cdot \binom{2m}{m}.$$

Hence it suffices to show that

$$\sum_{n=0}^{m} \binom{m}{n}^2 = \binom{2m}{m}.$$

We do this combinatorially: consider $2m$ balls numbered from 1 to $2m$.Balls 1 to m are blue, balls $m+1$ to $2m$ are red. We can choose m balls from the $2m$ balls in $\binom{2m}{m}$ ways. On the other hand, we can also first choose n blue balls, $0 \leq n \leq m$, and then choose $m-n$ red balls. Equivalently, wecan choose n blue balls to include, and n red balls to not include. Hence the number of ways in which one can choose m balls is also equal to

$$\sum_{n=0}^{m} \binom{m}{n}^2,$$

therefore this sum equals $\binom{2m}{m}$, as desired.

356. Subtract the fourth equation from the first and the third equation from the second:

$$(b-d)(a-1) = -3, (b-d)(c-1) = 3.$$

We see that $b - d \neq 0$. Adding these equations, we get $a + c = 2$. Plugging this value in the sum of the first and the second equation, we obtain $b + d = 3$. Now, $c = 2 - a$, $d = 3 - b$ and the given system becomes

$$\begin{cases} ab - a - b = -2 \\ ab - 2a - b = -4 \end{cases}$$

with solution $(a, b) = (2, 0)$.

The solution of the given system is $(a, b, c, d) = (2, 0, 0, 3)$.

357. Putting $\varepsilon_i = a_i - 1$, $1 \le i \le k$, gives

$$p(z) - z = \sum_{i=1}^{k} \varepsilon_i x_i 4^{i-1}.$$

Let the number of solutions of $p(z) - z = 0, p(z) - z = -1$ be $f(\varepsilon_k, \varepsilon_{k-1}, \ldots, \varepsilon_1)$, $g(\varepsilon_k, \varepsilon_{k-1}, \ldots, \varepsilon_1)$, respectively, where $z \in \{0, 1, 2, \ldots, 4^k - 1\}$. We will prove by induction on k that $f(\varepsilon_k, \varepsilon_{k-1}, \ldots, \varepsilon_1)$ and $f(\varepsilon_k, \varepsilon_{k-1}, \ldots, \varepsilon_1) + g(\varepsilon_k, \varepsilon_{k-1}, \ldots, \varepsilon_1)$ are powers of 2. For $k = 1$, it holds true, since $f(-1) = f(1) = f(2) = 1, f(0) = 4$; $g(-1) = 1$, $g(0) = g(1) = g(2) = 0$. Suppose that it holds true for all integers greater than 1 and less than k. Then

$$\varepsilon_1 = 0 \Rightarrow f(\varepsilon_k, \ldots, \varepsilon_2, 0) = 4f(\varepsilon_k, \ldots, \varepsilon_2), g(\varepsilon_k, \ldots, \varepsilon_2, 0) = 0;$$

$$\varepsilon_1 = \pm 1 \Rightarrow f(\varepsilon_k, \ldots, \varepsilon_2, \pm 1) = f(\varepsilon_k, \ldots, \varepsilon_2), g(\varepsilon_k, \ldots, \varepsilon_2, -1) = f(\varepsilon_k, \ldots, \varepsilon_2),$$
$$g(\varepsilon_k, \ldots, \varepsilon_1, 1) = g(\varepsilon_k, \ldots, \varepsilon_2);$$

$$\varepsilon_1 = 2 \Rightarrow f(\varepsilon_k, \ldots, \varepsilon_2, 2) = f(\varepsilon_k, \ldots, \varepsilon_2) + g(\varepsilon_k, \ldots, \varepsilon_2), g(\varepsilon_k, \ldots, \varepsilon_2, 2) = 0$$

$$\left(\begin{array}{c} x_1 = 0 \Rightarrow \varepsilon_1 x_1 = 0, \sum_{i=2}^{k} \varepsilon_i x_i 4^{i-2} = 0, \\ x_1 = 2 \Rightarrow \varepsilon_1 x_1 = 4, \sum_{i=2}^{k} \varepsilon_i x_i 4^{i-2} = -1 \end{array} \right).$$

To sum up, $f(\varepsilon_k, \varepsilon_{k-1}, \ldots, \varepsilon_1)$ and $f(\varepsilon_k, \varepsilon_{k-1}, \ldots, \varepsilon_1) + g(\varepsilon_k, \varepsilon_{k-1}, \ldots, \varepsilon_1)$ are powers of 2 for all k.

358. Suppose that the assertion is false. Then

$$-1 < a^i - P(i) < 1, 0 \le i \le n+1 \Rightarrow -\binom{n+1}{i} < (-1)^i \binom{n+1}{i} (a^i - P(i)) < \binom{n+1}{i}.$$

Sum from $i = 0$ to $n + 1$ and use the fact that the $(n + 1)$th partial difference of a degree n polynomial is 0 to find

$$-2^{n+1} < \sum_{i=0}^{n+1} (-1)^i \binom{n+1}{i} a^i = (1-a)^{n+1} < 2^{n+1},$$

which implies $|1 - a| < 2$, contradiction with $a \geq 3$.

359. Apply induction on n. For $n = 0$ there are nothing to prove, since the relation $u - v|P(u) - P(v)$ is always true. Suppose that

$$u^{2^n} - v^{2^n} \Big| u^{2^{n+1}} - v^{2^{n+1}} | P(u) - P(v)$$

for all nonnegative integers u, v. By the induction hypothesis, there is a polynomial Q' with integer coefficients, such that $P(x) = Q'\left(x^{2^n}\right)$. Then $a - b|Q'(a) - Q'(b)$ for all integers a, b. Taking $a = u^{2^n}, b = -v^{2^n}$, we get $u^{2^n} + v^{2^n} | Q'\left(u^{2^n}\right) - Q'\left(-v^{2^n}\right)$. On the other hand, $u^{2^n} + v^{2^n} \Big| u^{2^{n+1}} - v^{2^{n+1}} | Q'\left(u^{2^n}\right) - Q'\left(v^{2^n}\right)$ for all positive integers u, v. But then $u^{2^n} + v^{2^n} | Q'\left(v^{2^n}\right) - Q'\left(-v^{2^n}\right)$ for all positive integers u, v. Fix v and take $u \to \infty$, then we see that $Q'\left(v^{2^n}\right) = Q'\left(-v^{2^n}\right)$ for infinitely many values v, hence $Q'(x) = Q'(-x)$, that is, there exists a polynomial Q with integer coefficients such that $Q'(x) = Q(x^2)$. Then $P(x) = Q'\left(x^{2^n}\right) = Q\left(x^{2^{n+1}}\right)$, as desired.

360. Answer: $K = 1$.

The case $|x| = 2$ is obvious. If $|x| \neq 2$, then

$$xy + y^2 + z^2 + 1 - z(xz + yz + y) = \left(x + y - \frac{x + z^2}{2}\right)^2$$
$$+ \frac{4 - x^2}{4}\left(1 - \frac{z(xz + 2y)}{4 - x^2}\right)^2 + \frac{(4 - x^2 - y^2 - z^2 - xyz)z^2}{4 - x^2} \geq 0.$$

Equality holds when $x = \frac{1+\sqrt{5}}{2}, y = z = \frac{1-\sqrt{5}}{2}$.

361. Answer: $k = 13$.

For $x = y = t, z = \frac{1}{t^2} t > 0, t \neq 1,$ we have the equivalences

$$t^2 + \frac{2}{t} + \frac{k}{2t + \frac{1}{t^2} + 1} \geq 3 + \frac{k}{4} \Leftrightarrow t^2 + \frac{2}{t} - 3 \geq k\left(\frac{1}{4} - \frac{t^2}{2t^3 + t^2 + 1}\right) \Leftrightarrow k$$
$$\leq 4\left(t^2 + 2t + \frac{2}{t} - \frac{3}{2t + 1}\right).$$

Choose $t = \frac{2}{3}$, then $k \leq \frac{880}{63} < 14$, hence $k \leq 13$.

Now, we need to show that $abc = 1$ implies

$$\frac{1}{a}+\frac{1}{b}+\frac{1}{c}+\frac{13}{a+b+c+1} \geq \frac{25}{4}.$$

Replace (a, b, c) by $\left(\frac{a^2}{bc}, \frac{b^2}{ca}, \frac{c^2}{ab}\right)$, then

$$\frac{bc}{a^2}+\frac{ca}{b^2}+\frac{ab}{c^2}+\frac{13abc}{a^3+b^3+c^3+abc} \geq \frac{25}{4} \Leftrightarrow \frac{a^3b^3+b^3c^3+c^3a^3-3a^2b^2c^2}{a^2b^2c^2}$$
$$\geq \frac{13\left(a^3+b^3+c^3-3abc\right)}{4\left(a^3+b^3+c^3+abc\right)}.$$

If $c = \max\{a, b, c\}$, then

$$a^3b^3+b^3c^3+c^3a^3-3a^2b^2c^2=(ab+bc+ca)\left[c^2(a-b)+ab(c-a)(c-b)\right]$$
$$\geq (ab+bc+ca)\cdot ab\cdot\left[(a-b)^2+(c-a)(c-b)\right].$$

It suffices to prove

$$\frac{(ab+bc+ca)\left[(a-b)^2+(c-a)(c-b)\right]}{abc^2}$$
$$\geq \frac{13(a+b+c)\left[(a-b)^2+(c-a)(c-b)\right]}{4\left(a^3+b^3+c^3\right)},$$

which is equivalent to

$$4(ab+bc+ca)\left(a^3+b^3+c^3+abc\right) \geq 13abc^2(a+b+c).$$

But $(ab + bc + ca)^2 \geq 3abc(a + b + c)$, therefore we need to prove that

$$12\left(a^3+b^3+c^3+abc\right) \geq 13c(ab+bc+ca) \Leftrightarrow 12\left(a^3+b^3+c^3\right) \geq 13c^2(a+b)$$
$$+abc.$$

We have

$$\frac{18}{\sqrt[3]{2}}\cdot c^2(a+b) \geq 13c^2(a+b)+abc \Leftrightarrow \left(\frac{18}{\sqrt[3]{2}}-13\right)c(a+b) \geq ab,$$

which is true.

3.43 2014

362. Denote

$$H_n = \sum_{i=1}^{n} \frac{1}{i}.$$

Then we have the equivalences

$$\frac{1}{n+1}\left(H_{2n} - \frac{1}{2}H_n\right) > \frac{1}{2n}H_n \Leftrightarrow \frac{2n}{n+1}H_{2n} > \frac{2n+1}{n+1}H_n \Leftrightarrow 2n\sum_{i=n+1}^{2n}\frac{1}{i} > H_n,$$

which is true.

363. *a*), *b*). Simple calculations show that the values of the expressions from *a*), *b*) are respectively, 1, − 1.

c) Rewrite the inequality in the form

$$\frac{(1+ab)^2+(1-ab)^2}{(a-b)^2}+\frac{(1+bc)^2+(1-bc)^2}{(b-c)^2}+\frac{(1+ca)^2+(1-ca)^2}{(c-a)^2} \geq 3.$$

For real numbers x, y, z, we have

$$x^2+y^2+z^2 \geq xy+yz+zx, x^2+y^2+z^2 \geq -2(xy+yz+zx),$$

hence

$$\sum\left(\frac{1+ab}{a-b}\right)^2 \geq \sum\frac{1+ab}{a-b}\cdot\frac{1+bc}{b-c} = 1,$$

$$\sum\left(\frac{1-ab}{a-b}\right)^2 \geq -2\left(\sum\frac{1-ab}{a-b}\cdot\frac{1-bc}{b-c}\right) = 2.$$

Adding these inequalities, we get the desired inequality.

364. Answer: $\lambda(2k) = \frac{k}{2}, \lambda(2k+1) = \frac{2k+1}{\left(2\sin\frac{k\pi}{2k+1}\right)^2}$, k positive integer.

Case 1. n even. Let $z_j = a_j + b_j i$, a_j, $b_j \in \mathbb{R}$, $1 \leq j \leq n$. Then (indices are taken mod n)

$$\sum_{j=1}^{n}|z_j|^2 = \sum_{j=1}^{n}\left(a_j^2+b_j^2\right) \geq \frac{1}{2}\left(\sum_{j=1}^{n}\left(a_j^2+b_j^2-a_ja_{j+1}-b_jb_{j+1}\right)\right) = \frac{1}{4}$$

$$\times\sum_{j=1}^{n}\left|z_j-z_{j+1}\right|^2 \geq \frac{n}{4}\min_{1\leq j\leq n}\left\{\left|z_j-z_{j+1}\right|^2\right\}.$$

Equality holds when $z_j + z_{j+1} = 0$ for all j.

Case 2. n odd. The equality is attained when z_j are the vertices of a regular star polygon, where every "spoke" of the star has measure $\frac{\pi}{2k+1}$. Consider a configuration when $\left|z_{j+1}-z_j\right|$ are fixed and minimize $\sum_{j=1}^{2k+1}\left|z_j\right|^2$. The configuration can be translated so that its center is at the origin.

Now, consider a configuration of z_j when $|z_j|$ are fixed and we want to maximize

$$\min_{1\leq j\leq 2k+1}\left\{\left|z_{j+1}-z_j\right|^2\right\}.$$

Assume that m differences $|z_{j+1}-z_j|$ attain this maximum. Then there is a sequence z_j, z_{j+1}, z_{j+2} such that $|z_{j+1}-z_j|^2$ attains the maximum and $|z_{j+2}-z_{j+1}|^2$ does not. We can smooth z_{j+1} by rotating it around the origin such that

$$\left|z_{j+1}-z_j\right|=\left|z_{j+2}-z_{j+1}\right|,$$

which means that now $m-1$ of these differences attain the maximum. We continue this process until 0 differences attain the maximum, contradiction. We conclude that $|z_{j+1}-z_j|$ has the same value x for each j.

Let $\theta_j=\angle\,(z_j, z_{j+1})$. Then

$$\min_{1\leq j\leq 2k+1}\left\{\left|z_{j+1}-z_j\right|^2\right\}$$

is maximized when each angle is obtuse and

$$\sum_{j=1}^{2k+1}\theta_k=2k\pi.$$

By the law of cosines,

$$x^2=\left|z_j\right|^2+\left|z_{j+1}\right|^2-2\left|z_j\right|\left|z_{j+1}\right|\cos\theta_j\leq\left(\left|z_j\right|^2+\left|z_{j+1}\right|^2\right)(1-\cos\theta_j).$$

By the Jensen's inequality,

$$\sum_{j=1}^{2k+1}\left|z_j\right|^2\geq\frac{x^2}{2}\sum_{j=1}^{2k+1}\frac{1}{1-\cos\theta_j}\geq\frac{x^2}{2}\left(\frac{2k+1}{1-\cos\frac{2k\pi}{2k+1}}\right)=\frac{(2k+1)x^2}{\left(2\sin\frac{k\pi}{2k+1}\right)^2}.$$

365. The condition $a+b+c=1$ implies $a, b, c\in(0,1)$, hence

$$b^3 < b, c^4 < c, b^3 + c^4 + 1 < b + c + 1 = 2 - a, \frac{a^2}{b^3 + c^4 + 1} > \frac{a^2}{2 - a} = -2 - a + \frac{4}{2 - a}.$$

Adding with two similar inequalities, we get

$$E := \frac{a^2}{b^3 + c^4 + 1} + \frac{b^2}{c^3 + a^4 + 1} + \frac{c^2}{a^3 + b^4} > -7 + 4\left(\frac{1}{2 - a} + \frac{1}{2 - b} + \frac{1}{2 - c}\right).$$

The AM–HM inequality yields

$$\frac{1}{2 - a} + \frac{1}{2 - b} + \frac{1}{2 - c} \geq 3 \cdot \frac{3}{(2 - a) + (2 - b) + (2 - c)} = \frac{9}{5}.$$

Finally,

$$E > -7 + 4 \cdot \frac{9}{5} = \frac{1}{5}.$$

366. Answer: $t \in \left[\frac{2}{3}, 2\right]$.
If $a = \varepsilon < 1$, $b = c = 1$, then

$$a^2 + bct < b^2 + cat + c^2 + abt \Longrightarrow \varepsilon^2 + t \leq 2 + 2\varepsilon t, \forall \varepsilon > 0 \Longrightarrow t \leq 2.$$

If $a = 2 - \varepsilon$, $\varepsilon > 0$, $b = c = 1$, then

$$b^2 + cat + c^2 + abt - \left(a^2 + bct\right) = (3 - 2\varepsilon)t + 2 - (2 - \varepsilon)^2 > 0, \forall \varepsilon > 0 \Longrightarrow t \geq \frac{2}{3}.$$

Now, choose $t \in \left[\frac{2}{3}, 2\right]$ and denote $2x = a - b + c, 2y = a + b - c, 2z = -a + b + c$. Then x, y, z are positive, $a = x + y, b = y + z, c = z + x$ and

$$b^2 + cat + c^2 + abt - \left(a^2 + bct\right) = (2 - t)z^2 + (x - y)^2 t + (3t - 2)xy + z(x + y) \times (2 + t) > 0$$

367. Since $a_1 = 1$ and $1 + 2 = 3 \cdot 1$, hence $a_2 \neq 2$. Thus, $a_2 = 3$ and $a_3 = 4$.
Since $5 + 4 = 6 + 3 = 3 \cdot 3$, hence $a_4 \neq 5, 6$, so $a_4 = 7$.
Since $8 + 1 = 3 \cdot 3$, $9 + 3 = 3 \cdot 4$, hence $a_5 \neq 8, 9$, so $a_5 = 10$.
Similar reasoning gives the first twelve terms of the sequence:

n	1	2	3	4	5	6	7	8	9	10	11	12
a_n	1	3	4	7	10	12	13	16	19	21	22	25

From above, we guess that for $r = 0, 1, 2, \ldots,$

$$a_{4r+1} = 9r + 1, a_{4r+2} = 9r + 3, a_{4r+3} = 9r + 4, a_{4r+4} = 9r + 7.$$

We will prove this by induction. The statements are true for $r = 0, 1, 2$. Assume that they are true for $r = 0, 1, 2, \ldots, m$ $(m \geq 2)$. This means that among $a_1, a_2, \ldots, a_{4m+4}$ none of terms are $\equiv 2 \pmod 3$, further, in the interval $[1, 9m + 7]$ all numbers $\equiv 1 \pmod 3$ appear in the sequence.

(a) Since $9m + 8 + 4 = 3(3m + 4)$, hence $a_{4m+5} \neq 9m + 8$. Similarly, since $9m + 9 + + 3 = 3(3m + 4)$, hence $a_{4m+5} \neq 9m + 9$. Actually, $a_{4m+5} = 9m + 10$, because if

$$(9m + 10) + y = 3z,$$

then $y \equiv 2 \pmod 3$ and y is not in the sequence.

(b) Since $(9m + 11) + 1 = 3(3m + 4)$, hence $a_{4m+6} \neq 9m + 11$. Actually, $a_{4m+6} = 9m + 12$, because $(9m + 12) + y = 3z$, then $y \equiv 0 \pmod 3$ and y is not in the sequence.

(c) Next, $a_{4m+7} = 9m + 13$, because $(9m + 13) + y = 3z$ implies $y \equiv 2 \pmod 3$ and y is not in the sequence.

(d) Since $(9m + 14) + 7 = 3(3m + 7)$, and $3m + 7 \equiv 1 \pmod 3$, is in the sequence, hence $a_{4m+8} \neq 9m + 14$. Since $(9m + 15) + (9m + 15) = 3(6m + 10)$ and $6m + 10 \equiv 1 \pmod 3$, hence $a_{4m+8} \neq 9m + 15$. Actually, $a_{4m+8} = 9m + 16$, because

$$(9m + 16) + y = 3z$$

implies $y \equiv 2 \pmod 3$ and y is not in the sequence.

We conclude that the proposition is true for all positive integers, hence

$$a_{2015} = a_{4 \cdot 503+3} = 9 \cdot 503 + 4 = 4531.$$

368. We will prove a stronger inequality:

$$(1 + a)^4 (1 + b)^4 \geq 64ab(a + b)^2.$$

Denote $a + b = 2u$, $ab = v^2$. Then

$$\left(1 + 2u + v^2\right)^4 \geq 256u^2v^2 \Leftrightarrow 1 + 2u + v^2 \geq 4\sqrt{uv} \Leftrightarrow 2\left(\sqrt{u} - \sqrt{v}\right)^2 + (v - 1)^2 \geq 0,$$

obvious.

369. If $x = 0$, then $1 - xy = 1^2$. If $x \neq 0$, denote $xy = a \in \mathbb{Q}$. Then

$$x^5 + \frac{a^5}{x^5} = 2a^2,$$

so that x^5, y^5 are solutions of the quadratic equation $t^2 - 2a^2t + a^5 = 0$. Since x^5, y^5 are both rational, then $\Delta = 4a^4(1 - a)$ must be the square of a rational number, therefore $1 - a = 1 - xy$ is the square of a rational number.

370. If $x + y = 0$, then from the second equation we get $xy = 1$, which is impossible. So, $x + y \neq 0$ and

$$z = \frac{1 - xy}{x + y}.$$

Replacing this in the first equation, we obtain

$$b^2y^2 + \left[(a^2 + b^2 - c^2)bx - 1\right]y + a^2x^2 + x - c^2 = 0.$$

The quadratic equation in y has a unique solution if and only if the discriminant

$$\Delta = \left[(a^2 + b^2 - c^2)^2 - 4a^2b^2\right]x^2 - 2(a^2 + b^2 - c^2)x + 1 - 4b^2c^2 = 0.$$

Also, the quadratic equation in x has a unique solution if and only if its discriminant vanishes:

$$\Delta' = a^4 + b^4 + c^4 - 2(a^2b^2 + b^2c^2 + c^2a^2) + 1 = 0,$$

whence it follows that there is a triangle with sidelengths equal to a, b, c, whose area is equal to 1 (see the Heron's formula).

371. If $x_n > \frac{1}{3}$, then

$$\frac{1}{3} \leq x_1 + \ldots + x_{n-1} = 1 - x_n < \frac{2}{3}$$

and $k = n - 1$.

If $x_n \leq \frac{1}{3}$, then let k be the smallest integer such that $s_k = x_1 + \ldots + x_k \geq \frac{2}{3}$. We deduce that s_{k-1} satisfies the required condition or $s_{k-1} < \frac{1}{3}$. But then

$$x_k = s_k - s_{k-1} > \frac{1}{3} \geq x_n,$$

contradiction.

372. The given equality implies $ab + b^2 + a^2 + 2abc = abc^2$. We cannot have $c = 1$, otherwise $a + b = 0$, contradiction. Then

$$\sqrt{ab} = \frac{a+b}{c-1}$$

is rational and

$$\sqrt{(c-3)(c+1)} = \sqrt{(c-1)^2 - 4} = \sqrt{\frac{(a+b)^2}{ab} - 4} = \frac{|a-b|}{\sqrt{ab}}$$

is rational.

373. Answer: $(-1, b)$, $b \in (-\infty, 0] \cup (2, +\infty)$.

We eliminate y:

$$(a^3+1)x^3 + 3a^2bx^2 + 3ab^2x + b^3 - 2 = 0.$$

If $a^3 + 1 \neq 0$, this is an equation of third degree, hence it has at least one real root x and $y = ax + b$ is also real. It follows that $a^3 + 1 = 0$, whence $a = -1$, and the equation becomes

$$3bx^2 - 3b^2x + b^3 - 2 = 0.$$

If $b = 0$, then the equation has no real roots. If $b \neq 0$, this is an equation of second degree without real roots if and only if the discriminant is negative:

$$\Delta = (-3b^2)^2 - 4 \cdot 3b \cdot (b^3 - 2) = 3b(2-b)(b^2 + 2b + 4),$$

that is, for $b < 0$ or $b > 2$.

374. Answer: $x_{\max} = \frac{5}{2}$.

Case 1. $x \leq 0$. Then $x_{\max} \leq 0$.

Case 2. $x > 0$. Then

$$x^2 + y^2 + z^2 + (x+y+z)^2 = 9 \Leftrightarrow x(x+2y+2z)^2 + 4\left(\frac{15}{32} - xyz\right) + \left(x - \frac{5}{2}\right) \times \left(3x^2 + \frac{15}{2}x + \frac{9}{4}\right) = 0,$$

so we have $x \leq \frac{5}{2}$ and $x = \frac{5}{2}$ if and only if $y + z = \frac{5}{4}, yz = \frac{3}{16}$, whence

$$(x, y, z) = \left(\frac{5}{2}, \frac{-5+\sqrt{13}}{8}, \frac{-5-\sqrt{13}}{8}\right).$$

375. Let $a_{n+1} = 0$ and $b_k = \sqrt{k}(a_k - a_{k+1})$, then $b_k \geq 0$ and

$$a_k = \frac{b_k}{\sqrt{k}} + \frac{b_{k+1}}{\sqrt{k+1}} + \ldots + \frac{b_n}{\sqrt{n}},$$

therefore

$$\text{LHS}^2 = \sum_{k=1}^{n} \left(\sum_{j=k}^{n} \frac{b_j}{\sqrt{j}} \right) = b_1^2 + \ldots + b_n^2 + \sum_{k<l} \sqrt{\frac{k}{l}} b_k b_l \leq \sum_k b_k^2 + 2 \sum_{k<l} b_k b_l$$
$$= (b_1 + \ldots + b_n)^2 = \text{RHS}^2.$$

Equality holds only when $n - 1$ of $b_1, \ldots, b_n$ are zero, hence $a_1 = a_2 = \ldots = a_n$.

376. For $x = 0$, we get $P(0) = 0$. For $x = -1$, we get $P(-1) = 0, \ldots$ It follows that $0, -1, -2, \ldots, -2013$ are roots of P. Now we will prove that deg $(P) = 2014$. Let

$P(X) = a_n X^n + a_{n-1} X^{n-1} + \ldots + a_1 X + a_0$, $a_n \neq 0$. Identifying the coefficients of X^n, we have $2014 a_n + a_{n-1} = n a_n + a_{n-1}$, hence $n = 2014$ and

$$P(X) = cX(X+1)\ldots(X+2014), c \in \mathbb{R}.$$

Clearly, these polynomials satisfy the given condition.

377. The Jensen's inequality applied for $f(x) = \sqrt[3]{x}$ get us

$$8\sqrt[3]{abc} + \sqrt[3]{\frac{a^3+b^3+c^3}{3}} \leq 9 \sqrt[3]{\frac{8abc + \frac{a^3+b^3+c^3}{3}}{9}} = 3\sqrt[3]{a^3+b^3+c^3+24abc}$$
$$\leq 3\sqrt[3]{(a+b+c)^3} = 3(a+b+c).$$

378. Let $x_1, \ldots, x_n$ be the roots of the polynomial. Using the Viète's relations, we have

$$\sum x_i^2 = \left(\sum x_i\right)^2 - 2\sum_{i<j} x_i x_j = -2a_{n-2}.$$

By the Cauchy–Schwarz inequality,

$$(n-1)\sum_{j \neq i} x_j^2 \geq \left(\sum_{j \neq i} x_j \right)^2 = x_i^2$$

for all i, hence $(n-1)\sum x_j^2 \geq n x_i^2$. We deduce

$$x_i^2 \leq \frac{n-1}{n} \sum x_j^2 = -\frac{2(n-1)}{n} a_{n-2},$$

as desired.

379. For $1 \geq a \geq b \geq c \geq 0$, we have

$$\sqrt{b+c} \geq 5abc, \sqrt{a+b} + \sqrt{c+a} \geq 2.$$

Indeed, $\sqrt{b+c} \geq \sqrt{2\sqrt{bc}} = \sqrt[4]{4bc} \geq 5abc$. The last inequality is a consequence of

$$\left(a^4b^3c^3\right)^2 = 4^4 \cdot 3^3 \cdot 3^3 \cdot \left(\frac{a^2}{4}\right)^4 \left(\frac{b^2}{3}\right)^3 \left(\frac{c^2}{3}\right)^3 \leq 4^4 \cdot 3^3 \cdot 3^3$$
$$\cdot \left(\frac{4 \cdot \frac{a^2}{4} + 3 \cdot \frac{b^2}{3} + 3 \cdot \frac{c^2}{3}}{10}\right)^{10} = \frac{4^4 \cdot 3^6}{10^{10}} < \left(\frac{4}{5^4}\right)^2.$$

Also, $(a+b)(c+a) = a^2 + ab + bc + ca \geq 1$, hence $\sqrt{a+b} + \sqrt{c+a} \geq 2$. Adding the two proved inequalities, we get the required inequality.

3.44 2015

380. We will prove the inequalities

$$(a+b+c)^5 \geq E = \frac{27}{8}(a+b)(b+c)(c+a)(a+b+c)^2 \geq 81abc\left(a^2+b^2+c^2\right).$$

The first inequality follows from the AM–GM inequality, the second is a consequence of

$$E - 81abc\left(a^2+b^2+c^2\right) = \frac{27}{8}\sum_{cyc} a[(b-c)(3a-b-c)]^2 \geq 0.$$

381. We have

$$\frac{a+1}{a(a+2)} + \frac{b+1}{b(b+2)} + \frac{c+1}{c(c+2)} = \frac{1}{2}\left(\frac{1}{a} + \frac{1}{b} + \frac{1}{c} + \frac{1}{a+2} + \frac{1}{b+2} + \frac{1}{c+2}\right) \geq \frac{1}{2}$$
$$\times \left(\frac{9}{a+b+c} + \frac{9}{a+b+c+6}\right) \geq 2.$$

Equality holds for $a = b = c = 1$.

382. We have

$$\frac{1}{x_1} \geq \frac{2}{x_2} \geq \ldots \geq \frac{n}{x_n}.$$

By the AM–GM and Chebyshev's inequalities,

$$\frac{n(n+1)}{2} = \sum_{k=1}^{n} x_k \cdot \frac{k}{x_k} \geq \frac{1}{n} \sum_{k=1}^{n} x_k \sum_{k=1}^{n} \frac{k}{x_k} \geq \frac{1}{n} \sum_{k=1}^{n} x_k \cdot n \sqrt[n]{\frac{n!}{x_1 x_2 \ldots x_n}}$$
$$= \frac{(x_1 + x_2 + \ldots + x_n)\sqrt[n]{n!}}{\sqrt[n]{x_1 x_2 \ldots x_n}},$$

hence

$$\frac{x_1 + x_2 + \ldots + x_n}{\sqrt[n]{x_1 x_2 \ldots x_n}} \geq \frac{n+1}{2\sqrt[n]{n!}}.$$

383. Let d be the common difference of the arithmetic progression. We will show that $d = 0$. Assume that $a_1 \leq a_2 \leq \ldots \leq a_n$ and $d \geq 0$. Then $a_1 + a_2 \leq a_1 + a_3$ are the least sums, hence $d = a_3 - a_2$. Analogously, $a_{n-2} + a_n \leq a_{n-1} + a_n$ are the greatest sums, so that $d = a_{n-1} - a_{n-2}$.

If $n \geq 6$, then 2, 3, $n - 2$, $n - 1$ are different indices, hence $a_2 + a_{n-1}$, $a_3 + a_{n-2}$ are elements of the arithmetic progression and

$$a_3 + a_{n-2} = (a_2 + d) + (a_{n-1} - d) = a_2 + a_{n-1},$$

therefore $d = 0$.

If $n = 5$, then as above, $a_3 = a_2 + d$, $a_4 = a_3 + d$. Assume that $d > 0$. Then the sums $a_1 + a_2$, $a_1 + a_3$, $a_1 + a_4$ are the smallest and the sums $a_2 + a_5$, $a_3 + a_5$, $a_4 + a_5$ are the greatest among all pairwise sums. Observe that $a_2 + a_3$, $a_2 + a_4$, $a_3 + a_4$ are consecutive sums, hence there are only two possibilities of ordering all sums:

$$a_1 + a_2 < a_1 + a_3 < a_1 + a_4 < a_1 + a_5 < a_2 + a_3 < a_2 + a_4 < a_3 + a_4 < a_2 + a_5$$
$$< a_3 + a_5 < a_4 + a_5,$$

$$a_1 + a_2 < a_1 + a_3 < a_1 + a_4 < a_2 + a_3 < a_2 + a_4 < a_3 + a_4 < a_1 + a_5 < a_2 + a_5$$
$$< a_3 + a_5 < a_4 + a_5.$$

In the first case, $a_5 - a_4 = d$ implies $a_2 + a_5 = a_3 + a_4$, contradiction. In the second case, $a_2 - a_1 = d$ implies $a_1 + a_4 = a_2 + a_3$, contradiction. Hence $d = 0$.

384. Answer: $f(x) = 0$; $f(x) = x^2$.

Inverting the roles of x, y, we get $f(f(y) + f(x)) = f(y^2) + 2y^2 f(x) + (f(x))^2$, hence $f(x^2) + 2x^2 f(y) + (f(y))^2 = f(y^2) + 2y^2 f(x) + (f(x))^2$. Taking $x = 0$, we get

$$f(x^2) + 2ax^2 + a^2 = a + (f(x))^2,$$

where $a = f(0)$. Instead, taking $y = 1$, we get

$$f(x^2) + 2bx^2 + b^2 = b + 2f(x) + (f(x))^2,$$

where $b = f(1)$. Subtracting these equations, we obtain

$$2(b - a)x^2 + b^2 - a^2 = b - a + 2f(x).$$

Taking $x = 0$, we have $b^2 - a^2 = b + a$.

Case 1. $b + a = 0$. Then choose $(x, y) = (0, 1)$ in the given equation, to obtain

$$f(0) = f(a + b) = a + b^2,$$

whence $b = 0$ and so $a = 0$. The function is $f(x) = 0$ for all x.

Case 2. $b - a = 1$. Then $b = a + 1$, $2f(x) = 2x^2 + b + a - 1 = 2x^2 + 2a$, and so $f(x) = x^2 + a$. Choose $x = 2a$, then $f(2a) = 4a^2 + a$. Choose $(x, y) = (0, 0)$ in the given equation, to obtain $f(2a) = a + a^2$, so $a = 0$ and $f(x) = x^2$ for all x.

Direct calculations show that both functions satisfy the given equation.

385. With the notations $x = \sqrt{a}, y = \sqrt{b},$ the given inequality becomes

$$\left(x^{2n} - y^{2n}\right)\left(\frac{1}{y^{2n-2}} - \frac{1}{x^{2n-2}}\right) > 4n(n-1)(x-y)^2.$$

Applying the AM–GM inequality, we have

$$\begin{aligned}\left(x^{2n} - y^{2n}\right)\left(\frac{1}{y^{2n-2}} - \frac{1}{x^{2n-2}}\right) &> (x-y) \cdot 2n \sqrt[2n]{x^{2n-1} \cdot x^{2n-2}y \cdot \ldots \cdot y^{2n-1}} \\ &\times \left(\frac{1}{y} - \frac{1}{x}\right) \cdot (2n-2) \sqrt[2n-2]{\frac{1}{y^{2n-3}} \cdot \frac{1}{y^{2n-4}x} \cdot \ldots \cdot \frac{1}{x^{2n-3}}} = 4n(n-1)(x-y)^2 \\ &\times (xy)^{\frac{2n-1}{2}} \cdot \cdot \frac{1}{xy} \cdot \frac{1}{(xy)^{\frac{2n-3}{2}}} = 4n(n-1)(x-y)^2.\end{aligned}$$

386. Replacing y by $-y$ in

$$f(xy) \le yf(x) + f(y) \quad (1)$$

we get

$$f(-xy) \le -yf(x) + f(-y). \quad (2)$$

Adding (1) and (2), we get $f(xy) + f(-xy) \le f(y) + f(-y)$ (3). Setting $y = 1$, we get

$$f(x) + f(-x) \le f(1) + f(-1). \quad (4)$$

In (3), using $\frac{1}{y}$ with $y \ne 0$ to replace x, we get $f(1) + f(-1) \le f(y) + f(-y)$ (5). By (4) and (5), for $y \ne 0$ we have $f(y) + f(-y) = f(1) + f(-1)$. Let $c = f(1) + f(-1)$. Then (2) becomes $c - f(xy) \le -yf(x) + c - f(y)$, hence $f(xy) \ge yf(x) + f(y)$ (6). By (1) and (6), for all $x, y \ne 0, f(xy) = yf(x) + f(y)$ (7). Setting $x = y = 1$, we get $f(1) = 0$. Interchanging x, y in (7), we get $f(yx) = xf(y) + f(x)$ (8). Subtracting (8) from (7), we get $(y-1)f(x) = (x-1)f(y)$. Then for $x, y \ne 0, 1$, we have

$$\frac{f(x)}{x-1} = \frac{f(y)}{y-1}.$$

Since $f(1) = 0$, we see that there exists a such that $f(x) = a(x - 1)$ for all $x \ne 0$. Setting $x = 0$ in (1), we get $f(y) \ge (1 - y)f(0)$. Then for $y \ne 0$, we get

$$a(y-1) \ge (1-y)f(0),$$

that is, $(y - 1)(a + f(0)) \ge 0$. Then $a = -f(0)$ and $f(x) = a(x - 1)$ for all x, where a is an arbitrary constant.

387. Answer: $n = 2014$.

We can take $a_{2k} = 2014$, $a_{2k+1} = 2015$, so $P(-1) = 0$.

Assume that there is such polynomial of degree $2n < 2 \cdot 2014$. Then $P(x) \ge 2014$ for $x \ge 0$, so if P has a real root, then it must be negative, so $Q(x) = P(-x)$ has a positive real root. But for $x > 0$ we have

$$Q(x) \ge 2014x^{2n} - 2015x^{2n-1} + 2014x^{2n-2} - \ldots - 2015x + 2014 = R(x),$$

and it suffices to show that $R(x) > 0$ for $x > 0$. We have

$$R(1) = 2014(n+1) - 2015n = 2014 - n > 0.$$

If $x \ne 1$, then

$$R(x) > 0 \Leftrightarrow 2014\left(x^{2n} + x^{2n-2} + \ldots + 1\right) > 2015\left(x^{2n-1} + \ldots + x\right) \Leftrightarrow \frac{x^{2n+2} - 1}{x(x^{2n} - 1)} \geq 1 + \frac{1}{2014}.$$

Since $n < 2014$, it suffices to prove that

$$\frac{x^{2n+2} - 1}{x(x^{2n} - 1)} \geq 1 + \frac{1}{n},$$

or, equivalently,

$$n\left(x^{2n} + x^{2n-2} + \ldots + 1\right) \geq (n+1)\left(x^{2n-1} + \ldots + x\right). \quad (1)$$

By the AM–GM inequality,

$$nx^{2k+1} \leq \frac{n}{2}\left(x^{2k} + x^{2k+2}\right). \quad (2)$$

Moreover,

$$x^k + x^{2n-k} \leq x^{2n} + 1 \quad (3)$$

holds, since it is equivalent to $(x^k - 1)(x^{2n-k} - 1) \geq 0$, which is true. Adding (2) for $k = 0, \ldots, n - 1$ and (3) for $k = 1, \ldots, 2n - 1$, we get (1).

388. Denote these sets by S. As f has odd degree, S is not bounded, so g has also odd degree. We can find N such that for all $x \geq N$, and $y < x$, we have $f(y) < f(x)$.

Choose $t \in S$ such that $t > \max(f(N), g(N))$. Let $t = f(m) = g(n)$ and consider the function $h(x) = g(x + n - m)$. Then h has the same range as g, $h(m) = g(n) = f(m)$. As h, f are strictly increasing after m and all values are greater than all previous values, it follows that $h(m + 1) = f(m + 1)$, $h(m + 2) = f(m + 2)$, …, so $h = f$.

Hence $g(x) = f(x + k)$, $k = m - n$.

389. Answer: $P = X^4$, $Q = 1$ or $P = 1$, $Q = X^4$.

The following decompositions hold

$$(P + Q)\left(P^2 + Q^2 - PQ\right) = \left(X^4 + 1\right)\left(X^8 - X^4 + 1\right).$$

The polynomials $X^4 + 1$ and $X^8 - X^4 + 1$ are irreducible, being cyclotomic. Moreover,

$$\gcd\left(P + Q, P^2 + Q^2 - PQ\right) = \gcd\left(X^4 + 1, X^8 - X^4 + 1\right) = 1.$$

Case 1. $P + Q = X^4 + 1$. Then $P^2 + Q^2 - PQ = X^8 - X^4 + 1$, so

$$PQ = \frac{\left(X^4+1\right)^2 - \left(X^8 - X^4 + 1\right)}{3} = X^4,$$

hence $P = X^4$, $Q = 1$ or $P = 1$, $Q = X^4$.

Case 2. $P + Q = X^8 - X^4 + 1$. Then $P^2 + Q^2 - PQ = X^4 + 1$, so

$$P(x)Q(x) = \frac{\left(x^8 - x^4 + 1\right)^2 - \left(x^4 + 1\right)}{3} \leq \left(\frac{x^8 - x^4 + 1}{2}\right)^2$$

for all real x, contradiction.

390. By the Cauchy–Schwarz inequality,

$$\left(\left(\frac{x}{\sqrt{x+y}}\right)^2 + \left(\frac{y}{\sqrt{y+z}}\right)^2 + \left(\frac{z}{\sqrt{z+w}}\right)^2 + \left(\frac{w}{\sqrt{w+x}}\right)^2\right)$$
$$\times \left(\left(\sqrt{x+y}\right)^2 + \left(\sqrt{y+z}\right)^2 + \left(\sqrt{z+w}\right)^2 + \left(\sqrt{w+x}\right)^2\right) \geq (x+y+z+w)^2,$$

with equality only when $x = y = z = w$. This simplifies to

$$\frac{x^2}{x+y} + \frac{y^2}{y+z} + \frac{z^2}{z+w} + \frac{w^2}{w+x} \geq \frac{x+y+z+w}{2}.$$

Applying the AM–GM inequality and using the given condition, it follows that

$$\frac{x^2}{x+y} + \frac{y^2}{y+z} + \frac{z^2}{z+w} + \frac{w^2}{w+x} \geq 2\sqrt[4]{xyzw} = 2\sqrt[4]{16} = 4,$$

with equality only when $x = y = z = w$.

391. We have

$$a^3 + b^3 + c^3 + \frac{ab}{a^2+b^2} + \frac{bc}{b^2+c^2} + \frac{ca}{c^2+a^2} \geq \frac{a^3+b^3+c^3}{2}$$
$$+ \sum_{cyc}\left(\frac{a^2+b^2}{4} + \frac{ab}{a^2+b^2}\right) \geq \frac{a^3+b^3+c^3}{2} + \sum_{cyc}\sqrt{ab} \geq \frac{3}{2} + 3 = \frac{9}{2}.$$

392. The polynomial $Q(X) = (X+1)P(X-1) - (X-1)P(X)$ is constant. Clearly, constant polynomials are solutions. Suppose that $P(X) = aX^n + bX^{n-1} + R(X)$, with $\deg(R) \leq n - 2$. The coefficient of X^n in $Q(X)$ is $(n-2)a = 0$, hence $n = 2$ and

$$P(X) = aX^2 + bX + c.$$

Setting this in the given condition, we get $a = b$, therefore $P(X) = aX^2 + aX + c$, with a, c real numbers.

393. Answer: $n = 1$ or n even.

For $n = 1$, choose $x_1 = 0$.

For $n = 2k$, choose

$$x_1 = \ldots = x_k = \sqrt{1 - \frac{1}{k^2}}, x_{k+1} = \ldots = x_{2k} = -\sqrt{1 - \frac{1}{k^2}}.$$

If $n = 2k + 1, k \geq 1$, then $x_i \neq 0$ for all i, otherwise $\sqrt{1 - x_1^2} + \ldots + \sqrt{1 - x_n^2} > 1$. It follows that at least $k + 1$ of $x_1, \ldots, x_n$ have the same sign. Without loss of generality, suppose that $x_1, \ldots, x_{k+1} > 0$.Since $x > 1 - \sqrt{1 - x^2}$ for $x \in (0, 1)$, we deduce

$$x_1 + \ldots + x_{k+1} > k + 1 - \sum_{i=1}^{k+1} \sqrt{1 - x_i^2} > k,$$

hence

$$x_1 + x_2 + \ldots + x_n > k + x_{k+2} + \ldots + x_n > 0,$$

contradiction.

394. From

$$\left(1 + \frac{1}{k(k-1)}\right)^k > 1 + \frac{1}{k-1},$$

it follows that

$$\sqrt[k]{\frac{k}{k-1}} < 1 + \frac{1}{k-1} - \frac{1}{k},$$

whence

$$\sum_{k=2}^{n} \sqrt[k]{\frac{k}{k-1}} < \sum_{k=2}^{n} \left(1 + \frac{1}{k-1} - \frac{1}{k}\right) = n - \frac{1}{n}.$$

395. By the homogeneity, we can suppose that

$$\frac{1}{a} + \frac{1}{b} + \frac{1}{c} = 1.$$

Then we have $1 < c < 10$ and

$$a+b=10-c,\ \frac{1}{a}+\frac{1}{b}=1-\frac{1}{c}.$$

The AM–HM inequality gets us

$$(10-c)\left(1-\frac{1}{c}\right)=(a+b)\left(\frac{1}{a}+\frac{1}{b}\right)\geq 4,$$

whence $c^2-10c+7\leq 0$, therefore $2\leq c\leq 5$. Similarly, $2\leq a,b\leq 5$. Now,

$$(a-2)(b-2)(c-2)\geq 0.$$

Expanding the expression from the left and using the conditions $a+b+c=10$, $abc=ab+bc+ca$, we get $abc\leq 32$.

Similarly, from $(a-5)(b-5)(c-5)\leq 0$, we get $abc\geq\frac{25}{4}$. Denote

$$t=abc=ab+bc+ca.$$

We have

$$T=\left(a^2+b^2+c^2\right)\left(\frac{1}{a^2}+\frac{1}{b^2}+\frac{1}{c^2}\right)=(100-2t)\cdot\frac{t^2-20t}{t^2}=140-\left(2t+\frac{2000}{t}\right),$$

where $\frac{125}{4}\leq t\leq 32$. Now, $T_{min}=\frac{27}{2}$, achieved for $t=32$ or $t=\frac{125}{4}$, that is, when $(a,b,c)=(2,4,4)$ or $\left(5,\frac{5}{2},\frac{5}{2}\right)$ and their permutations, $T_{max}=140-40\sqrt{10}$, achieved when $t=10\sqrt{10}$, that is, when a,b,c are the roots of the polynomial

$$X^3-10X^2+10\sqrt{10}X-10\sqrt{10}.$$

396. Answer: $P=0$; $P=1$; $P=(X^2-X+2)^n$, $n\in\mathbb{N}^*$.

If P is constant, then $P=0, 1$. Suppose that $\deg(P)\geq 1$. Let x be a real root of P, then $x^2+2>x$ is also a root of P, so P has infinitely many roots, contradiction. We conclude that P has no real roots, in particular, $\deg(P)$ is even. Denote by c the leading coefficient of P. Then $c^2=c$, hence $c=1$, since $c\neq 0$.

Direct calculations show that the unique polynomial of second degree that satisfies the given condition is $P=X^2-X+2$. We see that $P=(X^2-X+2)^n$, $n\in\mathbb{N}^*$, is also a solution and we will prove that no other solutions exist.

Let P, Q be polynomials of same degree which are solutions and let $R=P-Q$. Then $\deg(R)<\deg(P)=\deg(Q)$, since P, Q are monic. For all real x we have

$$(R(x)+Q(x))(R(x+1)+Q(x+1))=R(x^2)+Q(x^2),$$

hence

$$R(x)R(x+1)+R(x)Q(x+1)+R(x+1)Q(x)=R(x^2+2).$$

This equality holds for infinitely many values of x, therefore these functions coincide as polynomials. But the degrees of the polynomial functions in the left-hand side and the right-hand sides are different, contradiction.

397. Answer: $n=5$.

The 5-tuple $(7,4,2,-6,-6)$ satisfies the given conditions. Let $n=4$ and $a_1 \geq a_2 \geq \geq a_3 \geq a_4$. Then it is easy to see that $a_1 > 0 > a_4$.

Case 1. $a_1 \geq a_2 > 0 > a_3 \geq a_4$. Denote $a_1 = a, a_2 = b, a_3 = -c, a_4 = -d$. Then

$$a+b>c+d, a^3+b^3<c^3+d^3, a^5+b^5<c^5+d^5.$$

Now,

$$a \geq b > 0, d \geq c > 0 \Rightarrow a > d, > b; a^3+b^3-c^3<d^3<(a+b-c)^3 \Rightarrow 3(a+b) \times (a-c)(b-c)>0 \Rightarrow c>a \geq b.$$

By the AM–GM inequality,

$$a^5+b^5=a^3(a^2-b^2)+b^2(a^3+b^3)<c^3(a^2-b^2)+b^2(c^3+d^3) < \frac{2a^5+2b^5+3c^5+3d^5}{5} \Rightarrow a^5+b^5<c^5+d^5,$$

contradiction.

Case 2. $0 > a_2 \geq a_3$. Then $a_1 > -(a_2+a_3+a_4)$ from a), hence

$$a_1^3+a_2^3+a_3^3+a_4^3 > -3(a_2+a_3)(a_2+a_4)(a_3+a_4)>0,$$

contradiction.

Case 3. $a_2 \geq a_3 > 0$. Then $a_1^3+a_2^3+a_3^3 < -a_4^3$, thus $0<a_3 \leq a_2 \leq a_1 < -a_4$, so

$$a_1^5+a_2^5+a_3^5+a_4^5<a_4^2(a_1^3+a_2^3+a_3^3+a_4^3)<0,$$

contradiction.

If $n<4$, then we can include $4-n$ zeros and we will have a 4-tuple.

3.45 2016

398. Answer: $P(x) = 0, \pm 1$; $P(x) = \pm x$.
Changing the variables, the equation becomes

$$(P(x))^3 + (P(y))^3 + (P(z))^3 = 3P(xyz)$$

and the condition $x + y + z = 0$ remains unchanged.
Choose $y = -x$, $z = 0$, then $(P(x))^3 + (P(-x))^3 = 3P(0) - (P(0))^3 = c$.
Choose $y = 1 - x$, $z = -1$, then

$$(P(x))^3 + (P(1-x))^3 + (P(-1))^3 = 3P(x(x-1)) = (P(x))^3 + c - (P(x-1))^3 + (P(-1))^3.$$

Let $n = \deg(P)$, then $\deg((P(x))^3 - (P(x-1))^3) = 3n - 1$, $\deg(P(x(x-1))) = 2n$, so $n = 1$. For $n = 0$, we have $P = 0, \pm 1$. For $n = 1$, we have $P(x) = ax + b$. Simple calculations show that $a = \pm 1$, $b = 0$.

399. Denote $a^2 = x$, $b^2 = y$, $c^2 = z$. Now, we have to prove the inequality

$$\left(x + \frac{y}{z}\right)\left(y + \frac{z}{x}\right)\left(z + \frac{x}{y}\right) \geq (x+1)(y+1)(z+1).$$

By the Cauchy–Schwarz inequality

$$\left(x + \frac{y}{z}\right)\left(y + \frac{z}{x}\right) \geq \left(\sqrt{xy} + \sqrt{\frac{y}{x}}\right)^2 = \frac{y}{x}(x+1)^2.$$

Multiplying with two similar inequalities and extracting square roots, we get the desired result.

400. Answer: $f(x) = -1$ for all x; $f(x) = x + 1$ for all x.
Both functions satisfy the given equation (easy verification!).
Let f be an arbitrary solution. Taking $(x, y) = (0, f(0))$ in the equation, we find $f(z) = -1$, for $z = -f(f(0))$. Choose $y = z$, then

$$f(x+1) = f(f(x)), \forall x \in \mathbb{Z}. \quad (1)$$

The given equation becomes

$$f(x - f(y)) = f(x+1) - f(y) - 1. \quad (2)$$

Choose $y = x$, then $f(x + 1) - f(x) = f(x - f(x)) + 1 = f(f(x - 1. - f(x))) + 1$. From (2) we have $f(x - 1 - f(x)) = f(x) - f(x) - 1 = -1$, and we obtain

$$f(x+1)=f(x)+c,$$

where $c=f(-1)+1$ is a constant.

By induction we can show that $f(x) = Ax + B$ for all integers x, where $B=f(0)$. From (1) we obtain

$$Ax+A+B=A^2x+AB+B$$

for all integers x. If $x=0, 1$, respectively, we have $A+B=AB+B, A^2=A$.

Case 1. $A=0$. Then f is constant and the given equation shows that this constant is -1.

Case 2. $A=1$. Then $B=1$, hence f is the succesor function.

401. Answer: $f(n)=kn^2$, $k\in\mathbb{N}$.

The function $f(n)=kn^2$, $k=f(1)$, satisfies the condition.

We have $f(0)=0$. Denote by $P(a,b,c,d)$ the given equation. Then

$$P(n,n,n,n)\Rightarrow f(2n)=4f(n)\Rightarrow f(2)=4k,$$
$$P\left(n^2,1,n,n\right)\Rightarrow f\left(n^2+1\right)=f\left(n^2\right)+k+2f(n),$$
$$P\left(n^2,2,2n,0\right)\Rightarrow f\left(n^2+2\right)=f\left(n^2\right)+4k+f(2n)=f\left(n^2\right)+4k+4f(n),$$
$$P\left(n^2+1,1,n+1,n-1\right)\Rightarrow f\left(n^2+2\right)=f\left(n^2+1\right)+k+f(n+1)+f(n-1),$$

so $f(n+1)=2f(n)-f(n-1)+2k$. By induction, it follows that $f(n)=kn^2$ for all nonnegative integers n.

402. Answer: $\lambda=1$.

Let $z_1\to 1$, $z_2\to\omega$, $z_3\to\omega^2$, where $\omega^3=1$, $\omega\neq 1$. Then $\lambda\geq 1$.

Define $|z_1|^2=r_1, |z_2|^2=r_2, |z_3|^2=r_3$. Then

$z_1\bar{z}_2+\bar{z}_1z_2=(z_1+z_2)(\overline{z_1+z_2})-|z_1|^2-|z_2|^2=r_3-r_1-r_2, z_2\bar{z}_3+\bar{z}_2z_3=r_1-r_2-r_3, z_3\bar{z}_1+\bar{z}_3z_1=r_2-r_3-r_1, |z_1z_2+z_2z_3+z_3z_1|^2+|z_1z_2z_3|^2=r_1^2+r_2^2+r_3^2-r_1r_2-r_2r_3-r_3r_1+r_1r_2r_3<r_1+r_2+r_3-r_1r_2-r_2r_3-r_3r_1+r_1r_2r_3=(r_1-1)(r_2-1)(r_3-1)+1<1.$

403. Answer: $f(x)=0$, $x\in\mathbb{R}$; $f(x)=x$, $x\in\mathbb{R}$.

Clearly, $f(x)=0$ for all x is a solution. Let

$$P(x,y):f\left(x^2\right)+xf(y)=f(x)f(x+f(y))$$

and let f be a nonconstant solution. If $f(f(x))=1$ for all x, then $f(1)=1$ and $P(1,1)$ gives $f(2)=2$, contradiction. Now,

$$P(0,x) \Rightarrow f(0)=0; P(x,0) \Rightarrow (f(x))^2 = f(x^2);$$
$$f(x_0)=0, P(x_0,x) \Rightarrow x_0=0; P(x,x) \Rightarrow f(x+f(x)) = x+f(x);$$
$$P(-f(x),x) \Rightarrow f\left((f(x))^2\right) = (f(x))^2;$$
$$P(-f(x), x+f(x)) \Rightarrow f \text{ surjective} \Rightarrow f(x)=x, \forall x>0;$$
$$P(x,y), x>0 \Rightarrow f(x+z) = x+z, \forall z.$$

404. Answer: $n=3$.

The following equivalences hold:

$$(3x_i - x_j)(x_i - 3x_j) \geq (1 - x_ix_j)^2 \Leftrightarrow 3x_i^2 - 10x_ix_j + 3x_j^2 \geq 1 - 2x_ix_j + x_i^2x_j^2 \Leftrightarrow 3x_i^2$$
$$- 6x_ix_j + 3x_j^2 \geq 1 + 2x_ix_j + x_i^2x_j^2 \Leftrightarrow 3(x_i - x_j)^2 \geq (1 + x_ix_j)^2 \Leftrightarrow \left|\frac{x_i - x_j}{1 + x_ix_j}\right| \geq \frac{1}{\sqrt{3}}.$$

Assume that $0 < x_1 < x_2 < \ldots < x_n$. We can find real numbers $\theta_i \in \left(0, \frac{\pi}{2}\right)$ such that $x_i = \tan\theta_i$ for $1 \leq i \leq n$. Now,

$$\left|\frac{x_i - x_j}{1 + x_ix_j}\right| \geq \frac{1}{\sqrt{3}} \Leftrightarrow \tan(\theta_j - \theta_i) \geq \frac{1}{\sqrt{3}} \Leftrightarrow \theta_j - \theta_i \geq \frac{\pi}{6}$$

for $1 \leq i < j \leq n$. If $n \geq 4$, by the pigeonhole principle, there exists $1 \leq i \leq 3$ $\theta_{i+1} - \theta_i < \frac{\pi}{6}$, which contradicts the above inequality.

For $n=3$, we can choose $x_i = \tan\frac{(2i-1)\pi}{12}, 1 \leq i \leq 3$, which satisfy the given condition.

405. By the AM – GM inequality, $a^3 + 2 = a^3 + 1 + 1 \geq 3a$. Denote by E the given expression and let $x = \frac{1}{a}, y = \frac{1}{b}, z = \frac{1}{c}$. Applying the Cauchy–Schwarz inequality,

$$E = \sum_{cyc} \frac{a^3+8}{a^3(b+c)} \geq \sum_{cyc} \frac{3a+6}{a^3(b+c)} = 3\sum_{cyc} \frac{x}{y+z} + 6\sum_{cyc} \frac{x^2}{y+z}$$
$$\geq 3\left[\left(\sum x\right)\left(\sum_{cyc} \frac{1}{y+z}\right) - 3\right] + 6\left[\frac{\left(\sum x\right)^2}{2\sum x}\right] \geq 3\left(\frac{9}{2} - 3\right) + 6 \cdot \frac{\sum x}{2} \geq 3 \cdot \frac{3}{2}$$
$$+ 6 \cdot \frac{3}{2} = \frac{27}{2}.$$

The minimum value occurs for $a = b = c = 1$.

406. By the given conditions,

$$(a+1)(b+1)(c+1)(d+1)=(a-1)(b-1)(c-1)(d-1).$$

Then

$$\sum \frac{2}{a-1}=\sum \frac{a+1}{a-1}-4>4-4=0,$$

by the AM–GM inequality, since

$$\frac{a+1}{a-1}=\frac{a^2-1}{(a-1)^2}>0.$$

Equality cannot hold, because $a + 1 > a - 1$.

407. The given equality implies

$$a_k \geq \frac{k}{a_{k+1}}-\frac{k-1}{a_k}.$$

Summing it for $k = 1, \ldots, n$, we have

$$a_1+\ldots+a_n \geq \frac{n}{a_{n+1}}.$$

We will prove the proposed inequality by induction. If $n = 2$, then by definition,

$$a_1+a_2 \geq a_1+\frac{1}{a_1} \geq 2.$$

Assume that the inequality holds for $n \geq 2$. If $a_{n+1} \geq 1$, then

$$(a_1+\ldots+a_n)+a_{n+1} \geq n+1,$$

otherwise,

$$(a_1+\ldots+a_n)+a_{n+1} \geq \frac{n}{a_{n+1}}+a_{n+1}=\frac{n-1}{a_{n+1}}+\left(\frac{1}{a_{n+1}}+a_{n+1}\right)>(n-1)+2=n+1.$$

Equality holds only when $a_1 = \ldots = a_n = 1$.

408. Let E be the given expression. Since it is linear in each variable, it attains the maximum value when $x_i = -1$ or 1, so that it suffices to consider $x_i \in \{-1, 1\}$, for $1 \leq i \leq 2n$. Define

$$y_i = \sum_{r=1}^{i} x_r - \sum_{r=i+1}^{2n} x_r.$$

Squaring, we get

$$y_i^2 = \sum_{r=1}^{2n} x_r^2 + \sum_{r<s\leq i} 2x_r x_s + \sum_{i<r<s} 2x_r x_s - \sum_{r\leq i<s} 2x_r x_s = 2n + \sum_{r<s\leq i} 2x_r x_s$$
$$+ \sum_{i<r<s} 2x_r x_s - \sum_{r\leq i<s} 2x_r x_s.$$

For every $r < s$, the coefficient of $x_r x_s$ is 2 if $i = 1, \ldots, r-1, s, \ldots, 2n$ and this coefficient is -2 if $i = r, \ldots, s-1$. We deduce that the coefficient of $x_r x_s$ in

$$\sum_{i=1}^{2n} y_i^2$$

is $2(2n - s + r) - 2(s - r) = 4(n - s + r)$. Summing it yields

$$\sum_{i=1}^{2n} y_i^2 = 4n^2 + \sum_{1\leq r<s\leq 2n} 4(n - s + r)x_r x_s = 4n^2 - 4E.$$

Now, we are looking for the minimum of LHS. We see that y_i is an even integer and $y_i - y_{i-1} = 2x_i = \pm 2$, so that y_{i-1}, y_i are consecutive even integers for $2 \leq i \leq 2n$. It follows that $y_{i-1}^2 + y_i^2 \geq 4$, hence

$$\sum_{i=1}^{2n} y_i^2 = \sum_{j=1}^{n} \left(y_{2j-1}^2 + y_{2j}^2\right) \geq 4n.$$

In conclusion, $4n^2 - 4E \geq 4n$, whence $E \leq n(n-1)$.

The maximum value is attained, for example, taking $x_i = (-1)^{i-1}$, $1 \leq i \leq 2n$.

409. Denote $b + 2c = x$, $c + 2a = y$, $a + 2b = z$. Then

$$a+b+c=\frac{x+y+z}{3}, 2a+b=\frac{-x+2(y+z)}{3}, 2b+c=\frac{-y+2(z+x)}{3}, 2c+a=\frac{-z+2(x+y)}{3}.$$

Now, we have to prove the inequality

$$\frac{1}{3}\left(\frac{-x+2(y+z)}{x}+\frac{-y+2(z+x)}{y}+\frac{-z+2(x+y)}{z}\right)\geq 3,$$

which is equivalent to

$$\left(\frac{x}{y}+\frac{y}{x}\right)+\left(\frac{y}{z}+\frac{z}{y}\right)+\left(\frac{z}{x}+\frac{x}{z}\right)\geq 6.$$

This follows from the AM−GM inequality.

410. Answer: $k=\frac{1}{3}$.

For $a_0=1$, $a_i=2^{i-1}$, $1\leq i\leq n$, we have

$$\frac{1}{2}+\frac{1}{4}+\ldots+\frac{1}{2^n}<k\left(1+1+\frac{1}{2}+\ldots+\frac{1}{2^{n-1}}\right),$$

that is,

$$k>\frac{1-\frac{1}{2^n}}{3-\frac{1}{2^{n-1}}}.$$

Taking $n\to\infty$, we get $k\geq\frac{1}{3}$.

By the Cauchy–Schwarz inequality,

$$\frac{1}{a_0}+\frac{1}{a_1}\geq\frac{4}{a_0+a_1}, \frac{1}{a_0+a_1}+\frac{1}{a_2}\geq\frac{4}{a_0+a_1+a_2},\ldots,\frac{1}{a_0+\ldots+a_{n-1}}+\frac{1}{a_n}\geq\frac{4}{a_0+\ldots+a_n}.$$

Summing it, we obtain

$$\frac{1}{a_0}+\frac{1}{a_1}+\ldots+\frac{1}{a_n}\geq 3\left(\frac{1}{a_0+a_1}+\ldots+\frac{1}{a_0+\ldots+a_{n-1}}\right)+\frac{4}{a_0+\ldots+a_n}$$
$$>3\left(\frac{1}{a_0+a_1}+\ldots+\frac{1}{a_0+\ldots+a_n}\right).$$

411. We have

$$\sqrt{x+\sqrt{4x+\sqrt{16x+\sqrt{\ldots+\sqrt{4^n x+3}}}}}=1+\sqrt{x},$$

$$x+\sqrt{4x+\sqrt{16x+\sqrt{\ldots+\sqrt{4^n x+3}}}}=1+2\sqrt{x}+x,$$

$$4x+\sqrt{16x+\sqrt{\ldots+\sqrt{4^n x+3}}}=1+4\sqrt{x}+4x,$$

$$\vdots$$

$$4^n x+3=1+2^{n+1}\sqrt{x}+4^n x,$$

$$x=\frac{1}{4^n}.$$

412. We have $P = \sin\alpha + \cos\alpha + \tan\alpha + \cot\alpha + \sec\alpha + \csc\alpha + 2$. Denote

$$t+1=\sqrt{2}\sin\left(\alpha+\frac{\pi}{4}\right).$$

Then

$$t\in\left(0,\sqrt{2}-1\right],P(t)=t+\frac{2}{t}+3,P'(t)=1-\frac{2}{t^2}\leq 0,$$

hence P is decreasing on $\left(0,\sqrt{2}-1\right]$. The minimum value is attained at $t=\sqrt{2}-1$, that is,

$$P_{\min}=P\left(\sqrt{2}-1\right)=4+3\sqrt{2}.$$

413. The AM–GM inequality yields

$$a+\sqrt{ab}+\sqrt[3]{abc}=a+\sqrt{\frac{a}{2}\cdot 2b}+\sqrt[3]{\frac{a}{4}\cdot b\cdot 4c}\leq a+\frac{1}{2}\left(\frac{a}{2}+2b\right)+\frac{1}{3}$$
$$\times\left(\frac{a}{4}+b+4c\right)\leq\frac{4}{3}(a+b+c),$$

with equality for $a = 4b = 16c$.

414. Denote $a = x^2$, $b = y^2$, $c = z^2$, then

$$a^2<16bc\Rightarrow x^4<16y^2z^2\Rightarrow x<2\sqrt{yz}<y+z,$$

so x, y, z are the sidelengths of a triangle with area A. By the Heron's formula,

$$16A^2 = 2\left(x^2y^2 + y^2z^2 + z^2x^2\right) - \left(x^4 + y^4 + z^4\right) > 0,$$

whence the conclusion.

415. We will prove first by induction that $P_n(2) = 0$ for all even n. For $n = 0$ this is true. We have

$$P_{n+2}(2) = P_{n+1}(3)P_{n+1}(-1) - 1 = [P_n(4)P_n(2) - 1][P_n(2)P_n(0) - 1] - 1 = 1 - 1 = 0,$$

which ends the proof.

Now, we will prove by induction on k that x^{2k} divides $P_{2k}(x)$. Express P_{2k+2} in function of P_{2k}:

$$P_{2k+2}(x) = P_{2k+1}(1-x)P_{2k+1}(1+x) - 1 = [P_{2k}(2-x)P_{2k}(x) - 1] \times [P_{2k}(x+2)P_{2k}(-x) - 1] - 1 = -P_{2k}(2-x)P_{2k}(x) - P_{2k}(x+2)P_{2k}(-x) + P_{2k}(x+2)P_{2k}(-x)P_{2k}(x)P_{2k}(2-x).$$

By the induction hypothesis, $x^{2k+2} | P_{2k}(x+2)P_{2k}(-x)P_{2k}(x)P_{2k}(2-x)$. Now, we will prove that $x^{2k+2} | Q(x) + Q(-x)$, where $Q(x) = P_{2k}(x)P_{2k}(2-x)$. We have

$$Q(x) = P_{2k}(x)P_{2k}(2-x) = \left(a_n x^n + \ldots + a_{2k}x^{2k}\right)\left(a_n 2^n + \ldots + a_{2k}2^{2k} + xR(x)\right),$$

with $R(x) \neq 0$. The coefficient of x^{2k} in $Q(x) + Q(-x)$ is $2a_{2k}P_{2k}(2) = 0$. Thus, $x^{2k} | P_{2k}(x)$ implies $x^{2k+2} | P_{2k+2}(x)$.

For $k = 1008$, we obtain the desired result.

416. For $x = 0$ and $x = 2$, we have $P(-2) = P(0) = P(2)$, hence

$$P(x) = x\left(x^2 - 4\right)Q(x) + a.$$

Replacing in the given equality, we deduce

$$(x-4)Q(x-2) + (x+4)Q(x+2) = 2xQ(x).$$

If there is $s < 4$ such that $Q(s) = Q(s+2) = k$, then

$$Q(s-2) = \frac{2sk - (s+4)k}{s-4} = k,$$

hence $Q(s - 2n) = k$ by induction, therefore Q is constant and

$$P(x) = bx\left(x^2 - 4\right) + a, a, b \in \mathbb{R}.$$

417. Put $x = y = 0$ in (1), then $(f(0))^2 \leq 0$, hence $f(0) = 0$. Assume that there exists $x > 0$ such that $f(x) > x^2$. Define the sequence $(a_n)_{n \geq 0}$ by $a_n = 2^n - 2n - 1$. Then $a_{n+1} = 2a_n + 2n - 1$ for all n. We will prove by induction that

$$f\left(\frac{x}{2^n}\right) \geq 2^{a_n} x^2.$$

The base case is our assumption. Suppose that the inequality holds for some n and *put* $\frac{x}{2^n}$ for x, y in (1) to get

$$\left(f\left(\frac{x}{2^n}\right)\right)^2 \leq 2\left(\frac{x}{2^n}\right)^2 f\left(\frac{x}{2^{n+1}}\right) \Rightarrow f\left(\frac{x}{2^{n+1}}\right) \geq \frac{\left(f\left(\frac{x}{2^n}\right)\right)^2}{\frac{x^2}{2^{2n-1}}}.$$

Using the induction hypothesis, we obtain

$$f\left(\frac{x}{2^{n+1}}\right) \geq \frac{\left(2^{a_n}x^2\right)^2}{\frac{x^2}{2^{2n-1}}} = 2^{a_{n+1}}x^2,$$

which ends the proof by induction.

For n sufficiently large, we have $\frac{x}{2^n} < 1$, and $2^{a_n}x^2 = 2^{2^n - 2n - 1}x^2 > 2016$, which contradicts (2).

3.46 2017

418. Answer: $f(x) = 0$ for all x; $f(x) = x$ for all x.
Clearly, $f(0) = 0$. Let $P(x, y)$ be the given equation. If $f(x_0) = 0, x_0 \neq 0$, then $P(x, x_0)$ and $P(x_0, x)$ show that $f(x) = 0$ for all x. Also, $P\left(\sqrt{x^2+1}+1, x\right)$ shows that $f(x) = x$ for all x.

419. We have

$$\frac{1}{2}\sum_{\text{cyc}}\left[\frac{(x-1)^2}{y} + \frac{(y-1)^2}{z}\right] \geq \frac{1}{2}\sum_{\text{cyc}}\frac{(x+y-2)^2}{y+z} = \frac{1}{4}\sum_{\text{cyc}}\frac{2z^2}{y+z} = \frac{1}{4}\sum_{\text{cyc}}\frac{x^2+y^2}{x+y},$$

since

$$\sum_{\text{cyc}}\frac{2z^2-(y^2+z^2)}{y+z} = \sum_{\text{cyc}}\frac{z^2-y^2}{y+z} = \sum_{\text{cyc}}(z-y) = 0.$$

420. Answer: $C = \frac{1}{2}$.

The sequence 1, 1, 1, $\frac{1}{2}, x$, with arbitrarily small x, shows that $C \geq \frac{1}{2}$. Suppose that $C > \frac{1}{2}$ and $a_1 \leq a_2 \leq a_3 \leq a_4 \leq a_5$. Then since $\frac{a_3}{a_4} \geq \frac{a_2}{a_5}$, we have also $\frac{a_3}{a_4} \geq \frac{a_1}{a_2}$, hence $\frac{a_1}{a_2} < \frac{1}{2}$. Similarly, $\frac{a_4}{a_5} < \frac{1}{2}$, contradiction.

421. It suffices to consider $x \in (1,2)$, since the case $x \geq 2$ is obvious.

Lemma 1. For any positive integer k and $x \in (1,2)$, it holds

$$\sum_{ix<k} \{ix\} < \frac{k}{2}.$$

Proof. Observe that LHS is increasing as a function of x, excepting the cases when ix is an integer less than k, that is, $x = \frac{p}{q}$, $\gcd(p,q) = 1, p \leq k$.

Apply strong induction on k. Take $x = \frac{p}{q} - \varepsilon$, with $\varepsilon > 0$ arbitrary. If $p < k$, then

$$\sum_{ix<k} \{ix\} = \sum_{ix<p} \{ix\} + \sum_{p \leq ix < k} \{ix\} = \sum_{ix<p} \{ix\} + \sum_{jx<k-p} \{jx+p\} = \sum_{ix<p} \{ix\} + \sum_{jx<k-p} \{jx\},$$

and the bound follows from the induction hypothesis on p and $k-p$. If $p = k$, as x approaches $\frac{p}{q}$, the $q-1$ numbers $\{x\}, \{2x\}, \ldots, \{(q-1)x\}$ form a permutation of $\frac{1}{q}, \frac{2}{q}, \ldots \frac{q-1}{q}$, so

$$\sum_{ix<q} \{ix\} < \frac{1}{q} + \frac{2}{q} + \ldots + \frac{q-1}{q} = \frac{q-1}{2} < \frac{p}{2}.$$

Lemma 2. For all positive integers k and $x \in (1,2)$, it holds

$$\sum_{ix<k-1} \{ix\} + \frac{1}{2} \sum_{k-1 \leq ix < k} \{ix\} < \frac{k-1}{2}$$

Proof. The proof is analogous. For $p = k$, observe that the second term in the LHS is $\frac{1}{2}$ and that $\{x\}, \{2x\}$, $\{(q-1)x\}$ approaches a permutation of $\frac{1}{q}, \frac{2}{q}, \ldots, \frac{q-1}{q}$, so

$$\sum_{x \leq q-1} \{ix\} + \frac{\{iq\}}{2} < \left(\frac{1}{q} + \frac{2}{q} + \ldots + \frac{q-1}{q}\right) + \frac{1}{2} = \frac{q}{2} \leq \frac{p-1}{2}.$$

Back to the problem, consider the sequence $(a_n)_{n \geq 1}$ defined by

$$a_n = \begin{cases} \{kx\}, n = \lfloor kx \rfloor, \\ 0, n \neq \lfloor kx \rfloor \end{cases}.$$

Let $s_k = a_1 + \ldots + a_k$. By the lemmas above,

$$s_k < \frac{k}{2}, \frac{s_{k-1} + s_k}{2} < \frac{k-1}{2}.$$

Using the inequality $\frac{2}{i} < \frac{1}{i-1} + \frac{1}{i+1}$, we have

$$\sum_{i=1}^{2n-1} \frac{a_i}{i} \leq \frac{1}{2}\left[a_1\left(\frac{1}{1}+\frac{1}{1}\right) + a_2\left(\frac{1}{1}+\frac{1}{3}\right) + a_3\left(\frac{1}{3}+\frac{1}{3}\right) + a_4\left(\frac{1}{3}+\frac{1}{5}\right) + \ldots\right] = \frac{1}{2}$$
$$\times\left(\frac{a_1}{1}+\frac{a_2}{1}+\frac{a_3}{3}+\frac{a_4}{3}+\ldots\right) + \frac{1}{2}\left(\frac{a_1}{1}+\frac{a_2}{3}+\frac{a_3}{3}+\frac{a_4}{5}+\ldots\right) = \frac{1}{2}$$
$$\times\left[\left(1-\frac{1}{3}\right)s_2 + \left(\frac{1}{3}-\frac{1}{5}\right)s_4 + \ldots + \left(\frac{1}{2n-3} - \frac{1}{2n-1}\right)s_{2n-2} + \frac{s_{2n-1}}{2n-1}\right] + \frac{1}{2}$$
$$\times\left[\left(1-\frac{1}{3}\right)s_1 + \left(\frac{1}{3}-\frac{1}{5}\right)s_3 + \ldots + \left(\frac{1}{2n-3} - \frac{1}{2n-1}\right)s_{2n-3} + \frac{s_{2n-1}}{2n-1}\right]$$
$$= \left(1-\frac{1}{3}\right)\cdot\frac{s_1+s_2}{2} + \left(\frac{1}{3}-\frac{1}{5}\right)\cdot\frac{s_3+s_4}{2} + \ldots + \frac{2s_{n-1}}{2n-1} < \frac{2}{1\cdot 3}\cdot\frac{1}{2} + \frac{2}{3\cdot 5}\cdot\frac{3}{2}$$
$$+ \ldots + \frac{2}{(2n-3)(2n-1)}\cdot\frac{2n-3}{2} + 1 = 1 + \frac{1}{3} + \frac{1}{5} + \ldots + \frac{1}{2n-1},$$

as desired.

422. Answer: -125.

Let a, b, c be the roots of the polynomial f. Then $f(x) = (x-a)(x-b)(x-c)$ and $abc = 64$. We have

$$f(-1) \leq -125 \Leftrightarrow (1+a)(1+b)(1+c) \geq 125.$$

By the Hölder's inequality

$$(1+a)(1+b)(1+c) \geq \left(\sqrt[3]{1} + \sqrt[3]{abc}\right)^3 = (1+4)^3 = 125.$$

423. Inductively, we find

$$a_n + b_n = \frac{a_1+b_1}{2^{n-1}}, a_n b_n = \sqrt[2^{n-1}]{a_1 b_1}, n \geq 2.$$

By the AM–GM inequality,

$$\left(\frac{a_1+b_1}{2^{n-1}}\right)^2 \geq 4\sqrt[2^{n-1}]{a_1b_1},$$

hence $(a_1+b_1)^2 \geq 2^{2n}\sqrt[2^{n-1}]{a_1b_1}$. This inequality holds for all $n \geq 2$, therefore $a_1b_1 = 0$, whence $b_1 = 0$. Now,

$$0 = \sqrt[2^{2015}]{a_1b_1} = a_{2016}b_{2016} = a_{2016}, \frac{a_1}{2^{2015}} = \frac{a_1+b_1}{2^{2015}} = a_{2016} + b_{2016} = 1,$$

which implies $a_1 = 2^{2015}$.

This value can be obtained, for example, choosing $a_n = 2^{2016-n}$, $b_n = 0$ for odd n and $a_n = 0$, $b_n = 2^{2016-n}$ for even n.

424. Proof by induction: the first step is obvious, assume that $P(i) = a_i$, $0 \leq i \leq N$, for a polynomial P. Then we prove that $P(n) = a_n$ implies $P(n+1) = a_{n+1}$. For

$n \leq N — 1$ this is clear, for $n \geq N$ we have

$$\sum_{k=0}^{n+1} P(k)(-1)^k\binom{n+1}{k} = 0, \sum_{k=0}^{n+1} a_k(-1)^k\binom{n+1}{k} = 0,$$

so $P(n+1) = a_{n+1}$.

425. *a*) If $P_c(r) = Q_c(r) = 0$, then $(P_c - Q_c)(r) = (a-c)(r^3 - r)$. Since $r \neq 0$, it follows that $r = \pm 1$. From $P_c(-1) = P_c(1) = 0$ and $Q_c(-1) = Q_c(1) = 0$, we deduce $a = -c$, $b = -2$, so $P_c(x) = (x^2-1)(x^2-cx-1)$, $Q_c(x) = (x^2-1)(x^2+cx-1)$. For any c there is an unordered pair $\{P_c(x), Q_c(x)\}$, hence the number of unordered pairs is 2017.

b) Let x_1, x_2 be the roots of the polynomial $P(x) = x^2 - cx - 1$, then the polynomial $Q(x) = x^2 + cx - 1$ has the roots $-x_1$, $-x_2$, therefore $A_c\Delta B_c = \{x_1, x_2, -x_1, -x_2\}$, with sum of the elements 0, since x_1, x_2, $-x_1$, $-x_2 \neq 0$.

426. More generally,

$$\sum_{cyc}\frac{a^n}{(a-b)(a-c)} > \binom{n}{2}, n \geq 3.$$

For $n \geq 3$,

$$\sum_{cyc}\frac{a^n}{(a-b)(a-c)} = \sum_{i+j+k=n-2} a^ib^jc^k.$$

Since there are $\binom{n}{2}$ triples (i,j,k) of nonnegative integers such that $i+j+k = n-2$, by the AM – GM inequality and the condition $abc = 1$,

$$\sum_{i+j+k=n-2} a^i b^j c^k \geq \binom{n}{2}.$$

The equality case $a = b = c$ cannot hold, whence the required result.

427. We will prove first the following result:

If $p \in \mathbb{R}[x]$ is the square of a nonconstant polynomial and $a \in \mathbb{R}^*$, then $p - a$ is not a square. Indeed, if $p = q^2, p - a = h^2$ are nonconstant, then $a = (q - h)(q + h)$, contradiction, since $q - h$, $q + h$ are nonconstant.

Let $f(x) = (l(x))^2(x - b_1)\ldots(x - b_k)$, $l \in \mathbb{R}[x]$, $b_1, \ldots, b_k \in \mathbb{R}$. Then

$$f(g(x)) = (l(g(x)))^2(g(x) - b_1)\ldots(g(x) - b_k), (g(x) - b_1)\ldots(g(x) - b_k) = (h(x))^2.$$

If $g(t) = b_1$ for some t, then $h(t) = 0$, so $h(x) = (x - t)j(x)$. Therefore $(x - t)^2 | g(x) - b_1$, which implies that $g(x) - b_i$ is a square, $i = 1, , , , , k$. By the above result, $k = 1$, so $f(x) = (l(x))^2(x - t)$ and $g(x) = (u(x))^2 + t$. We deduce

$$g(f(x)) = (u(f(x)))^2 + t,$$

which is not a square, as we have shown.

428. *a*) From $f_n(1)=1-n<0$, $f_n(2)=1>0$, by the Intermediate Value Theorem, it follows that f_n has a root $\alpha_n \in (1,2)$. Denote $g_n(x)=x^n+x^{n-1}+\ldots+x-1$. Then $x_0>0$ is a root of f_n if and only if $\frac{1}{x_0}$ is a root of g_n. Since g_n is increasing on $(0,\infty)$, $g_n(0)=-1$, $g_n(1)=n-1$, we deduce that g_n has a unique positive root, hence f_n has a unique positive root.

b) For $x \neq 1$, we have

$$f_n(x) = \frac{x^{n+1} - 2x^n + 1}{x - 1}.$$

The derivative of the numerator vanishes at

$$x_n = \frac{2n}{n + 1}$$

and is positive on $(x_n, 2)$. At this point f_n is negative:

$$f_n(x_n) < 0 \Leftrightarrow x_n^{n+1} - 2x_n^n + 1 < 0 \Leftrightarrow x_n^n > \frac{1}{2 - x_n} \Leftrightarrow \frac{2^{n+1}}{n + 1} > \left(1 + \frac{1}{n}\right)^n.$$

This inequality is true, because

$$\frac{2^{n+1}}{n+1} > 3 > \left(1 + \frac{1}{n}\right)^n.$$

The function f_n is increasing on $(x_n, 2)$ and

$$0 = f_n(\alpha_n) < f_n(\alpha_{n+1}) = \alpha_{n+1}^n (2 - \alpha_{n+1}),$$

so $\alpha_n < \alpha_{n+1}$.
c) The sequence (α_n) is bounded and increasing, so it is convergent. From

$$\frac{2n}{n+1} < \alpha_n < 2$$

we deduce

$$\lim_{n \to \infty} \alpha_n = 2.$$

429. Observe that the first and the last term of the given inequality are equal:

$$\left(\frac{b^2}{a+b} + \frac{c^2}{b+c} + \frac{a^2}{c+a}\right) - \left(\frac{a^2}{a+b} + \frac{b^2}{b+c} + \frac{c^2}{c+a}\right) = (b-a) + (c-b) + (a-c) = 0.$$

Now, we have the equivalences:

$$\frac{b^2}{a+b} + \frac{c^2}{b+c} + \frac{a^2}{c+a} = \frac{c^2}{a+b} + \frac{a^2}{b+c} + \frac{b^2}{c+a} \Leftrightarrow a^2b^2 + b^2c^2 + c^2a^2 - a^4 - b^4 - c^4 = 0 \Leftrightarrow \left(a^2 - b^2\right)^2 + \left(b^2 - c^2\right)^2 + \left(c^2 - a^2\right)^2 = 0 \Leftrightarrow a = b = c.$$

430. We have

$$\begin{aligned} a_n &= \sqrt{(n+2)\sqrt[3]{(n+2)(n+3)\sqrt[4]{(n+2)(n+3)(n+4)\sqrt[5]{(n+2)(n+3)(n+4)(n+5)}}}} \\ &= \sqrt[120]{(n+2)^{86}(n+3)^{26}(n+4)^{6}(n+5)} \\ &\le \frac{86(n+2) + 26(n+3) + 6(n+4) + (n+5) + 1}{120} = \frac{119n}{120} + \frac{7}{3} \end{aligned}$$

and the equality is impossible.

431. If P has a root z with $|z| > 1$, then

$$0=|(z-1)P(z)|=\left|a_n z^{n+1}+(a_{n-1}-a_n)z^n+\ldots+(a_0-a_1)z-a_0\right|\geq\left|a_n z^{n+1}\right|$$
$$-\left|(a_{n-1}-a_n)z^n\right|-\ldots-|a_0-a_1|z-|a_0|>a_n\left|z^{n+1}\right|-(a_n-a_{n-1})\left|z^{n+1}\right|-\ldots$$
$$-(a_1-a_0)\left|z^{n+1}\right|-a_0\left|z^{n+1}\right|=0,$$

contradiction.

432. Answer: $n = 3125$.

Define $f = xg$, where $g\colon \mathbb{Z}_+ \to \mathbb{Q}$. Then

$$i)\ g(p)=\frac{1}{p} \text{ for all prime } p;\ ii)\ g(xy)=g(x)+g(y) \text{ for all } x, y.$$

From $ii)$ it follows that $g(1)=2g(1)$, hence $g(1)=0$. Let $n=\prod_{i=1}^{k} p_i^{a_i}$ the canonical decomposition of n, then

$$g(n)=\sum_{i=1}^{k} a_i g(p_i)=\sum_{i=1}^{k}\frac{a_i}{p_i}$$

and we need to solve the equation $g(n) = 1$, that is,

$$\sum_{i=1}^{k}\frac{a_i}{p_i}=1.$$

From this we deduce $p_1|a_1$, hence $k = 1$, $a_1 = p_1 = p$ and $n = p^p$.

The relations $3^3 = 27 < 2016 < 3125 = 5^5$ show that $n = 3125$.

433. The equality $x^3 + y^3 + (-xy)^3 = 3(x)(y)(-xy)$ forces $x + y - xy = 0$ or $x = y= = -xy$. In the first case, we have

$$\frac{1}{x}+\frac{1}{y}=1.$$

In the second case, we have $x = y = -1$, hence

$$\frac{1}{x}+\frac{1}{y}=-2.$$

434. Denote $k = ab + bc + ca$ and consider the polynomial functions

$$f(x)=(x-a)(x-b)(x-c)=x^3-51x^2+kx-4000,$$

$$g(x) = (x - 10)(x - 16)(x - 25) = x^3 - 51x^2 + 810x - 4000.$$

Then the function $h(x) = f(x) - g(x) = (k - 810)x$ is linear and

$$h(c) = (k - 810)c = f(c) - g(c) = 0 - (c - 10)(c - 16)(c - 25) \leq 0,$$

since $c \geq 25$, whence $k - 810 \leq 0$. From $0 < a \leq 10$, it follows that

$$h(a) = (k - 810)a = f(a) - g(a) = 0 - (a - 10)(a - 16)(a - 25) \geq 0,$$

so that $k - 810 \geq 0$, therefore $k = 810$.

We deduce $\{a, b, c\} = \{10, 16, 25\}$. But $a \leq 10$, $c \geq 25$, hence $(a, b, c) = (10, 16, 25)$.

435. If $u_n = 2n^2 - 2n + 1$, then $4n^4 + 1 = u_n u_{n+1}$, hence

$$a_n = u_n + u_{n+1} - 2\sqrt{u_n u_{n+1}} \Rightarrow \sqrt{a_n} = \sqrt{u_{n+1}} - \sqrt{u_n} \Rightarrow \sqrt{a_1} + \ldots + \sqrt{a_{119}}$$
$$= \sqrt{u_{120}} - \sqrt{u_1} = 168.$$

436. There are no two consecutive negative values, otherwise $C_n < 0$. The sequence consists then in blocks containing nonnegative values and a negative value. Consider an arbitrary block of positive terms $a_k, a_{k+1}, \ldots, a_{k+l-1}$. We call *initial* the term a_k and *final* the term a_{k+l-1}. Let P be the sum of all initial terms, K be the sum of the final ones, N be the sum of all non − positive terms and S be the sum of positive terms which are not initial or final.

Summing the inequalities $a_{k+l-1} \leq a_{k+l} + C(n)a_{k+l+1}$ over all blocks yields

$$K \leq N + C(n)P.$$

Since $N = -K - S - P$, we get $2K \leq (C(n) - 1) - S$ (1). If $C(n) \leq 2$, then summing the inequalities $a_{k+2i} \leq a_{k+2i+1} + 2a_{k+2i+2}$ for $0 \leq i \leq \left\lfloor \frac{l-3}{2} \right\rfloor$ and adding the inequality $a_{k+l-2} \leq 2a_{k+l-1}$ if $2|l$, we obtain

$$a_k \leq a_{k+1} + a_{k+2} + \ldots + a_{k+l-2} + 2a_{k+l-1}.$$

Summing over all blocks then yields $P \leq S + 2K$ (2), with equality only if the length l of the block is odd. Indeed, if l is even, then all inequalities are equalities, in particular, $a_{k+l-2} = 2a_{k+l-1}$, which contradicts the condition $0 < a_{k+l-2} \leq a_{k+l-1}$.

Summing (1) and (2) gives us $0 \leq (C(n) - 2)P$, therefore $C(n) \geq 2$. If $C(n) = 2$, then all blocks have odd lengths, so n is even. The example $x_r = (-1)^r$ shows that $C(n) = 2$ for even n.

Remark. An example for $n = 4$ is $x_1 = 3a + 2b, x_2 = a, x_3 = a + b, x_4 = -5a - 3b$, where $a \in \mathbb{R}$, $b \geq 0$. It can be shown that

$$C(3)=3, C(5)=\frac{1+\sqrt{11}}{2}, \lim_{n\to\infty} C(n)=2.$$

437. We have

$$\sum_{\text{cyc}}\frac{c}{a+2b}+\sum_{\text{cyc}}\frac{a+2b}{c}+8=\sum_{\text{cyc}}\frac{a+2b+c}{a+2b}+\sum_{\text{cyc}}\frac{a+2b+c}{c}$$

$$=\sum_{\text{cyc}}\frac{(a+2b+c)^2}{(a+2b)c}\geq\frac{\left(\sum_{\text{cyc}}(a+2b+c)\right)^2}{\sum_{\text{cyc}}(a+2b)c}=\frac{8(a+b+c+d)^2}{ab+ac+ad+bc+bd+cd}.$$

438. Answer: $P=0$; $P=1$; $P=(x^2-2x+3)^n$, $n\in\mathbb{N}^*$.

It is easy to verify these solutions. If P is a nonconstant solution, then deg (P) is even. Indeed, if P has a real root t, then $t^2-t+3>t$ is also a real root, hence P has infinitely many real roots, contradiction. So, P is without real roots, in particular, P has even degree. As in the solution of problem **396**, we can prove that for a given degree the solution is unique.

3.47 2018

439. We will prove that

$$a_i(x+1)^{\frac{1}{a_i}}\leq x+a_i,$$

with equality if and only if $a_i=1$. If $a_i=1$, then the equality is true. If $a_i>1$, then the AM $-$ GM inequality applied to $x+1$, $1,\ldots,1$ (a_i-1 times) yields

$$\frac{(x+1)+(1+\ldots+1)}{a_i}\geq\sqrt[a_i]{x+1}\Longrightarrow a_i(x+1)^{\frac{1}{a_i}}\leq x+a_i.$$

Since $x+1>1$, the inequality is strict for $a_i>1$.

Multiplying these inequalities for $i=1,\ldots,n$ yields

$$\prod_{i=1}^{n}a_i(x+1)^{\frac{1}{a_i}}\leq\prod_{i=1}^{n}(x+a_i)\Leftrightarrow M(x+1)^{\sum_{i=1}^{n}\frac{1}{a_i}}-\prod_{i=1}^{n}(x+a_i)\leq 0\Leftrightarrow P(x)\leq 0$$

with equality if and only if $a_i=1$ for all $i\in\{1,2,\ldots,n\}$. This implies $M=1$, contradiction, hence $P(x)<0$ for all nonnegative x, and P has no positive roots.

440. We will use the following well-known estimates:

$$H(n) \leq 1 + \ln n, n! \geq \left(\frac{n}{e}\right)^n, e^x \geq \left(1 + \frac{x}{2}\right)^2, x \geq 0.$$

The given inequality holds for $C = 2\sqrt{e}$. Indeed, by the AM – GM inequality,

$$4e\left(\sum_{i=1}^{m} ia_i\right) \geq 4em\sqrt[m]{m!} \cdot \sqrt[m]{\prod_{i=1}^{m} a_i} \geq 4m^2\sqrt[m]{\prod_{i=1}^{m} a_i} \geq 4m^2\left(\frac{\ln\left(\prod_{i=1}^{n} a_i\right)}{2m} + 1\right)^2$$

$$> \left(\ln\left(\prod_{i=1}^{m} a_i\right) + m\right)^2 = \left(\sum_{i=1}^{m} (\ln a_i + 1)\right)^2 \geq \left(\sum_{i=1}^{m} H(a_i)\right)^2.$$

441. Take $x = a + 2b + c$, $y = a + b + 2c$, $z = a + b + 3c$. Then

$$K = \frac{2y}{x} + \frac{4x}{y} + \frac{4z}{y} + \frac{8y}{z} - 17.$$

By the AM–GM inequality,

$$K \geq 2\sqrt{8} + 2\sqrt{32} - 17 = 12\sqrt{2} - 17.$$

Equality holds for $4x^2 = 2y^2 = z^2$. Solving the corresponding linear system in a, b, c, we find $(a, b, c) = \left(a, \left(1 + \sqrt{2}\right)a, \left(4 + 3\sqrt{2}\right)a\right), a > 0$.

442. Answer: $f(x) = 0$ for all x or $f(x) = x$ for all x.

For $(x, y) = (x, 1)$, we get $f(x + f(x)) = 2xf(1)$; for $(x, y) = (1, x)$, we get $f(x + f(x)) = 2f(x)$. Hence $f(x) = xf(1)$. Now, it remains to find $f(1)$. Choose $x = 1$ in the equality $f(x + f(x)) = 2f(x)$, then $f(1 + f(1)) = 2f(1)$; choose $x = 1 + f(1)$ in the equality $f(x) = xf(1)$, then $f(1 + f(1)) = (1 + f(1))f(1)$. Hence $2f(1) = (1 + f(1))f(1)$, so that $f(1) = 0$ or $f(1) = 1$.

Both functions $f(x) = 0$ and $f(x) = x$ satisfy the given equation.

443. By definition, we have $b_n - b_{n-2} = a_n - a_{n-2} + b_{n-1}$, $n \geq 3$, hence $b_n - b_{n-2} = a_{n-1} + b_{n-1}$, so that $b_n = a_{n-1} + b_{n-1} + b_{n-2}$. We will prove the inequalities by induction on n.

The left-hand inequality is true for $n = 2$ and $n = 3$:

$$\frac{b_2}{2a_2} = \frac{a_2 + b_1}{2a_2} = \frac{a_1 + a_2}{2a_2} > \frac{a_2}{2a_2} = \frac{1}{2} > \frac{1}{3},$$

$$\frac{b_3}{3a_3} = \frac{a_3 + b_2}{3a_3} = \frac{2(a_1 + a_2)}{3(a_1 + a_2)} = \frac{2}{3} > \frac{1}{3}.$$

Assume $n \geq 4$ and suppose that the left – hand inequality is true for $n - 1$ and $n - 2$. Then

$$b_n = a_{n-1} + b_{n-1} + b_{n-2} > a_{n-1} + \frac{(n-1)a_{n-1}}{3} + \frac{(n-2)a_{n-2}}{3} = \frac{n(a_{n-1} + a_{n-2})}{3}$$
$$+ \frac{2(a_{n-1} - a_{n-2})}{3} > \frac{na_n}{3}.$$

The last inequality follows from $a_{n-1} = a_{n-2} + a_{n-3} > a_{n-2}$. We deduce that the inequality is true for all $n \geq 2$.

The right-hand inequality holds for $n = 3$ and $n = 4$:

$$\frac{b_3}{3a_3} = \frac{2}{3} < 1, \quad \frac{b_4}{4a_4} = \frac{a_4 + b_3 + b_1}{4a_4} < 1.$$

Assume $n \geq 5$ and suppose that the inequality is true for $n - 1$ and $n - 2$. Then

$$b_n = a_{n-1} + b_{n-1} + b_{n-2} < a_{n-1} + (n-1)a_{n-1} + (n-2)a_{n-2}$$
$$= n(a_{n-1} + a_{n-2}) - 2a_{n-2} < na_n.$$

We deduce that the inequality is true for all $n \geq 3$.

444. By the Hölder's and AM–GM inequalities, we have

$$\sum_{cyc} \frac{x}{\sqrt{x^2 + 2yz + 2}} = \sqrt{\frac{\left(\sum_{cyc} \frac{x}{\sqrt{x^2+2yz+2}}\right)^2 \left(\sum_{cyc} x(x^2 + 2yz + 2)\right)}{\sum_{cyc}(x^3 + 2xyz + 2x)}}$$
$$\geq \sqrt{\frac{(x+y+z)^3}{\sum_{cyc}(x^3 + 2xyz + 6xyz)}} \geq \sqrt{\frac{(x+y+z)^3}{\sum_{cyc}(x^3 + 2xyz + 3x^2y + 3x^2z)}} = 1.$$

Equality holds when $x = y = z = \frac{\sqrt{3}}{3}$.

445. Answer: $f(x) = x^{10} + x^9 + x^8 + x^7 + x^6 + x^5 + 1$.

Observe that f satisfies the required conditions.

446. Let $Q(x) = P(x + 1) - (x + 1)$, then $Q(\sqrt[3]{2}) = 0$. The minimal polynomial of $\sqrt[3]{2}$ over the rationals is $x^3 - 2$, hence $Q(x) = (x^3 - 2)R(x)$, with $R(x) \in \mathbb{Z}[x]$. We have

$$Q\left(\sqrt{5}\right)=\left(-2+5\sqrt{5}\right)R\left(\sqrt{5}\right)=\left(-2+5\sqrt{5}\right)\left(a+b\sqrt{5}\right)=1+2\sqrt{5}, a, b\in\mathbb{Z}.$$

This is a contradiction, since

$$\frac{1+2\sqrt{5}}{-2+5\sqrt{5}}=\frac{\left(1+2\sqrt{5}\right)\left(2+5\sqrt{5}\right)}{121}\notin\mathbb{Z}\left[\sqrt{5}\right].$$

447. Multiplying the given relations, we have

$$a_{n+1}b_{n+1}=1+a_nb_n+\frac{1}{4a_nb_n}\geq 2,$$

which implies $a_nb_n > n + 1$ for all n, hence

$$\max\{a_{2018}, b_{2018}\}>\sqrt{2019}>44.$$

448. Let $x_i = a_i + b_i$. Then

$$\sum_{i=1}^{n} x_i=2n, x_n\geq x_i\geq 1, x_n^2+x_i^2\leq 1+(x_n+x_i-1)^2,$$

$$n^2+3n=\sum_{k=1}^{n}\left(a_k^2+b_k^2\right)\leq\sum_{k=1}^{n}x_k^2\leq n-1+(x_1+\ldots+x_n-(n-1))^2=n-1$$
$$+(n+1)^2=n^2+3n,$$

so $x_1 = x_2 = \ldots = x_{n-1} = 1$, $x_n = n + 1$.
The only solution is

$$(a_1, a_2, \ldots, a_n)=(0, 0, \ldots, 0), (b_1, b_2, \ldots, b_n)=(1, 1, \ldots, n+1).$$

449. The given condition is homogenous, so we can suppose

$$\sum_{i=1}^{n} a_i=\frac{n}{2},$$

which implies

$$\sum_{i=1}^{n} a_i^2=\frac{n}{2}.$$

Let $b_i = 1 - a_i$, $1 \leq i \leq n$. Then

$$\sum_{i=1}^{n} b_i^2 = \sum_{i=1}^{n} (1 - a_i)^2 = \sum_{i=1}^{n} a_i^2 = \frac{n}{2}.$$

The sequence (a_i) is nondecreasing, (b_i) is a non – increasing sequence. Suppose that $b_m \geq 1$. Then $b_1 \geq b_2 \geq \ldots \geq b_m \geq 1$, hence

$$\sum_{i=1}^{n} b_i^2 \geq m + \sum_{i=m+1}^{n} b_i^2 \geq m.$$

so $\frac{n}{2} \geq \left\lfloor \frac{n}{2} \right\rfloor + 1$, contradiction. It follows that $1 - a_m = b_m < 1$, hence $a_m > 0$.

450. The condition is equivalent to

$$x_i^{\frac{1}{\lfloor x_i \rfloor}} = x_j^{\frac{1}{\lfloor x_j \rfloor}},$$

hence $x_i^{\frac{1}{\lfloor x_i \rfloor}}$ is constant. Consider the function $f(x) = x^{\frac{1}{\lfloor x \rfloor}}$. For n positive integer and $n \leq x < n + 1, f$ attains all value in the interval $I_n = \left[\sqrt[n]{n}, \sqrt[n]{n+1}\right)$.

The number $\frac{10}{7} \approx 1{,}428 \in I_k, 1 \leq k \leq 4$. For $n \geq 3$, the left endpoint of I_n is strictly larger than the right endpoint of $I_{n+2}, I_{n+3}, \ldots$.

Hence the maximum possible value is $n = 4$.

451. Let

$$a = \frac{x}{y+z+t}, b = \frac{y}{z+t+x}, c = \frac{z}{t+x+y}, d = \frac{t}{x+y+z},$$

then

$$\sum_{cyc} \sqrt[3]{abc} = \sum_{cyc} \sqrt[3]{\frac{xyz}{(y+z+t)(z+t+x)(t+x+y)}} \leq \sum_{cyc} \frac{\frac{x+y}{t+x+y} + \frac{y+z}{t+y+z} + \frac{z+x}{t+z+x}}{6}$$
$$= \frac{4}{3}.$$

452. Take $a = 2019x$, $b = 2019y$, $c = 2019z$. Then

$$\frac{1}{x+1} + \frac{1}{y+1} + \frac{1}{z+1} = 1.$$

This equality is equivalent to $xyz = x + y + z + 2$. By the AM–GM inequality,

$$xyz = x + y + z + 2 \geq 4\sqrt[4]{2xyz},$$

hence $xyz \geq 8$ and $abc = 2019^3 xyz \geq 2019^3 \cdot 2^3 = 4038^3$.

453. The expression from the left is less than

$$\int_0^1 (1 - x^2)dx = \left(x - \frac{x^3}{3}\right)\Big|_0^1 = \frac{2}{3}.$$

454. We have

$$\sum \frac{a}{a^2b + 2} = \sum \frac{1}{\frac{1}{c} + \frac{2}{a}} \leq \frac{1}{9} \sum (2a + c) = \frac{a + b + c}{3}.$$

455. *a)* Case 1. n odd. Then $k \in \phi_n$ if and only if $n - k \in \phi_n$ and $(-1)^k + (-1)^{n-k} = 0$, so we can choose $r = 1$.

Case 2. $4|n$. Then $i^k + i^{n-k} = 0$, for $k \in \phi_n$, so $P_n(i) = 0$ and we can choose $r = 2$.

Case 3. $4|n - 2$. Then there is an odd prime factor p of n. Let $r = 2^k$ be the maximal power of 2 dividing $p + 1$ and let s be the product of all prime divisors of n excepting 2 and p. We will show that $x^r + 1|P_n$.

By the Möbius inversion formula

$$xP_n(x) = (x^n - 1)\sum_{d|n} \frac{\mu(d)}{x^d - 1} = (x^n - 1)\sum_{d|s} \mu(d)$$
$$\times \left(\frac{1}{x^d - 1} - \frac{1}{x^{2d} - 1} - \frac{1}{x^{pd} - 1} + \frac{1}{x^{2pd} - 1}\right).$$

All denominators are not divisible by $x^r + 1$, since $4 \nmid n$. The polynomial $x^r + 1$ is irreducible over $\mathbb{Q}[x]$, so it suffices to prove that $x^r + 1$ divides each summand. Since $x^r + 1|x^{rd} + 1$, it suffices to prove this for $d = 1$, that is, $x^r + 1$ divides

$$\frac{1}{x - 1} - \frac{1}{x^2 - 1} - \frac{1}{x^p - 1} + \frac{1}{x^{2p} - 1}.$$

In the field $\frac{\mathbb{Q}[x]}{(x^r+1)}$, we have $x^{p+1} = -1$, hence

$$\frac{1}{x-1} - \frac{1}{x^2 - 1} - \frac{1}{x^p - 1} + \frac{1}{x^{2p} - 1} = \frac{x}{x^2 - 1} - \frac{x^p}{x^{2p} - 1} = \frac{x}{x^2 - 1} - \frac{x^{p+2}}{x^{2p+2} - x^2}$$
$$= \frac{x}{x^2 - 1} + \frac{x}{1 - x^2} = 0,$$

as desired.

b) The polynomial Q_n must be constant, hence $P_n = x^r + 1$. In particular, $\varphi(n) = P_n(1) = 2$. It is well known that this happens only for $n \in \{3, 4, 6\}$.

3.48 2019

456. Squaring, we get

$$a_{n+2}^2 = a_{n+1}^2 + \frac{2a_{n+1}}{a_n} + \frac{1}{a_n^2} > a_{n+1}^2 + 2.$$

This holds since $a_{n+1} \geq a_n$. Summing it, we get

$$a_{n+2}^2 > a_4^2 + 2(n + 2 - 4).$$

Since $a_4 = 3$, we have $a_{180}^2 > 9 + 2(180 - 4) = 361$. The sequence contains only positive terms, hence $a_{180} > 19$.

457. Let $p = ax^2 + bx + c$. Then

$$c \not\equiv 0 \pmod 3), \mathrm{a} + \mathrm{b} + \mathrm{c} \not\equiv 0 \pmod 3), \mathrm{a} + 2\mathrm{b} + \mathrm{c} \not\equiv 0 \pmod 3),$$

with solutions $\pm(a, b, c) \in \{(0, 0, 1), (1, 0, 1), (2, 1, 1), (2, 2, 1)\}$. It suffices to consider the "+" cases, since otherwise we can change all signs.

$$(a, b, c) = (0, 0, 1), f = x^6 + x^4 + x^2 + 1,$$
$$(a, b, c) = (1, 0, 1), f = x^4 + 1,$$
$$(a, b, c) = (2, 1, 1), f = 2x^4 + 2x^3 + 2x + 1,$$
$$(a, b, c) = (2, 2, 1), f = 2x^4 + x^3 + x + 1.$$

458. Let $S = \{x \in \mathbb{Z} | f(x) = x + 1\}$. Then

a) $S \neq \varnothing$. For $x \to f(x)$, we have

$$f(f(f(x))) = f\left(xf(x) - x^2 + 2\right) = f(x)f(f(x)) - (f(x))^2 + 2 = x(f(x))^2 - x^2 f(x)$$
$$+ 2f(x) - (f(x))^2 + 2.$$

Let $d \in \mathbb{Z}$ such that $f(d) = 1$, then $f(f(1)) = f(d - d^2 + 2) = d - d^2 + 3$, hence $d - d^2 + 2 \in S$.

b) $a \in S \Rightarrow a + 1 \in S$. Indeed, if $f(a) = a + 1$, then

$$f(a+1) = a(a+1) - a^2 + 2 = a+2,$$

hence $a + 1 \in S$

c) $a \in S \Rightarrow a - 1 \in S$. For $t \in S$, there exists c such that $f(c) = t$. From the given equation we get $t + 1 = ct - c^2 + 2$, whence $c = 1$ or $c = t - 1$. But $f(1)$ has only one value, therefore 1 is the only possible value not in S. The function f is surjective, so for $k \in \mathbb{Z} - \{1\}$ we have $f(k) = k + 1$. If $k + 1 = 2$, then $k = 1$, so $1 \in S$.

We conclude that $S = \mathbb{Z}$, hence $f(x) = x + 1$ is the only function that satisfies the given conditions.

459. For every positive integer n, $\sqrt{n-1} < \sqrt{n} < \sqrt{n+1}$, then

$$\sqrt{n-1} + \sqrt{n} < 2\sqrt{n} < \sqrt{n} + \sqrt{n+1},$$

$$2\left(\sqrt{n+1} - \sqrt{n}\right) < \frac{1}{\sqrt{n}} < 2\left(\sqrt{n} - \sqrt{n-1}\right).$$

Summing it from 2 to 2019, we get

$$2\left(\sqrt{2020} - \sqrt{2}\right) < \sum_{n=2}^{2019} \frac{1}{\sqrt{n}} < 2\left(\sqrt{2019} - 1\right).$$

From $\sqrt{2020} > 44$ and $\sqrt{2} < \frac{3}{2}$, it follows that $2\left(\sqrt{2020} - \sqrt{2}\right) > 2\left(44 - \frac{3}{2}\right) = 85$. Also, $\sqrt{2019} < 45$ implies $2\left(\sqrt{2019} - 1\right) < 88$. We deduce

$$85 < \sum_{n=2}^{2019} \frac{1}{\sqrt{n}} < 88.$$

Direct calculations show that $2\left(\sqrt{2020} - \sqrt{2}\right) > 87$, hence

$$88 < \sum_{n=1}^{2019} \frac{1}{\sqrt{n}} < 89.$$

Finally,

$$\left\lfloor \sum_{n=1}^{2019} \frac{1}{\sqrt{n}} \right\rfloor = 88.$$

460. It is known that $\tan A + \tan B + \tan C = \tan A \tan B \tan C$, if $A + B + C = 180°$. Let z be the fourth solution of the equation. By the Viète's formulae,

$$p = \tan A + \tan B + \tan C + z,$$
$$q = \tan A \tan B + \tan B \tan C + \tan C \tan A + (\tan A + \tan B + \tan C)z,$$
$$r = \tan A \tan B \tan C + (\tan A \tan B + \tan B \tan C + \tan C \tan A)z,$$
$$s = (\tan A \tan B \tan C)z.$$

The first equation and the above condition imply

$$\tan A \tan B \tan C = \tan A + \tan B + \tan C = p - z.$$

From the second and the fourth equations, we deduce

$$\tan A \tan B + \tan B \tan C + \tan C \tan A = q - s.$$

Replacing this in the third equation yields

$$r = (p - z) + (q - s)z,$$

hence

$$z = \frac{r - p}{q - s - 1}.$$

461. Denote $a = 2^x - 4$, $b = 4^x - 2$, then $a^3 + b^3 = (a + b)^3$, which is equivalent to

$$ab(a + b) = 0.$$

If $a = 0$, then $2^x = 4$, whence $x = 2$.
If $b = 0$, then $4^x = 2$, whence $x = \frac{1}{2}$.
If $a + b = 0$, then $4^x + 2^x - 6 = (2^x + 3)(2^x - 2) = 0$, whence $x = 1$.

462. The first equation represents the union of the angle bisectors of the coordinate axes; the second equation represents a circle with center $(a, 0)$ and radius 1.
Replacing y^2 from the first equation in the second, we get

$$(x - a)^2 + x^2 = 1,$$

with solutions

$$x_{1,2} = \frac{a \pm \sqrt{2 - a^2}}{2}.$$

Having in mind the geometric representation, we conclude that the system has:

0 solutions, for $|a| > \sqrt{2}$;
2 solutions, for $a = \pm\sqrt{2}$;
3 solutions for $a = \pm 1$;
4 solutions for $a \in \left(-\sqrt{2}, -1\right) \cup (-1, 1) \cup \left(1, \sqrt{2}\right)$.

There are no values of a for which the system has 1 or 5 solutions.

463. From $x + \frac{1}{x} = 3$ it follows that $x \in \{r_1, r_2\}$, where r_1, r_2 are the solutions of the equation $x^2 - 3x + 1 = 0$. Similarly, $y, z \in \{r_1, r_2\}$. There are eight possible values of the triple (x, y, z).

If $x = y = z$, then $\frac{x}{y} + \frac{y}{z} + \frac{z}{x} = 3$.

If $x \neq y = z$, then $\frac{x}{y} + \frac{y}{z} + \frac{z}{x} = \frac{r_1^2 + r_2^2}{r_1 r_2} + 1 = 3^2 - 2 \cdot 1 + 1 = 8$.

In conclusion, the given expression can take only two values, whose sum is

$$3 + 8 = 11.$$

464. Write the equation as

$$\sqrt{3(x+1)^2 + 4} + \sqrt{5(x+1)^2 + 9} = 5 - (x+1)^2.$$

We have LHS $\geq \sqrt{4} + \sqrt{9} = 5$ and RHS ≤ 5, with equality only for $x = -1$.

In conclusion, the given equation has the unique solution $x = -1$.

465. Answer: $f(x) = 0$ for all x; $f(x) = 2|x|$ for all x.

Clearly, the above functions are solutions. Let f be an arbitrary solution. Substituting $-x$ for x in the given equation, we get

$$f(x^2)f(y^2) + |x|f(xy^2) = 3|y|f(x^2y), \quad (1)$$

which combined with the original equation yields $|x|f(-xy^2) = |x|f(xy^2)$. Then $f(-y^2) = f(y^2)$, hence $f(-x) = f(x)$ for all real x. If $x = y = 0$ in the given equation, we get $f(0) = 0$. Interchanging in (1) x, y, we obtain

$$f(x^2)f(y^2) + |y|f(x^2y) = 3|x|f(xy^2). \quad (2)$$

Subtracting (2) from (1), we get $|x|f(xy^2) - |y|f(x^2y) = 3|y|f(x^2y) - 3|x|f(-xy^2)$. Simplifying and dividing by 4, we get

$$|x|f(xy^2) = |y|f(x^2y). \quad (3)$$

Then $y = 1$ gives us

$$f(x^2) = |x|f(x) \quad (4)$$

for all x. Substitution of (3) to (1) or (2) gives us

$$f(x^2)f(y^2) = 2|x|f(xy^2). \quad (5)$$

Substituting $f(x^2)$ from (4) to (5) gives us $|x|f(x)f(y^2) = 2|x|f(xy^2)$. For $x > 0$ and $y = \sqrt{x}$, we get

$$f(x^2) = \frac{1}{2}(f(x))^2. \quad (6)$$

Since f is an even function and $f(0) = 0$, equation (6) holds also for $x \leq 0$. Subtracting (4) from (6) gives $\frac{1}{2}(f(x))^2 - |x|f(x) = 0$, whence $f(x) = 0$ or $f(x) = 2|x|$. For each $x \neq 0$, only one of these equalities can hold.

Let $c \neq 0$ such that $f(c) = 0$. Taking $x = c$ in (4), we obtain $f(c^2) = 0$. Taking $x = c$ in (5) and simplifying, we get $f(cy^2) = 0$ for all y, therefore $f(-x) = 0$ for all x, as f is an even function. In conclusion, these are the only solutions of the given functional equation.

466. Let

$$B(x) = b_n x^n + b_{n-1}x^{n-1} + \ldots + b_1 x + b_0,$$

$$Q(x) = q_m x^m + q_{m-1}x^{m-1} + \ldots + q_1 x + q_0,$$

$$A(x) = a_{n+m}x^{n+m} + a_{n+m-1}x^{n+m-1} + \ldots + a_1 x + a_0.$$

By assumption, the coefficients a_i, b_j are all interesting and $b_0 = 1$. We will prove by induction that all coefficients q_k are interesting. As $A(x) = B(x)Q(x)$, we get that $a_0 = b_0 q_0 = q_0$ is interesting. It is easy to see that the sum and the product of two interesting numbers are also interesting. Suppose that $q_0, \ldots, q_{k-1}$ are interesting. Then the equality

$$q_k = a_k - b_k q_0 - b_{k-1}q_1 - \ldots - b_1 q_{k-1}$$

shows that q_k is interesting.

467. Taking the derivative gives

$$-f'(x) = \frac{a}{(x-1)^2} + \frac{1}{(x-2)^2} + \frac{1}{(x-6)^2}.$$

From $f'(4) = 0$, it follows that

$$0=\frac{a}{9}+\frac{1}{4}+\frac{1}{4},$$

whence $a=-\frac{9}{2}$.

468. Adding the inequalities

$$(a-3)^2(a+11)\geq 0\Leftrightarrow a^3+5a^2\geq 57a-99,$$
$$(b-4)^2(b+13)\geq 0\Leftrightarrow b^3+5b^2\geq 88b-208,$$
$$(c-5)^2(c+15)\geq 0\Leftrightarrow c^3+5c^2\geq 125c-375,$$

and taking account of the given condition, we get

$$a^3+b^3+c^3+5\left(a^2+b^2+c^2\right)\geq 466,$$

with equality only when $a=3$, $b=4$, $c=5$.

469. Answer: $f(x)=1$ for all x.

Using the substitutions $(x,y)=(f(a),b)$ and $(x,y)=(f(b),a)$, $a,b\in\mathbb{Q}_+$, in the functional equation, we get

$$(f(f(a)))^2f(b)=f\left((f(a))^2(f(b))^2\right)=(f(f(b)))^2f(a),$$

hence

$$\frac{(f(f(a)))^2}{f(a)}=\frac{(f(f(b)))^2}{f(b)},a,b\in\mathbb{Q}_+.$$

Then there is a constant $c\in\mathbb{Q}_+$ such that $(f(f(a)))^2=cf(a)$, $\forall\, a\in\mathbb{Q}_+$. Denote by f^n the nth iteration of f. We have

$$\frac{f(a)}{c}=\left(\frac{f^2(a)}{c}\right)^2=\left(\frac{f^3(a)}{c}\right)^4=\ldots=\left(\frac{f^{n+1}(a)}{c}\right)^{2^n}$$

for all positive integers n. We deduce that $\frac{f(a)}{c}$ is a 2^nth power for all n, therefore f is constant. The functional equation get us $c=c^3$, whence $c=1$.

470. Answer: $\frac{8}{\sqrt[3]{7}}$, attained at $(a,b,c,d)=(1,49,1,49)$ and cyclic permutations.

By the Hölder's inequality,

$$S^3=\left(\sum_{cyc}\frac{\sqrt[6]{a}\cdot\sqrt[6]{a}}{\sqrt[3]{b+7}}\right)^3\le\sum_{cyc}(\sqrt[6]{a})^3\cdot\sum_{cyc}(\sqrt[6]{a})^3\cdot\sum_{cyc}\left(\frac{1}{\sqrt[3]{b+7}}\right)^3$$
$$=\left(\sum_{cyc}\sqrt{a}\right)^2\sum_{cyc}\frac{1}{b+7}.$$

From

$$\frac{(x-1)^2(x-7)^2}{x^2+7}\ge 0\Leftrightarrow x^2-16x+71\ge\frac{448}{x^2+7},$$

we have

$$\sum\frac{1}{b+7}\le\frac{\sum(b-16\sqrt{b}+71)}{448}=\frac{384-16\sum\sqrt{b}}{448}=\frac{48-2\sum\sqrt{b}}{56}.$$

By the AM–GM inequality,

$$S^3\le\frac{(\sum\sqrt{a})^2(48-2\sum\sqrt{a})}{56}\le\frac{1}{56}\left(\frac{\sum\sqrt{a}+\sum\sqrt{a}+(48-2\sum\sqrt{a})}{3}\right)^3=\frac{512}{7},$$

whence the conclusion.

471. We will prove that $a_1=a_2=\ldots=a_n$ is the unique solution. Assume that there is a solution with $d=|a_1-a_2|\ne 0$. Then

$$\pm\frac{1}{1}\pm\frac{1}{2}\pm\ldots\pm\frac{1}{n}=0.$$

Choose a prime $p\in\left(\frac{n}{2},n\right)$, which exists, by the Bertrand's postulate. Thus

$$p|1\cdot 2\cdot\ldots\cdot(p-1)(p+1)\cdot\ldots\cdot n,$$

therefore $n\ge 2p$, contradiction.

472. The AM–GM inequality implies

$$\frac{1}{2}=\frac{1-x+x}{2}\ge\sqrt{x(1-x)},$$

so that $x(1-x)\le\frac{1}{4}$. Write two similar inequalities for y, z and multiply them:

$$\frac{1}{x(1-x)y(1-y)z(1-z)} \geq 4^3.$$

Now, by the AM–GM inequality, we have

$$\frac{1}{x(1-y)} + \frac{1}{y(1-z)} + \frac{1}{z(1-x)} \geq 3\sqrt[3]{\frac{1}{x(1-y)y(1-z)z(1-x)}} \geq 3 \cdot 4 = 12.$$

473. Denote $x^3 + y = a, x^2 + y^2 = b, x + y^3 = c, xy = p$. Then p is an integer. This is obvious when $x = 0$ or $y = 0$. Suppose that $xy \neq 0$. We have

$$a^2 + c^2 = b\left(x^4 - x^2y^2 + y^4 + 2xy + 1\right) \in \mathbb{Z},$$

hence $x^4 - x^2y^2 + y^4 + 2xy = b^2 - 3p^2 + 2p \in \mathbb{Q}$, therefore $3p^2 - 2p \in \mathbb{Q}$. Moreover, $ac - b^2 = p^3 - 2p^2 + p \in \mathbb{Z}$, so

$$\left(3p^2 - 2p + \frac{1}{3}\right)\left(\frac{1}{3}p - \frac{4}{9}\right) = p^3 - 2p^2 + p - \frac{4}{27} \in \mathbb{Q},$$

whence $\frac{1}{3}p - \frac{4}{9} \in \mathbb{Q}$, that is, $p \in \mathbb{Q}$.

Let $p = \frac{m}{n}$, with m, n integers, $n \neq 0$ and $\gcd(m, n) = 1$.Observe that $n \nmid \mathrm{m} - \mathrm{n}$, so $p^3 - 2p^2 + p = \frac{\mathrm{m}(\mathrm{m}-\mathrm{n})^2}{\mathrm{n}^3} \in \mathbb{Z}$ if and only if $n = 1$, that is, p is an integer.

Now, $a + c = (x + y)(b - p + 1) \in \mathbb{Z}$, $a - c = (x - y)(b + p - 1) \in \mathbb{Z}$, so $x + y$, $x - y$ are rationals. Also, $(x + y)^2 = b + 2p$, $(x - y)^2 = b - 2p$ are integers, therefore $x + y$ and $x - y$ are integers. Let $2x = u$, $2y = v$, with u, v integers. From $u^2 + v^2 \equiv 0$

(mod 4), it follows that u, v are even, hence x, y are integers.

474. We have $Q'(x) = (2n + 1)b_{2n+1}x^{2n} + \ldots + 2b_2x + b_1$. Choose

$$b_{2n+1}, b_{2n-1}, \ldots, b_1 \geq b_{2n}, b_{2n-2}, \ldots, b_2.$$

Case 1. $x \geq 0$. Then $Q'(x) > 0$.

Case 2. $x < 0$. Take $y = -x$, then

$$Q'(x) = Q'(-y) = (n+1)b_{2n+1}y^{2n+1} + \sum_{k=1}^{n} kb_{2k+1}y^{2k-2}\left[\left(y - \frac{b_{2k}}{b_{2k+1}}\right)^2 + \frac{b_{2k-1}b_{2k+1} - b_{2k}^2}{b_{2k+1}^2}\right] > 0,$$

since $b_{2k+1} \geq b_{2k}$, $b_{2k-1} \geq b_{2k}$, $k = 1, \ldots, n$.

In both cases Q is increasing, so it has a unique real root.

475. The collinearity condition is equivalent to

$$\frac{P(y)-P(x)}{y-x}=\frac{P(z)-P(x)}{z-x}.$$

Taking $x = 0$, we get $P(y) + P(-y) = 2P(0)$, which means that P is odd. Taking $y = 2x$, we have $z = -3x$ and $5P(x) = 4P(2x) + P(3x)$.

Looking at the leading coefficient of P, we have $5 = 2^{2n+3} - 3^{2n+1}$. The values $n = 0, 1$ are solutions. For $n \geq 2$, we have $3^{2n+1} > 2^{2n+3}$, hence $n \leq 1$.

Checking $P(x) = ax^3 + bx + c$, we see that this is a solution.

476. We need a lemma.

Lemma. Let $a_1, \ldots, a_n$ be nonnegative integers and let $G(x)$ be a nonzero polynomial with $\deg(G) \leq a_1 + \ldots + a_n$. Suppose that the polynomial $F(x_1, \ldots, x_n)$ satisfies

$$F(x_1,\ldots,x_n)=G(x_1+\ldots+x_n),\,(x_1,\ldots,x_n)\in\{0,1,\ldots,a_1\}\times\ldots\times\{0,1,\ldots,a_n\}.$$

Then $\deg(F) \geq \deg(G)$.

Proof of the statement. Take the unique polynomial $g(x)$ such that $g(x)=\left\lfloor\frac{x}{m}\right\rfloor$ for $x\in\{0,1,\ldots,n(m-1)\}$ and $\deg(g)\leq n(m-1)$. From $g(0)=g(1)=0, g(m)=1,$ we deduce $\deg(g)\geq 2$. Applying the lemma to $a_1=\ldots=a_n=m-1$ and $f, g,$ we conclude that $\deg(f)\geq\deg(g)$. Consider the polynomial $h(x)=g(x+m)-g(x)---1$. The degree of $g(x+m)-g(x)$ is $\deg(g)-1\geq 1$, so $\deg(h)=\deg(g)-1\geq 1$. The polynomial h is not the zero polynomial and vanishes at $0,1,\ldots,n\,(m-1)-m$, so h has at least $(n-1)(m-1)$ roots, hence

$$\deg(f)\geq\deg(g)=\deg(h)+1\geq(n-1)(m-1)+1\geq n.$$

Proof of the lemma. If $p(x)$ is a polynomial of a single variable, then define

$$(\Delta p)(x)=p(x+1)-p(x).$$

If $\deg(p) \geq 1$, then $\deg(\Delta p) = \deg(p) - 1$. For a polynomial $p(x_1, \ldots, x_n)$ with n variables, and $1 \leq k \leq n$, define

$$(\Delta_k p)(x_1,\ldots,x_n)=p(x_1,\ldots,x_{k-1},x_k+1,x_{k+1},\ldots,x_n)-p(x_1,\ldots,x_n).$$

Then $\Delta_k p$ is the zero polynomial, or $\deg(\Delta_k p) \leq \deg(p) - 1$.

Now we apply induction on $\deg(G)$. If $\deg(G) = 0$, then $F(0, \ldots, 0) = G(0) \neq 0$, so F cannot be the zero polynomial. Suppose that $\deg(G) \geq 1$ and the lemma is true for lower degrees. Since $a_1 + \ldots + a_n \geq \deg(G) > 0$, one of $a_1, \ldots, a_n$ is positive and we can suppose $a_1 \geq 1$. For the polynomials $F_1 = \Delta_1 F$, $G_1 = \Delta G$, we have

$$F_1(x_1,\ldots,x_n)=F(x_1+1,x_2,\ldots,x_n)-F(x_1,x_2,\ldots,x_n)=G(x_1+\ldots+x_n+1)$$
$$-G(x_1+\ldots+x_n)=G_1(x_1+\ldots+x_n)$$

on $\{0,\ldots,a_1-1\}\times\{0,\ldots,a_2\}\times\ldots\times\{0,\ldots,a_n\}$. Since G is nonconstant, it follows deg (G_1) = deg (G) − 1 ≤ $(a_1-1)+a_2+\ldots+a_n$, therefore we can apply the induction hypothesis to F_1, G_1 and conclude that F_1 is not the zero polynomial and deg (F_1) ≥ deg (G_1), hence deg (F) ≥ deg (F_1) + 1 ≥ deg (G_1+1) = deg (G).

477. Let $a_i = 2019x_i$, then

$$\sum_{i=1}^{2019}\frac{1}{x_i+1}=1.$$

For each $k\in\{1,2,\ldots,2019\}$,

$$\sum_{\substack{i=1\\ i\neq k}}^{2019}\frac{1}{x_i+1}=1-\frac{1}{x_k+1}=\frac{x_k}{x_k+1}.$$

By the AM−GM inequality,

$$\frac{x_k}{x_k+1}=\sum_{\substack{i=1\\ i\neq k}}^{2019}\frac{1}{x_i+1}\geq 2018\sqrt[2018]{\prod_{\substack{i=1\\ i\neq k}}^{2019}\frac{1}{x_i+1}}.$$

Multiplying all inequalities for each $k\in\{1,2,\ldots,2019\}$, we get

$$\prod_{k=1}^{2019}\frac{x_k}{x_k+1}\geq 2018^{2019}.$$

Cancelling denominators gives

$$x_1x_2\ldots x_{2019}\geq 2018^{2019}$$

and substituting by a_i yields

$$a_1a_2\ldots a_{2019}\geq(2018\cdot 2019)^{2019}.$$

Equality holds only when $a_1=a_2=\ldots=a_{2019}=2018\cdot 2019$.

478. Suppose that both assertions are false. The for every rational number $r \in (0,1)$, there is exactly one subset $F_r \subset S$ such that r is the sum of inverses of elements from F_r. in particular, if r, $s \in (0,1) \cap \mathbb{Q}$ and $x \in S$ such that $r = s + \frac{1}{x}$, then

$$x \notin F_s \Leftrightarrow x \in F_r = F_{s+\frac{1}{x}}.$$

Choose $r \in (0,1) \cap \mathbb{Q}, x \in S$ and define $n = \lfloor rx \rfloor$.From $0 < r - \frac{n}{x} < \frac{1}{x}$, it follows that $x \notin F_{r-\frac{n}{x}}$. By the above remark,

$$x \notin F_{r-\frac{n}{x}} \Leftrightarrow x \in F_{r-\frac{n-1}{x}} \Leftrightarrow x \notin F_{r-\frac{n-2}{x}} \Leftrightarrow \ldots$$

Finally, $x \in F_r$ if and only if n is odd.

Consider now $F_{\frac{2}{3}}$. If $x \in F_{\frac{2}{3}}$, then $\lfloor \frac{2x}{3} \rfloor$ is odd. But $\frac{2x}{3}$ is not an integer, so that there is $q_x \in (0,1) \cap \mathbb{Q}$ such that $\lfloor \frac{2x}{3} \rfloor = \lfloor x(\frac{2}{3} - q_x) \rfloor$. As $F_{\frac{2}{3}}$ is a finite set, there is

$$\varepsilon = \min\left\{ q_x \middle| x \in F_{\frac{2}{3}} \right\}.$$

Then $\lfloor \frac{2x}{3} \rfloor = \lfloor x(\frac{2}{3} - \varepsilon) \rfloor$ for any $x \in F_{\frac{2}{3}}$ and

$$x \in F_{\frac{2}{3}} \Rightarrow \left\lfloor \frac{2x}{3} \right\rfloor \text{ odd} \Rightarrow \left\lfloor x\left(\frac{2}{3} - \varepsilon\right) \right\rfloor \text{ odd} \Rightarrow x \in F_{\frac{2}{3}-\varepsilon},$$

hence $F_{\frac{2}{3}} \subset F_{\frac{2}{3}-\varepsilon}$, contradiction.

479. We can choose $\varepsilon = \frac{1}{n}$. Write RHS as the sum of $n-1$ arithmetic means and 1 harmonic meam and apply the AM – GM inequality:

$$(n-1)\frac{x_1 + \ldots + x_n}{n^2} + \frac{x_1 \ldots x_n}{\sum_{j=1}^{n} \prod_{i \neq j} x_i} \geq n \sqrt[n]{\left(\frac{x_1 + \ldots + x_n}{n^2}\right)^{n-1} \frac{x_1 \ldots x_n}{\sum_{j=1}^{n} \prod_{i \neq j} x_i}}.$$

We want to show that RHS is larger than $\sqrt[n]{x_1 \ldots x_n}$. This is equivalent to

$$\frac{x_1 + \ldots + x_n}{n} \geq \sqrt[n-1]{\frac{\sum_{j=1}^{n} \prod_{i \neq j} x_i}{n}},$$

which is the Maclaurin's inequality.

3.49 2020

480. Answer: $\frac{27\sqrt[4]{3}}{8}$.

By the AM–GM inequality,

$$\frac{8x^8 + 8(1-x^8)}{9} \geq \sqrt[9]{(8x^8)(1-x^8)^8} \Leftrightarrow x^8\left(1-x^8\right)^8 \leq \frac{8^8}{9^9} \Leftrightarrow x\left(1-x^8\right)$$

$$\leq \frac{8}{9\sqrt[4]{3}} \Leftrightarrow \frac{x^3}{1-x^8} \geq \frac{9\sqrt[4]{3}}{8}x^4.$$

Adding with two similar inequalities, we obtain the result.

481. Denote by A_n, Q_n the arithmetic and quadratic means, respectively. Then

$$Q_1^2 \geq A_1^2 \geq Q_2^2 \geq A_2^2 \geq \ldots$$

If the sequence (a_n) is nonconstant, then choose $p > q$ such that $|a_p - a_q| = d > 0$. For $n > p$, we have

$$Q_n^2 - A_n^2 = \frac{1}{n^2} \sum_{1 \leq i < j \leq n} \left(x_i - x_j\right)^2 \geq \frac{1}{n^2}$$

$$\times \left(\left(x_p - x_q\right)^2 + \sum_{\substack{1 \leq i \leq n \\ i \neq p,q}} \left(\left(x_i - x_p\right)^2 + \left(x_i - x_q\right)^2 \right) \right) \geq \frac{1}{n^2}\left(d^2 + (n-2)\frac{d^2}{2}\right)$$

$$= \frac{d^2}{2n}.$$

Then

$$\sum_{i>p}\left(Q_i^2 - A_i^2\right) \geq \frac{d^2}{2}\sum_{i>p}\frac{1}{i} = \infty,$$

contradiction.

482. Let $A = \{k | a_k \neq 0\}$, $B = \{k | b_k \neq 0\}$. It is well-known that

$$|b_1|^2 + \ldots + |b_n|^2 = n\left(|a_1|^2 + \ldots + |a_n|^2\right),$$

that is,

$$\sum_{k\in B}|b_k|^2 = n\sum_{k\in A}|a_k|^2.$$

Clearly,

$$|b_k| \le \sum_{k\in A}|a_k|.$$

By the Cauchy–Schwarz inequality,

$$|B| \ge \frac{n}{|A|} = \frac{n}{p}.$$

483. Answer: $f(x) = 0$ for all x or $f(x) = x^3$ for all x.
For $x = y = 0$, we have $3(f(0))^3 = 0$, whence $f(0) = 0$. (1)
For $x = 0$, taking account of (1), we get $f(y^3) = y^6 f(y)$. (2)
Replacing the first term in the RHS, the given equation becomes

$$f(x^3 + y^3) = x^6 f(x) + 3x^3 f(x) f(y) + 3f(x)(f(y))^2 + y^6 f(y). \ (3)$$

Interchanging variables and subtracting the result from (3), we obtain

$$f(x)f(y)(f(x) - x^3 - f(y) + y^3) = 0. \ (4)$$

If there are $x \neq y$ such that $f(x) = f(y)$, then

$$x^3 \neq y^3 \Rightarrow f(x) - x^3 - f(y) + y^3 \neq 0 \Rightarrow f(x) = f(y) = 0.$$

If $f(a) \neq 0$ for some a and $f(b) = 0$ for some $b \neq 0$, then choose $x = \sqrt[3]{a}, y = b$ in the given equation to obtain $f(a + b^3) = f(a)$, contradiction, as $b^3 \neq 0$. We deduce $f(x) = 0$ for all x.

If $f(x) \neq 0$ for any $x \neq 0$, then $y = 1$ in (4) yields $f(x) = x^3 + c$, for $x \neq 0$, $c = f(1) - 1$. Replacing in (2) and simplifying, we have $c = y^6 c$, for all y, so $c = 0$ and $f(x) = x^3$ for all x.

484. Answer: *a*) yes; *b*) no.
a) Let f be a function $2n$ – positive and $x_1, \ldots, x_n$ real numbers such that

$$x_1 + \ldots + x_n = 0.$$

Then

$$0 \le f(x_1) + \ldots + f(x_n) + f(x_1) + \ldots + f(x_n) = 2(f(x_1) + \ldots + f(x_n)),$$

hence f is n–positive.

b) Consider the function f defined by $f(x) = -1$ if x is an integer and $x \equiv 1$

$(\bmod 2n)$ and $f(x) = n$, otherwise.

The function f is n – positive: let $x_1, \ldots, x_n$ be real numbers with $x_1 + \ldots + x_n = 0$. If there is an integer k such that $f(x_k) = n$, then $f(x_i) \geq -1$, $i \neq k$, implies

$$f(x_1) + \ldots + f(x_n) \geq f(x_k) - (n-1) \geq 1.$$

Otherwise, $x_k \equiv 1(\bmod 2n)$, hence $x_1 + \ldots + x_n \equiv n \equiv 0(\bmod 2n)$, contradiction.

The function f is not $2n$ – positive: if $x_1 = 1 - 2n$, $x_2 = x_3 = \ldots = x_{2n} = 1$, then $x_1 + \ldots + x_{2n} = 0$, but $f(x_1) + \ldots + f(x_{2n}) = -2n < 0$.

Taking $n = 1010$, we obtain the required results.

485. Observe that $f(xf(y)) < xy + yz + zx$. For $z \to 0$, we get $f(xf(y)) \leq xy$. We have also two similar cyclic relations, so $f(xf(y)) = xy$. Taking $(x, y) = \left(\frac{x}{f(1)}, 1\right)$, we have $f(x) = \frac{x}{f(1)}$. If $x = 1$. then $(f(1))^2 = 1$, hence $f(1) = 1$ and $f(x) = x$ for all x.

486. We will prove the stronger inequality

$$\frac{a}{b} + \frac{b}{c} + \frac{c}{a} + \frac{\sqrt{3}(a+b+c)}{\sqrt{a^2+b^2+c^2}} \geq 6.$$

By the Cauchy–Schwarz inequality, we have

$$\frac{a}{b} + \frac{b}{c} + \frac{c}{a} \geq \frac{(a+b+c)^2}{ab+bc+ca}.$$

The AM–GM inequality get us

$$\frac{(a+b+c)^2}{ab+bc+ca} + \frac{\sqrt{3}(a+b+c)}{\sqrt{a^2+b^2+c^2}} \geq 2\sqrt[4]{\frac{3(a+b+c)^6}{(a^2+b^2+c^2)(ab+bc+ca)^6}}.$$

Again by the AM–GM inequality

$$\begin{aligned}(a+b+c)^2 &= (a^2+b^2+c^2) + 2(ab+bc+ca) \\ &\geq 3\sqrt[3]{(a^2+b^2+c^2)(ab+bc+ca)^2},\end{aligned}$$

which ends the proof.

487. Denote $t = \tan\frac{x}{2}$. Then we have

$$\begin{cases} \dfrac{2t+1-t^2}{1+t^2} = \dfrac{\sqrt{n}}{2} \\ t = \dfrac{\sqrt{n}-2}{3} \end{cases} \Leftrightarrow \left(1+\left(\frac{\sqrt{n}-2}{3}\right)^2\right)\cdot\frac{\sqrt{n}}{2} = 2\left(\frac{\sqrt{n}-2}{3}\right)+1$$

$$-\left(\frac{\sqrt{n}-2}{3}\right)^2 \Leftrightarrow (\sqrt{n}-2)(n-7)=0 \Leftrightarrow n\in\{4,7\}.$$

If $n = 4$, then $x = 0$.

If $n=7$, then $x=2\arctan\frac{\sqrt{7}-2}{3}$.

488. From

$$f(\sin x)+f(\cos x)=2020=f(\sin(-x))+f(\cos(-x)),$$

it follows that f is even. Put $f(x) = g(1-2x^2) + 1010$, with g: $[-1, 1] \to \mathbb{R}$. Then

$$g(\cos 2x)+g(-\cos 2x)=0,$$

whence we deduce that g is odd.

Conversely, each function f satisfying

$$f(x)=g(1-2x^2)+1010$$

for some odd function g: $[-1, 1] \to \mathbb{R}$ is a solution of the given functional equation.

489. Answer: $f(x) = 0, \forall x$; $f(x) = 2x, \forall x$; $f(x) = 2|x|, \forall x$.

For $x = 1$, we obtain $f(y) + 2f(y^2) = (1 + f(y))f(y)$, whence

$$2f(y^2) = (f(y))^2. \qquad (1)$$

Then the given equation becomes $f(x^2y) = x^2f(y)$. For $y = 1$, this implies $f(x^2)= =x^2f(1)$, and for $y = -1$, we have $f(-x^2) = x^2f(-1)$. Since x^2 takes all nonnegative values, we find

$$f(x)=\begin{cases} cx, x\geq 0 \\ dx, x<0 \end{cases}.$$

Taking $y = 1$ and $y = -1$ in (1), we have $2c = c^2$, and $2c = (-d)^2$, respectively.

Case 1. $c = d = 0$. Then $f(x) = 0$ for all x.

Case 2. $c = d = 2$. Then $f(x) = 2x$ for all x.

Case 3. $c = 2$, $d = -2$. Then $f(x) = 2|x|$ for all x.

490. Answer: $P(x) = 1, 2$; $P(x) = x^n + 1$, $n \geq 1$.

If $P(x)$ is constant, $P(x) = c$, then $c + 2c = c^2 + 2$, hence $c \in \{1, 2\}$.

If $P(x) = ax^n + Q(x)$, $a \neq 0$, $n \geq 1$, $\deg Q < n$, then

$$ax^{2n} + Q(x^2) + 2ax^n + 2Q(x) = a^2x^{2n} + (Q(x))^2 + 2ax^nQ(x) + 2,$$

therefore $a = 1$, $\deg Q = 0$. Let $P(x) = x^n + c$, then

$$x^{2n} + c + 2x^n + 2c = x^{2n} + c^2 + 2cx^n + 2,$$

which implies $c = 1$.

491. For $-2 \leq a \leq 10$ holds $(a + 2)(a - 10) \leq 0$, hence $a^2 \leq 8a + 20$. Therefore

$$\sum_{i=1}^{101} a_i^2 \leq \sum_{i=1}^{101} (8a_i + 20) = 20 \cdot 101 = 2020.$$

Equality holds if and only if k numbers are equal to -2 and $101 - k$ numbers are equal to 10. Then

$$0 = \sum_{i=1}^{101} a_i = -2k + 10(101 - k) = 10 \cdot 101 - 12k \Rightarrow k = \frac{10 \cdot 101}{12} \in \mathbb{Z},$$

contradiction. Finally,

$$\sum_{i=1}^{101} a_i^2 < 2020.$$

492. Answer: $Q = \pm 1$; $Q = \pm x^n$, $n \geq 1$.

The polynomial Q is a solution if and only if $-Q$ is a solution, so we may assume that the leading coefficient of Q is $c > 1$. If $Q \neq x^n$ is a solution of degree n, then there exists $R \in \mathbb{Z}[x]$, $\deg R \leq n - 1$, such that $Q(x) = cx^n + R(x)$, $c \neq 1$ or $R \neq 0$. There exists an integer N such that $a \geq N$ implies

$$a^{n-1} < Q(a) < a^{n+1}.$$

Also, we can assume that $Q(a) \neq a^n$, for all $a \geq N$. Then, if a is a prime number, then $Q(a)$ is divisible by some prime $p \neq a$. If b is the inverse of a modulo p, then $p|ab - 1$ and $p|Q(a)$, contradiction.

493. By the Cauchy–Schwarz inequality,

$$\left(\frac{a^2}{a+b}+\frac{b^2}{b+c}+\frac{c^2}{c+d}+\frac{d^2}{d+a}\right)[4(a+b)+(b+c)+4(c+d)+(d+a)]$$

$$\geq (2a+b+2c+d)^2 \Leftrightarrow \frac{a^2}{a+b}+\frac{b^2}{b+c}+\frac{c^2}{c+d}+\frac{d^2}{d+a} \geq \frac{(a+c+1)^2}{5}.$$

The AM–GM inequality get us

$$a+c+1=a+c+\frac{1}{3}+\frac{1}{3}+\frac{1}{3} \geq 5\sqrt[5]{\frac{ac}{27}},$$

hence

$$\frac{a^2}{a+b}+\frac{b^2}{b+c}+\frac{c^2}{c+d}+\frac{d^2}{d+a} \geq 5\left(\sqrt[5]{\frac{ac}{27}}\right)^2.$$

Raising to the fifth power, we obtain the desired inequality.

494. From

$$\sum_{i=1}^{2019} u_i = 0, \sum_{i=1}^{2019} u_i^2 = 1,$$

we deduce that u_i cannot be all zero, hence $a < 0$, $b > 0$. Let

$$P = \{i | u_i > 0\}, \quad N = \{i | u_i \leq 0\}.$$

and let $p = |P|$, $n = |N|$ be the sizes of these sets. Then $p + n = 2019$ and

$$0 = \sum_{i=1}^{2019} u_i = \sum_{i \in P} u_i - \sum_{i \in N} |u_i| \Rightarrow \sum_{i \in P} u_i = \sum_{i \in N} |u_i|.$$

It follows that

$$\sum_{i \in P} u_i^2 \leq \sum_{i \in P} b u_i = b \sum_{i \in P} u_i = b \sum_{i \in N} |u_i| \leq b \sum_{i \in N} |a| = -nab,$$

$$\sum_{i \in N} u_i^2 \leq \sum_{i \in N} |a| \cdot |u_i| = |a| \sum_{i \in N} |u_i| = |a| \sum_{i \in P} u_i \leq |a| \sum_{i \in P} b = -pab,$$

hence

$$1 = \sum_{i=1}^{2019} u_i^2 = \sum_{i \in P} u_i^2 + \sum_{i \in N} u_i^2 \leq -(p+n)ab = -2019ab,$$

which proves the required result.

495. Define the sets

$$P = \{i | a_i \leq -1\}, Q = \{i | -1 < a_i \leq 0\}, R = \{i | 0 < a_i < 1\}, S = \{i | a_i \geq 1\}$$

and let

$$p = \sum_{i \in P} a_i, q = \sum_{i \in Q} a_i, r = \sum_{i \in R} a_i, s = \sum_{i \in S} a_i, \ \ t_+ = \sum_{(i,j) \in A,\, a_i a_j \geq 0} a_i a_j, \ \ t_- = \sum_{(i,j) \in A,\, a_i a_j \leq 0} a_i a_j.$$

We know that $p + q + r + s = 0$ and we need to prove that $t_+ + t_- < 0$. Observe that $t_+ \leq \frac{p^2}{2} + pq + rs + \frac{s^2}{2}$, with equality only for $p = s = 0$, and $t_- \leq pr + ps + qs$, with equality if there do not exist $i \in Q, j \in R$ with $a_j - a_i > 1$, therefore

$$t_+ + t_- \leq \frac{(p+q+r+s)^2 - (q+r)^2}{2} = -\frac{(q+r)^2}{2} \leq 0.$$

If $A \neq \emptyset$ and $p = s = 0$, then there exist $i \in Q, j \in R$ with $|a_i - a_j| > 1$ and the equality cannot hold.

496. We have

$$\frac{xyz(x+y+z) + (xy+yz+zx)(1-xyz)}{xyz\sqrt{1-xyz}} \geq \frac{3\left(1 - xyz + \sqrt[3]{(xyz)^2}\right)}{\sqrt{(1-xyz)\sqrt[3]{(xyz)^2}}} \geq 6.$$

Equality holds when $x = y = z = t$ and $t^3 + t^2 - 1 = 0$.

Consider the function f: $(0, 1) \to \mathbb{R}, f(t) = t^3 + t^2 - 1 = 0$. It is continuous and

$$f\left(\frac{1}{2}\right) = -\frac{5}{8} < 0, f\left(\frac{4}{5}\right) = \frac{29}{125} > 0,$$

hence by the Intermediate Value Property, there is $t_0 \in (0, 1)$ such that $f(t_0) = 0$. In fact, this t_0 is unique, since f is strictly increasing: $f'(t) = 3t^2 + 2t > 0$ on $(0, 1)$.

3.50 2021

497. Answer: $C = \frac{1}{2}$.
We have to find the maximum M_n of the Vandermonde's determinant

$$V_n = \prod_{1 \le i < j \le n} (x_i - x_j),$$

when $x_i \in [-1, 1]$, $1 \le i \le n$. The maximum is attained at the Fekete points, that is, the roots of $(1 - x^2)P'(x)$, where P is the Legendre's polynomial of degree $n - 1$. For

$$c_n = M_n^{\frac{1}{\binom{n}{2}}},$$

we have $C = \inf\{c_n | n \in \mathbb{N}^*\}$. Now the result follows from M. Fekete, *Über die Verteilung der Wurzeln bei gewissen algebraischen Gleichungen mit ganzzahligen Koeffizienten.*

498. Answer: $P(x) = x + (x - 1)(x - 2)\ldots(x - 2021)R(x)$ or $P(x) = 2022 - x + (x - 1)(x - 2)\ldots(x - 2021)R(x)$, $R \in \mathbb{Z}[x]$.
The polynomial P has integer coefficients; hence, it satisfies $a - b | P(a) - P(b)$ for any distinct integers a, b. Then $2020 | P(2021) - P(1)$. From

$$\{P(1), \ldots, P(2021)\} = \{1, \ldots, 2021\},$$

it follows that $P(2021) - P(1) \le 2020$, so that $\{P(1), P(2021)\} = \{1, 2021\}$.
Case 1. $P(1) = 1$. Substituting $a = k \in \{2020, 2019, \ldots, 2\}$, $b = 1$ in the above relation, we get $k - 1 | P(k) - 1$, therefore $P(k) = k$, so that

$$P(x) - x = (x - 1)(x - 2)\ldots(x - 2021)R(x), R \in \mathbb{Z}[x].$$

Case 2. $P(1) = 2021$. Substituting $a = k \in \{2, \ldots, 2020\}$, $b = 2021$ into the divisibility relation, we get $2021 - k | P(k) - 1$, which implies $P(k) = 2022 - k$. This implies

$$P(x) - (2022 - x) = (x - 1)(x - 2)\ldots(x - 2021)R(x), R \in \mathbb{Z}[x].$$

499. The given inequality is equivalent to

$$\frac{a^2}{a(1+b)} + \frac{b^2}{b(1+c)} + \frac{c^2}{c(1+a)} \ge \frac{3}{2}.$$

By the Bergström's inequality,

$$\frac{a^2}{a(1+b)}+\frac{b^2}{b(1+c)}+\frac{c^2}{c(1+a)}\geq\frac{(a+b+c)^2}{a(1+b)+b(1+c)+c(1+a)},$$

hence it suffices to prove the inequality

$$\frac{(a+b+c)^2}{(a+b+c)+(ab+bc+ca)}\geq\frac{3}{2},$$

which is equivalent to

$$\frac{1}{a+b+c}+\frac{ab+bc+ca}{(a+b+c)^2}\leq\frac{2}{3}.$$

By the AM – GM inequality, $a+b+c\geq 3\sqrt[3]{abc}=3$. On the other hand,

$$\sqrt{\frac{ab+bc+ca}{3}}\leq\frac{a+b+c}{3},$$

whence it is easy to obtain the desired result.

500. Answer: $P(x, y) = (a + b)x^2 + axy + by^2$, $a, b \in \mathbb{R}$.

By the given equality

$$P(2x,2y)=2P(x+y,x-y)=4P(x,y).$$

The coefficients of x^iy^j are the same: $2^i2^ja_{ij}=4a_{ij}$. If $a_{ij}\neq 0$, then $i+j=2$, so

$$P(x,y)=cx^2+axy+by^2, a,b,c\in\mathbb{R}.$$

Now the given equality is

$$(c+a+b)x^2+2(c-b)xy+(c-a+b)y^2=2cx^2+2axy+2by^2,$$

whence $c = a + b$.

501. Answer: $f(n) = c\nu_p(n)$, where c is a nonnegative integer, p is a prime number, and $\nu_p(n)$ is the exponent of p in the decomposition of n as a product of primes.

The above functions are solutions (easy verification). Let f be a nonzero solution. Call a positive integer n *beautiful* if the equality $f(k) = f(n - k)$ holds for any k such that $1 \leq k \leq n - 1$. Observe that each divisor of a beautiful number is beautiful. Indeed, if $n = ad$, with a, d positive integers, then

$$f(k)=f(ak)-f(a)=f(n-ak)-f(a)=f(a(d-k))-f(a)=f(d-k)$$

for any integer k such that $1 \le k \le d-1$.

By the first condition,

$$\prod_i f(p_i^{\alpha_i}) = \sum_i \alpha_i f(p_i)$$

for any decomposition in prime factors. We have to prove that there is at most a prime number p for which $f(p) > 0$. Let p be the smallest p such that $f(p) > 0$. From the above formula, it follows that p is a prime number. Put $c = f(p)$.

If there is a beautiful integer $n \ge p$ not divisible by p, then $n = pq + r$, with q positive integer and $1 \le r \le p-1$. Since n is beautiful, we have

$$f(r)=f(n-pq)=f(pq)\ge f(p)>0,$$

contradiction. It follows that any beautiful number is of the form ap^b, with b nonnegative integer and $1 \le a \le p-1$. Conversely, since there are infinitely many beautiful numbers and a takes only finitely many values, it follows that any power of p divides a beautiful number, hence is beautiful itself. We deduce that for any prime number $q \ne p$, the number p^{q-1} is beautiful. Since q divides $p^{q-1}-1$, we deduce

$$0=f(1)=f\left(p^{q-1}-1\right)\ge f(q),$$

whence the conclusion.

502. Denote by A, B the given sums. From

$$\frac{n^{-3}-(n+1)^{-3}}{n^{-2}-(n+1)^{-2}}=\frac{(n+1)^3-n^3}{(n+1)^3 n-n^3(n+1)}=\frac{3n^2+3n+1}{n(n+1)(2n+1)}=\frac{1}{n}+\frac{1}{n+1}-\frac{1}{2n+1},$$

it follows that

$$A=\sum_{n=1}^{2019}(-1)^{n+1}\left(\frac{1}{n}+\frac{1}{n+1}-\frac{1}{2n+1}\right)=\sum_{n=1}^{2019}\frac{(-1)^{n+1}}{n}+\sum_{n=1}^{2019}\frac{(-1)^{n+1}}{n+1}-\sum_{n=1}^{2019}\frac{(-1)^{n+1}}{2n+1}=1+\frac{1}{2020}+\sum_{n=1}^{2019}\frac{(-1)^n}{2n+1}=1+\frac{1}{2020}-\frac{1}{3}+\frac{1}{5}-\dots-\frac{1}{4039}.$$

Then

$$B-A=-\frac{1}{2020}+\frac{1}{4041}-\frac{1}{4043}+\frac{1}{4045}-\dots+\frac{1}{5781}=-\frac{2021}{2020\cdot 4041}$$
$$-\frac{2}{4043\cdot 4045}-\frac{2}{4047\cdot 4049}-\dots-\frac{2}{5779\cdot 5781}<0.$$

503. By the AM–GM inequality,

$$\frac{a+1}{\sqrt{a+bc}}\geq\frac{a+1}{\sqrt{a+\left(\frac{b+c}{2}\right)^2}}=\frac{a+1}{\sqrt{a+\left(\frac{1-a}{2}\right)^2}}=2,$$

hence

$$\sum_{cyc}\frac{a+1}{\sqrt{a+bc}}\geq 2+2+2=6.$$

Now, it suffices to prove that

$$6\geq\frac{2}{a^2+b^2+c^2},$$

which is a consequence of the QM–AM inequality.

504. Add the first and the sixth equations, to get

$$\sum_{i=1}^{4}x_i=\sum_{i=1}^{4}x_i^2+6x_1x_2+6x_3x_4.$$

For any permutation σ: $\{1,2,3,4\}\to\{1,2,3,4\}$, we must then have

$$x_{\sigma(1)}x_{\sigma(2)}=x_{\sigma(3)}x_{\sigma(4)}.$$

Using $x_1x_2+x_3x_4=x_1x_4+x_2x_3$, we get $(x_1-x_4)(x_2-x_3)=0$. Assume $x_1=x_4$, then from $x_1x_2+x_3x_4=x_1x_4+x_2x_3$, we get $(x_1-x_3)(x_2-x_4)=0$, so $x_1=x_3=x_4$ or $x_2=x_3=x_4$, namely, at least three variables are equal. Adding all, we get

$$3(x_1+x_2+x_3+x_4)=3(x_1+x_2+x_3+x_4)^2,$$

so $x_1+x_2+x_3+x_4\in\{0,1\}$. By symmetry, we have two cases:

$$(x_1,x_2,x_3,x_4)\in\{(c,c,c,-3c),(c,c,c,1-3c)\}.$$

505. If $\frac{S_{n+1}}{S_n}$ is a rational function, then

$$\frac{S_{n+1}-S_n}{S_n}=\frac{2^{n+1}}{(n+1)^2S_n}$$

is a rational function, hence $\frac{S_n}{2^n}=r(n)$ is a rational function. We have

$$\frac{2}{(n+1)^2}=2r(n+1)-r(n).$$

This equality holds for any positive integer, thus it is an equality of rational functions. We see that $r(0)=0$, $r(1)=1$, hence r has a pole (root of the denominator) at $x=-1$. Then r has a pole at $x=-2,-3,\ldots$, so it has infinitely many poles, contradiction.

506. We have to prove that

$$\frac{x^4}{w}+\frac{y^4}{z}\geq 1.$$

By the Hölder's inequality,

$$\left(z^3+w^3\right)\left(\frac{x^4}{w}+\frac{y^4}{z}\right)^3\geq\left(x^3+y^3\right)^4,$$

so

$$\left(\frac{x^4}{w}+\frac{y^4}{z}\right)^3\geq 1,$$

whence the conclusion.

507. Choose ε_n such that

$$\sum_{i=1}^{n}\varepsilon_i=\lfloor\alpha n\rfloor, n\in\mathbb{N}^*.$$

We say that $(a_i)_{i\geq 1}\geq(b_i)_{i\geq 1}$ if

$$\sum_{i=1}^{n}a_i\geq\sum_{i=1}^{n}b_i, n\in\mathbb{N}^*.$$

If $(a_i)_{i\geq 1}\geq(b_i)_{i\geq 1}$, then it is easy to see that

$$\sum_{i=1}^{n} \frac{a_i}{(n+i-1)(n+i)} \geq \sum_{i=1}^{n} \frac{b_i}{(n+i-1)(n+i)}, \quad n \in \mathbb{N}^*.$$

Let $(\delta_i)_{i \geq 1}$ such that

$$\sum_{i=1}^{n} \delta_i = \lfloor \alpha n \rfloor + 1, n \in \mathbb{N}^*.$$

Then $(\delta_i)_{i \geq 1} \geq (\alpha_i)_{i \geq 1} \geq (\varepsilon_i)_{i \geq 1}$. Observe that

$$2n\sum_{i=1}^{n} \frac{\alpha}{(n+i-1)(n+i)} = 2n\alpha \sum_{i=1}^{n} \left(\frac{1}{n+i-1} - \frac{1}{n+i}\right) = 2n\alpha\left(\frac{1}{n} - \frac{1}{2n}\right) = \alpha,$$

so $\alpha - 2ns_n \geq 0$. Now, it is enough to prove that

$$2n\sum_{i=1}^{n} \frac{\delta_i}{(n+i-1)(n+i)} - 2n\sum_{i=1}^{n} \frac{\varepsilon_i}{(n+i-1)(n+i)} = \frac{2}{n+1}.$$

From $\varepsilon_1 = 0$, $\delta_1 = 1$ and $\varepsilon_i = \delta_i$, $i \geq 2$, it follows that this difference is

$$2n \cdot \frac{1}{n(n+1)} = \frac{2}{n+1},$$

as desired.

508. Answer: yes.

Let $n > 2021$ be a positive integer and define

$$P = (1-x)^n, Q = \prod_{i=0}^{n-1} \left(1 + x^{2^i}\right)^{n-i}.$$

By the Newton's binomial formula, both P, Q have coefficients greater than n. From

$$PQ = \left(1-x^2\right)\left(1-x^4\right)\left(1-x^8\right)\ldots\left(1-x^{2^n}\right),$$

it follows that all coefficients of PQ are ± 1, since any positive integer has a unique binary expansion.

509. Case 1. P constant. Let $P(z) = c \in \mathbb{R}$ for any $z \in \mathbb{C}$. Then $cz = 1$ if and only if $c^2 = 0$, which is impossible. No solution.

Case 2. $\deg P = 1$. Let $P(z) = az + b$, $a, b \in \mathbb{R}$, $a \neq 0$. Then $az^2 + bz = 1$ if and only if $(a(z-1) + b)(a(z+1) + b) = 0$. Comparing the coefficients, we have $ab = 2ab$ and $-a = b^2 - a^2$, whence $b = 0$, $a = 1$, so $P(z) = z$ for all z.

Case 3. $\deg P \geq 2$. If $Q(z) = (z+1)P(z+1) - (z-1)P(z-1)$, then $\deg Q \leq \deg P$ and Q has the same roots as P, therefore Q is a multiple of P, since P has no multiple roots. Indeed, if z_0 is a root of P, then taking $z = z_0 \pm 1$, we obtain

$$(z_0-1)P(z_0-1) = (z_0+1)P(z_0+1) = 1,$$

so $Q(z_0) = 0$. Let

$$P(z) = \sum_{k=0}^{n} a_k z^k,$$

then

$$Q(z) = \sum_{k=0}^{n} b_k z^k = \sum_{k=0}^{n} a_k \left((z+1)^{k+1} - (z-1)^{k+1}\right).$$

Comparing coefficients, we find $b_n = 2(n+1)a_n$, $b_{n-1} = 2na_{n-1}$, contradiction, since Q is a multiple of P. No solution.

510. We will prove the inequality by induction on n. If $n \leq 16$, then the given sum is at most $n \leq 4\sqrt{kn}$.
Assume that the inequality holds for $1, 2, \ldots, n-1$, with $n \geq 17$.
Case 1. n even. Define

$$x_l = \frac{a_l}{\sqrt{a_1^2 + \ldots + a_l^2}}.$$

Let $n = 2t$. Then

$$e^{-x_{t+1}^2 - \ldots x_{2t}^2} \geq \left(1 - x_{t+1}^2\right)\ldots\left(1 - x_{2t}^2\right) = \frac{a_1^2 + \ldots + a_t^2}{a_1^2 + \ldots + a_{2t}^2} \geq \frac{1}{1+4^k},$$

since $a_{t+i} \leq 2^k a_i$, $1 \leq i \leq t$. We deduce

$$x_{t+1}^2 + \ldots + x_{2t}^2 \leq \ln\left(1 + 4^k\right) \leq 2k \Longrightarrow x_{t+1} + \ldots + x_{2t} \leq \sqrt{2kt},$$

by the AM–QM inequality. Using the induction hypothesis for $n = t$, we obtain

$$\sum_{l=1}^{2t} x_l \leq 4\sqrt{kt} + \sqrt{2kt} \leq 4\sqrt{2kt}.$$

Case 2. n odd. For $n = 2t+1$, we have analogously

$$x_{t+2}^2 + \ldots + x_{2t+1}^2 \le \ln\left(1 + 4^k\right) \le 2k, \sum_{l=1}^{2t+1} x_l \le 4\sqrt{k(t+1)} + \sqrt{2kt} \le 4\sqrt{k(2t+1)}.$$

511. Answer: $n \equiv 0, 4, 8 \pmod 9$.

We will prove that

$$\frac{x_1^2 + x_2^2 + \ldots + x_n^2}{(x_1 + 2x_2 + \ldots + nx_n)^2} \le \frac{27}{4n(n+1)(2n+1)} = \frac{9}{8\left(1^2 + 2^2 + \ldots + n^2\right)},$$

which is equivalent to

$$8\left(x_1^2 + x_2^2 + \ldots + x_n^2\right)\left(1^2 + 2^2 + \ldots + n^2\right) \le (3x_1 + 6x_2 + \ldots + 3nx_n)^2.$$

From the given condition, $(x_i - i)(2i - x_i) \ge 0$, $\forall\, i \in \{1, 2, \ldots, n\}$. Then

$$3ix_i \ge x_i^2 + 2i^2, 1 \le i \le n \Rightarrow (3x_1 + 6x_2 + \ldots + 3nx_n)^2 \ge$$
$$\ge \left(x_1^2 + x_2^2 + \ldots + x_n^2 + 2\left(1^2 + 2^2 + \ldots + n^2\right)\right)^2.$$

Denote $A = x_1^2 + x_2^2 + \ldots + x_n^2$, $B = 1^2 + 2^2 + \ldots + n^2$. It suffices to show that

$$8AB \le (A + 2B)^2 \Leftrightarrow (A - 2B)^2 \ge 0,$$

which is true. Equality holds if and only if $(x_i - i)(2i - x_i) = 0$ for all i. Then

$$\begin{aligned} &3ix_i = x_i^2 + 2i^2, 1 \le i \le n \Rightarrow x_1^2 + x_2^2 + \ldots + x_n^2 = 3(x_1 + 2x_2 + \ldots + nx_n) \\ &- 2\left(1^2 + 2^2 + \ldots + n^2\right) \Rightarrow 3(x_1 + 2x_2 + \ldots + nx_n) \\ &= 4\left(1^2 + 2^2 + \ldots + n^2\right) \Rightarrow 3 \mid 1^2 + 2^2 + \ldots + n^2, \end{aligned}$$

so 9 divides $n(n + 1)(2n + 1)$, whence $n \equiv 0, 4, 8 \pmod 9$.

We will construct inductively the set $\{x_1, x_2, \ldots, x_n\}$. Denote by S_n the set of indices i for which $x_i = 2i$. We have

$$S_4 = \{1, 3\}, S_8 = \{2, 8\}, S_9 = \{1, 2, 3, 9\}, S_{13} = \{2, 10, 13\}, S_{17} = \{1, 7, 16, 17\},$$
$$S_{18} = \{3, 9, 17, 18\}, S_{k+18} = S_k \cup \{k+1, k+4, k+11, k+12, k+14, k+15\}.$$

512. Denote $p = a + b + c$, $q = ab + bc + ca$, $r = abc$. Then from

$$0 \le q = 2p^2 - 5p \le \frac{p^2}{3},$$

it follows $p = 0$ or $\frac{5}{2} \le p \le 3$. We need to prove that $7r \le 8p^2 - 30p + 25$. We have

$$7r \le \frac{7pq}{9} = \frac{7p^2(2p-5)}{9} = 8p^2 - 30p + 25 + \frac{(2p-5)(7p-15)(p-3)}{9} \le 8p^2 - 30p + 25.$$

Equality holds only when $a = b = c = 1$.

3.51 2022

513. We can take $b = 1$ and denote $k = \sqrt{ac}$, then

$$2k^5 + 5k^4 - 5k^2 + 1 = (2k+1)\left(k^2 + k - 1\right)^2 \ge 0,$$

with equality only when $k = \frac{\sqrt{5}-1}{2}$.

514. Answer: $80 + 40\sqrt{3}$.

Notice that

$$z_1 + z_2 + \ldots + z_{240} = \int_0^{2\pi} F(\theta) e^{i\theta} d\theta,$$

where $F(\theta)$ is the number of the z_is in the arc $\left(\theta - \frac{\pi}{6}, \theta + \frac{\pi}{6}\right)$. The integral equals

$$\frac{1}{6} \int_0^{2\pi} \left(F(\theta) e^{i\theta} + F\left(\theta + \frac{\pi}{3}\right) e^{i\left(\theta + \frac{\pi}{3}\right)} + \ldots + F\left(\theta + \frac{5\pi}{3}\right) e^{i\left(\theta + \frac{5\pi}{3}\right)} \right) d\theta.$$

We can suppose that $z_1 + \ldots + z_{240}$ is real. The maximum is attained when

$$z_1 = \ldots = z_{40} = i, z_{41} = \ldots = z_{80} = -i, z_{81} = \ldots = z_{160} = 1,$$
$$z_{161} = \ldots = z_{200} = \omega, z_{201} = \ldots = z_{240} = \overline{\omega},$$

where $\omega = e^{\frac{i\pi}{3}}$.

515. Answer: $\{x, y\} = \{4, 9\}$.

Set $s = x + y, p = xy$. Then $s^2 - p = 133, s + \sqrt{p} = 19$, hence $s - \sqrt{p} = \frac{133}{19} = 7$. This implies $s = 13, \sqrt{p} = 6$, so that $p = 36$ and x, y are solutions of the quadratic equation

$$z^2 - 13z + 36 = 0.$$

516. Answer: constant functions.

If $P(x, y)$ is the assertion and $a = f(0)$, then

$$P(x, 0) \Rightarrow f(x) - a \leq 0, \forall x \in \mathbb{R};$$
$$P(x, -x) \Rightarrow f(x) - f(-x) \leq f(-x^2) - a \leq 0;$$
$$P(-x, x) \Rightarrow f(-x) - f(x) \leq f(-x^2) \leq 0,$$

so $f(-x^2) = a$ and $f(x) = f(-x)$, $\forall\, x \in \mathbb{R}$.

517. Answer: 11.

More generally, the minimal value of the sum

$$\left\lfloor \frac{a_1}{1} \right\rfloor + \left\lfloor \frac{a_2}{2} \right\rfloor + \ldots + \left\lfloor \frac{a_n}{n} \right\rfloor$$

over all permutations $(a_1, a_2, \ldots, a_n)$ of $(1, 2, \ldots, n)$ is $\lfloor \log_2 n \rfloor + 1$. We will prove first the inequality

$$\left\lfloor \frac{a}{b} \right\rfloor \geq \log_2 \frac{a+1}{b}$$

for all positive integers a, b. Indeed, we have

$$\left\lfloor \frac{a}{b} \right\rfloor \leq \frac{a}{b} < \frac{a+1}{b} \leq \left\lfloor \frac{a}{b} \right\rfloor + 1, 2^{\left\lfloor \frac{a}{b} \right\rfloor} \geq \left\lfloor \frac{a}{b} \right\rfloor + 1,$$

hence

$$\left\lfloor \frac{a}{b} \right\rfloor \geq \log_2 \left(\left\lfloor \frac{a}{b} \right\rfloor + 1 \right) \geq \log_2 \frac{a+1}{b}.$$

Applying this inequality to each term of the sum, we get

$$\sum_{i=1}^{n} \left\lfloor \frac{a_i}{i} \right\rfloor \geq \sum_{i=1}^{n} \log_2 \frac{a_i + 1}{i} = \log_2 (n+1),$$

since $(a_1 + 1, \ldots, a_n + 1)$ is a permutation of $(2, \ldots, n + 1)$ and $\log_2 1 = 0$. If $2^k \le n < < 2^{k+1}$, then the above sum is at least $k + 1$.

The lower bound is attained for the permutation containing $k + 1$ cycles

$$(a_1) = (1), (a_2, a_3) = (3, 2), (a_4, a_5, a_6, a_7) = (7, 4, 5, 6), \ldots, (a_{2^{k-1}}, a_{2^k - 1}) = \left(2^k - 1, 2^{k-1}, 2^{k-1} + 1, 2^k - 2\right), (a_{2^k}, \ldots, a_n) = \left(n, 2^k, 2^k + 1, \ldots, n - 1\right),$$

since the sum of terms in every cycle is 1.

518. For all $k \le n$, let

$$s_k = a_1 + a_2 + \ldots + a_k, \quad b_k = \frac{a_k s_{k-1}^2}{1 - a_k},$$

where $s_0 = 0$. We will prove that

$$b_k < \frac{s_k^3 - s_{k-1}^3}{3}. \quad (*)$$

Indeed,

$$(*) \Leftrightarrow 0 < (1 - a_k)\left((s_{k-1} + a_k)^3 - s_{k-1}^3\right) - 3a_k s_{k-1}^2 \Leftrightarrow 0 < (1 - a_k) \times \left(3s_{k-1}^2 + 3s_{k-1}a_k + a_k^2\right) - 3s_{k-1}^2 \Leftrightarrow 0 < -3a_k s_{k-1}^2 + 3(1 - a_k)s_{k-1}a_k + (1 - a_k)a_k^2 \Leftrightarrow 0 < 3(1 - a_k - s_{k-1})s_{k-1}a_k + (1 - a_k)a_k^2,$$

which is true, since $a_k + s_{k-1} = s_k \le 1$ and $a_k \in (0, 1)$.

Adding $(*)$ for $k = 1, \ldots, n$, we get

$$b_1 + b_2 + \ldots + b_n < \frac{s_n^3 - s_0^3}{3} = \frac{1}{3}.$$

519. Answer: $M = 2$.

Let $S_n = a_0 + a_1 + \ldots + a_n$. We will prove by induction that $S_n > 2S_{n-1}$, $n \ge 1$. We have $S_0 = 1$, $S_1 = 4$, so $4 = S_1 > 2S_0 = 2$. Suppose that $S_k > 2S_{k-1}$ for some $k \ge 1$. Then from $a_0 + \ldots + a_{k-1} \ge 3a_k - a_{k+1}$, we have $S_{k-1} \ge 3(S_k - S_{k-1}) - (S_{k+1} - S_k)$, hence $S_{k+1} \ge 4(S_k - S_{k-1}) > 2S_k$, which ends the proof.

From this we deduce $a_{n+1} - 2a_n \ge a_n - S_{n-1} = S_n - 2S_{n-1} > 0$, hence

$$\frac{a_{n+1}}{a_n} > 2.$$

Now, we will prove that $M = 2$. We have $S_0 = 1$, $S_1 = 4$, $S_{n+1} = 4(S_n - S_{n-1})$, whence $S_n = 2^n(n+2)$, $a_n = S_n - S_{n-1} = 2^{n-1}(n+3)$ and

$$\lim_{n\to\infty}\frac{a_{n+1}}{a_n} = \lim_{n\to\infty}\frac{(n+4)2^n}{(n+3)2^{n-1}} = 2,$$

therefore $M = 2$.

520. For each distinct integers m, n holds $m - n | P(m) - P(n)$. Define $b_n = a_{n+1} - a_n$ for every nonnegative integer n. Then $b_n | b_{n+1}, n \in \mathbb{N}$.

Let $m \in \mathbb{N}^*$ such that $a_m = 0$, then $a_m = a_0$, hence $a_{m+1} = a_1$ and so, $b_m = b_0$. If $b_0 = 0$, then $a_0 = a_1 = \ldots = a_m = 0$. If $b_0 \neq 0$, then $b_0 | b_1 | \ldots | b_m = b_0$ implies $b_k = \pm b_0$ for $1 \leq k \leq m$. But $b_0 + b_1 + \ldots + b_{m-1} = a_m - a_0 = 0$, hence half of the numbers $b_0, b_1, \ldots, b_{m-1}$ are positive, so that there exists $k \in \{1, 2, \ldots, m-1\}$ for which $b_{k-1} = -b_k$. We deduce $a_{k+1} = a_{k-1}$ and $a_{n+1} = a_n$, for each $n \geq k - 1$. Then $a_{m+2} = a_m$ and $a_2 = P(P(a_0)) = P(P(a_m)) = a_{m+2} = a_m = 0$.

521. The function P^2 is convex and increasing on $\mathbb{R}^+$. By the Jensen's and AM–GM inequalities, we have

$$\sum P^2\left(\frac{x_i}{x_{i+1}}\right) \geq nP^2\left(\frac{1}{n}\sum\frac{x_i}{x_{i+1}}\right) \geq nP^2(1).$$

Equality holds only when P is constant or $x_1 = x_2 = \ldots = x_n$.

522. Answer: $\frac{4}{9}$, at $\left(\frac{1}{3}, \frac{1}{3}\right)$ and $\left(\frac{2}{3}, \frac{2}{3}\right)$.

Setting $(x, y) \to (1 - x, 1 - y)$ results the same three numbers, so we can assume that $x + y \geq 1$.

Case 1. $x + y \geq \frac{4}{3}$. Then

$$\max\{xy, (x-1)(y-1), x+y-2xy\} = \max\{xy, x+y-2xy\}$$
$$\geq \frac{xy + xy + x + y - 2xy}{3} \geq \frac{4}{9},$$

with equality for $(x, y) = \left(\frac{2}{3}, \frac{2}{3}\right)$.

Case 2. $x + y < \frac{4}{3}$. Then

$$\max\{xy, (x-1)(y-1), x+y-2xy\} \geq x+y-2xy \geq x+y-\frac{(x+y)^2}{2} = \frac{4}{9}+\frac{1}{2}$$
$$\times\left(x+y-\frac{2}{3}\right)\left(\frac{4}{3}-x-y\right) \geq \frac{4}{9},$$

with equality for $(x, y) = \left(\frac{1}{3}, \frac{1}{3}\right)$.

523. Answer: $n - 1$.

Rewrite the condition as

$$\sum \frac{x_i^2 - n + 1}{1 + x_i^2} = 0.$$

Note that if $x_i \geq x_j$, then

$$\frac{x_i^2 - n - 1}{1 + x_i^2} - \frac{x_j^2 - n - 1}{1 + x_j^2} = \frac{(n+2)\left(x_i^2 - x_j^2\right)}{(1 + x_i^2)\left(1 + x_j^2\right)} \geq 0.$$

For any x_i, x_j,

$$\frac{1}{1 + x_i^2} + \frac{1}{1 + x_j^2} \leq 1 \Rightarrow x_i^2 x_j^2 \geq 1 \Rightarrow x_i x_j \geq 1,$$

then for $x_i \geq x_j$,

$$\frac{1 + x_i^2}{x_i} - \frac{1 + x_j^2}{x_j} = \frac{(x_i x_j - 1)(x_i - x_j)}{x_i x_j} \geq 0.$$

If the sequence (x_i) is increasing, then by the Chebyshev's inequality,

$$n \sum \frac{x_i^2 - n + 1}{x_i} \geq \left(\sum \frac{x_i^2 - n + 1}{1 + x_i^2}\right)\left(\sum \frac{1 + x_i^2}{x_i}\right) = 0,$$

thus,

$$\sum \frac{x_i^2 - n + 1}{x_i} \geq 0 \Rightarrow \sum \left(x_i - \frac{n-1}{x_i}\right) \geq 0 \Rightarrow \sum x_i \geq (n-1) \sum \frac{1}{x_i} \Rightarrow \frac{\sum x_i}{\sum \frac{1}{x_i}}$$

$$\geq n - 1.$$

The minimum is attained when $x_1 = x_2 = \ldots = x_n = \sqrt{n-1}$.

524. 1) We will prove by induction that the roots of g_n are distinct. For $n = 1$ we have $g_1(x) = f(x)$, which has the roots $x = 0$ and $x = 2$. Assume that the polynomial g_n has 2^n distinct roots, $x_1, x_2, \ldots, x_{2^n}$. Then

$$g_n(x) = c_n(x - x_1)\ldots(x - x_{2^n}),$$

$$g_{n+1}(x) = g_n(f(x)) = c_n(f(x) - x_1)\ldots(f(x) - x_{2^n}).$$

If $i \neq j$, the the sets of roots of $f(x) - x_i$ and $f(x) - x_j$ are disjoint. If the polynomial $f(x) - x_i$ has a double root, then this root must be 1. From

$$g_1(1) = f(1) = -5, g_2(1) = f(-5) = 175,$$

it follows that $g_n(1) > 0$, $\forall\, n \geq 2$, hence $x = 1$ cannot be a root of g_n for any n. This shows that g_{n+1} has distinct roots.

The set of roots of g_n can be partitioned in 2^{n-1} sets having two elements with sum 2, so the sum of all roots is $2^{n-1} \cdot 2 = 2^n$.

2) The polynomial $cx(x - 2) + d$ has two distinct real roots if and only if $c > d$ (1). in this case, the roots are

$$x = 1 \pm \sqrt{1 - \frac{d}{c}}.$$

The polynomial $f(x) - r$ has two positive roots for $r < 0$ and has two real roots of opposite signs for $r > 0$. Define the sequence $(a_n)_{n \geq 1}$ by

$$a_1 = 2, a_{n+1} = 1 + \sqrt{1 + \frac{a_n}{c}}, n \geq 1.$$

Then $g_n(a_n) = 0$, $a_2 > a_1$ and by induction, $(a_n)_{n \geq 1}$ is strictly increasing. Also by induction it follows that

$$a_n < 2 + \frac{1}{c}, \forall n.$$

The sequence $(a_n)_{n \geq 1}$ is convergent and

$$\lim_{n \to \infty} a_n = 2 + \frac{1}{c}.$$

Now, we will prove by induction that $a_n < c + 2$ for all n. Indeed, $a_1 = 2 < c + 2$. Assume that $a_n < c + 2$, then

$$a_{n+1} = 1 + \sqrt{1 + \frac{a_n}{c}} < 1 + \sqrt{2 + \frac{2}{c}} < c + 2 \Leftrightarrow 2 + \frac{2}{c}$$
$$< (c+1)^2 \Leftrightarrow (c-1)^2 (c+2) > 0,$$

which is true. It follows that

$$\lim_{n\to\infty} a_n = 2 + \frac{1}{c} \leq c + 2,$$

therefore $c \geq 1$.

525. Suppose that $a \geq b$ and let $a = bx$. Then $x \geq 1$ and we have to prove that

$$\sqrt{\frac{x^2+1}{2}} + \frac{2x}{x+1} \geq \frac{x+1}{2} + \sqrt{x}, x \geq 1.$$

This is equivalent to

$$\sqrt{\frac{x^2+1}{2}} - \sqrt{x} = \frac{\frac{x^2+1}{2} - x}{\sqrt{\frac{x^2+1}{2}} + \sqrt{x}} = \frac{(x-1)^2}{2\left(\sqrt{\frac{x^2+1}{2}} + \sqrt{x}\right)} \geq \frac{x+1}{2} - \frac{2x}{x+1} = \frac{(x-1)^2}{2(x+1)}.$$

If $x = 1$, then the inequality becomes an inequality, else we have to prove that

$$x + 1 \geq \sqrt{\frac{x^2+1}{2}} + \sqrt{x}.$$

Now, we have the equivalences:

$$x - \sqrt{x} + 1 \geq \sqrt{\frac{x^2+1}{2}} \Leftrightarrow x^2 + x + 1 - 2x\sqrt{x} + 2x - 2\sqrt{x} \geq \frac{x^2+1}{2} \Leftrightarrow x^2 - 4x\sqrt{x}$$
$$+ 6x - 4\sqrt{x} + 1 = \left(\sqrt{x} - 1\right)^4 \geq 0,$$

which is obvious.

526. By the Hölder's inequality,

$$\text{LHS}^2\left(a\left(b^2 + \lambda bc + c^2\right) + b\left(c^2 + \lambda ca + a^2\right) + c\left(a^2 + \lambda ab + b^2\right)\right) \geq (a+b+c)^3,$$

so it suffices to prove the inequality

$$\frac{(a+b+c)^3}{\sum a^2b + 3\lambda abc} \geq \frac{9}{\lambda + 2}.$$

This is equivalent to

$$2\left(a^3 + b^3 + c^3\right) + 12abc + \lambda\left(a^3 + b^3 + c^3 + 3\sum a^2b - 21abc\right) \geq \sum a^2b.$$

The coefficient of λ is positive, by the AM – GM inequality, so it suffices to prove the inequality for $\lambda = \frac{1}{4}$. After calculations, we obtain

$$a^3 + b^3 + c^3 + 3abc \geq \sum a^2 b,$$

which is true by the Schur's inequality.

527. Answer: $P = 0$.

For $x = n$, we get $P(n) = 0$, so $P(x) = (x - n)^r Q(x)$, with r positive integer, $Q \in \mathbb{R}[x]$ and $Q(n) \neq 0$. Then

$$\left(x^2 + x - n - n^2\right)^r Q\left(x^2 + x - n^2\right) = (x - n)^{2r} Q^2(x) + (x - n)^r Q(x).$$

Dividing by $(x - n)^r$, we obtain

$$(x + n + 1)^r Q\left(x^2 + x - n^2\right) = (x - n)^r Q^2(x) + Q(x).$$

For $x = n$, it follows that $Q(n) = 0$, contradiction.

528. Answer: $3\sqrt[3]{36}$.

By the AM–GM inequality,

$$xy + \frac{1}{3x} + \frac{1}{2y} \geq \frac{\sqrt[3]{36}}{2}, yz + \frac{3}{2y} + \frac{3}{z} \geq \frac{3\sqrt[3]{36}}{2}, zx + \frac{2}{z} + \frac{2}{3x} \geq \sqrt[3]{36}.$$

Adding these inequalities, we find the minimum value of the given expression. Equality holds when

$$(x, y, z) = \left(\frac{\sqrt[3]{6}}{3}, \frac{\sqrt[3]{6}}{2}, \sqrt[3]{6}\right).$$

3.52 2023

529. Answer: $n = k + 4$.

Let $a_1 < a_2 < \ldots < a_n$ be the ordered set of n numbers. Since a_1, a_n are sums of k distinct elements of the set, it follows that

$$a_1 \geq a_2 + \ldots + a_{k+1}, a_n \leq a_{n-1} + \ldots + a_{n-k}.$$

If $n = k + 1$, then $a_1 \geq a_2 + \ldots + a_{k+1} > a_1 + \ldots + a_k \geq a_{k+1}$, contradiction.

If $n = k + 2$, then $a_1 \geq a_2 + \ldots + a_{k+1} \geq a_{k+2}$, contradiction.

If $n = k + 3$, then $a_1 \geq a_2 + \ldots + a_{k+1}$, $a_{k+3} \leq a_{k+2} + \ldots + a_3$. Adding inequalities, we get $a_1 + a_{k+2} \geq a_2 + a_{k+3}$, contradiction.

Now, we will give an example of a set with $n = k + 4$ elements satisfying the required property.

Case 1. k even. Let $k = 2l$, $l \geq 1$, $A_i = \{-i, i\}$. The set $A_1 \cup \ldots \cup A_{l+2}$ has $k + 4$ elements and satisfies the desired property. Observe that i is expressible if and only if $-i$ is expressible, so it suffices to consider $1 \leq i \leq l + 2$. If $i < l + 2$, then we sum numbers from some $l - 1$ sets $A_j, j \neq 1, i+1$ with $i+1$ and -1. If $i = l+2$, then we sum numbers from some $l - 1$ sets A_j, $j \neq 1$, $i + 1$ with $l + 1$ and 1.

Case 2. k odd. Let $k = 2l + 1, l \geq 1$. We will add 0 to the previous set and to each previous expression. The number 0 can be expressed taking 1, 2, -3 and all numbers from the remaining $l - 1$ sets $A_4, A_5, \ldots, A_{l+2}$.

530. For $a \geq b \geq c$, the inequality becomes

$$\frac{a+b+c}{3} - \sqrt[3]{abc} \leq \left(\sqrt{a}-\sqrt{c}\right)^2 \Leftrightarrow a+b+c-3\sqrt[3]{abc} \leq 3a+3c-6\sqrt{ac} \Leftrightarrow 2a + 2c + 3\sqrt[3]{abc} \geq b + 6\sqrt{ac}.$$

The inequality is homogenous, so we can take $ac = 1$, whence

$$2a + 2c + 3\sqrt[3]{b} - b \geq 6.$$

The derivative of $3\sqrt[3]{b} - b$ is $b^{-\frac{2}{3}} - 1$, whence it follows that the function $3\sqrt[3]{b} - b$ is increasing on $(0, 1]$ and decreasing on $[1, \infty)$. From this we deduce that it suffices to prove the inequality for $b = a$ and $b = c$.

If $b = a$, then $a + 2c + 3\sqrt[3]{a} = a + \frac{2}{a} + 3\sqrt[3]{a} \geq 6$. The derivative of the LHS is

$$1 - \frac{2}{a^2} + \frac{1}{\sqrt[3]{a^2}},$$

so it decreases until $a = 1$, then it increases. The minimum value is attained at $a = 1$.

If $b = c$, then $2a + c + 3\sqrt[3]{c} = \frac{2}{c} + c + 3\sqrt[3]{c} \geq 6$, which is the same inequality as above.

531. Answer: $\lambda = \frac{n-1+\sqrt{n-1}}{\sqrt{n}}$.

We will use the inequality $\sqrt{|a|} + \sqrt{|b|} \geq \sqrt{|a+b|}, a, b \in \mathbb{R}$. Then

$$\sum_{i=1, i\neq j}^{n} \sqrt{|a_i - b|} + \sqrt{n-1}\sqrt{|a_j - b|} + \sqrt{\left|\sum_{i=1}^{n} a_i\right|} \geq \sqrt{\left|\sum_{i=1, i\neq j}^{n} a_i - (n-1)b\right|}$$

$$+ \sqrt{|(n-1)b - (n-1)a_j|} + \sqrt{\left|\sum_{i=1}^{n} a_i\right|} \geq \sqrt{\left|\sum_{i=1, i\neq j}^{n} a_i - (n-1)a_j\right|}$$

$$+ \sqrt{\left|\sum_{i=1}^{n} a_i\right|} \geq \sqrt{n}\sqrt{|a_j|}, 1 \leq j \leq n.$$

Summing over all j, we get the desired inequality. Equality holds for

$$b = a_1 = \ldots = a_{n-1} = -\frac{a_n}{n-1} \neq 0$$

and permutations.

532. Answer: no.

We will prove that for any positive integer k, there exists a positive integer which is not $k-$ good. Choose an integer $m > k$ and consider all expressions of the form

$$a_1^2 + a_2^4 + \ldots + a_k^{2^k},$$

where $a_1, a_2, \ldots, a_k$ are nonnegative integers and all summands are less than m^{2^k}. Then

$$0 \leq a_i \leq m^{2^{k-i}}, 1 \leq i \leq k,$$

that is, there are $m^{2^{k-i}}$ possible values of a_i. It follows that the number of such expressions is

$$m^{2^{k-1}} \cdot m^{2^{k-2}} \cdot \ldots \cdot m = m^{2^k - 1}.$$

If $n \leq m^{2^k}$ is $k-$good, then n is just counted or contains a unique summand m^{2^k}. There are $m^{2^k-1} + k$ such numbers. From

$$m^{2^k-1} + k < m^{2^k-1} + m \leq m^{2^k-1} + (m-1)m^{2^k-1} = m^{2^k},$$

we deduce that there are numbers less than m^{2^k} that cannot be represented in the required form.

533. Let $P(x) = (x - n^2)Q(x) + 2022 = (x - n^2)\left(a_0 + a_1 x + \ldots + a_d x^d\right) + 2022$, with $a_0, a_1, \ldots, a_d$ integers. Let $r = \frac{p}{q}$, $\gcd(p, q) = 1$. From $P(r^2) = 2024$, it follows that

$$\left(p^2 - q^2 n^2\right)\left(a_0 q^{2d} + a_1 p^2 q^{2d-2} + \ldots + a_d p^{2d}\right) = 2q^{2d+2}.$$

Since $\gcd(p^2 - q^2 n^2, q) = 1$, we deduce $p^2 - q^2 n^2 | 2$, impossible, because the difference of two distinct nonzero perfect squares is at least 3.

534. By homogeneity, we can suppose that $x + y + z = 1$, so

$$\frac{|x| + |x + 4y| + |y + 7z| + 2|z|}{|x + y + z|} = |x| + 2|z| + |4 - 3x - 4z| + |1 - x + 6z|.$$

Define

$$f(z) = |x| + 2|z| + |4 - 3x - 4z| + |1 - x + 6z|.$$

Then

$$f(z) \geq f\left(\frac{x-1}{6}\right),$$

so

$$|x| + 2|z| + |4 - 3x - 4z| + |1 - x + 6z| \geq |x| + \frac{|x - 1| + |11x - 14|}{3}.$$

Define

$$g(x) = |x| + \frac{|x - 1| + |11x - 14|}{3}.$$

Then $g(x) \geq g\left(\frac{14}{11}\right)$, so the minimum value is $\frac{15}{11}$, attaind when $x = -4y = 28z$.

535. Assume that

$$2a > 2 + \frac{1}{b}, 2b > 2 + \frac{1}{c}, 2c > 2 + \frac{1}{a},$$

then multiplying, we get

$$8abc = 16 > \left(2 + \frac{1}{a}\right)\left(2 + \frac{1}{b}\right)\left(2 + \frac{1}{c}\right).$$

By the AM–GM inequality, we have

$$\left(2+\frac{1}{a}\right)\left(2+\frac{1}{b}\right)\left(2+\frac{1}{c}\right)\geq 8\sqrt{\frac{8}{abc}}=16,$$

contradiction.

536. For positive integers k, n, let $e_{k,\,n}(x_1, \ldots, x_k)$ be the nth elementary symmetric polynomial in variables $x_1, \ldots, x_k$. Define inductively a sequence of sets: $S_3 = \{1, 2, 3\}$, $S_k = \{a_1, \ldots, a_k\}$, with elements written in increasing order,

$$S_{k+1} = \{a_1, \ldots, a_{k-1}, a_k + 1, e_{k,k-1}(a_1, \ldots, a_{k-1}, a_k + 1)\}.$$

By induction, we will prove that

$$(1) e_{k,k}(a_1, \ldots, a_k) = e_{k,k-2}(a_1, \ldots, a_k);$$

$$(2)\ e_{k,k}(a_1, \ldots, a_{k-1}, a_k + 1) = e_{k,k-2}(a_1, \ldots, a_{k-1}, a_k + 1) + 1.$$

It is easy to verify these statements for $k = 3, 4$. Suppose that $S_{k\,-\,1} = \{b_1, \ldots, b_{k\,-\,1}\}$, $b_i = a_i$ for $i < k - 1$, $b_{k\,-\,1} = a_{k\,-\,1} - 1$. By definition, $a_k = e_{k\,-\,1,\,k\,-\,2}(a_1, \ldots, a_{k\,-\,1})$.

We have

$$\begin{aligned}
e_{k,k-2}(a_1, \ldots, a_k) &= e_{k-1,k-2}(a_1, \ldots, a_{k-1}) + a_k e_{k-1,k-3}(a_1, \ldots, a_{k-1}) \\
&= a_k(1 + e_{k-1,k-3}(a_1, \ldots, a_{k-1})) = a_k(1 + e_{k-1,k-3}(b_1, \ldots, b_{k-2}, b_{k-1} + 1)) \\
&= a_k e_{k-1,k-1}(b_1, \ldots, b_{k-2}, b_{k-1} + 1) = a_k e_{k-1,k-1}(a_1, \ldots, a_{k-1}) \\
&= e_{k,k}(a_1, \ldots, a_k).
\end{aligned}$$

which shows (1). Further,

$$\begin{aligned}
e_{k,k}(a_1, \ldots, a_{k-1}, a_k + 1) &= e_{k,k}(a_1, \ldots, a_k) + e_{k-1,k-1}(a_1, \ldots, a_{k-1}) \\
&= e_{k,k-2}(a_1, \ldots, a_k) + e_{k-1,k-1}(b_1, \ldots, b_{k-2}, b_{k-1} + 1) = e_{k,k-2}(a_1, \ldots, a_k) \\
&+ e_{k-1,k-3}(b_1, \ldots, b_{k-2}, b_{k-1} + 1) = e_{k,k-2}(a_1, \ldots, a_k) + e_{k-1,k-3}(a_1, \ldots, a_{k-1}) \\
&+ 1 = e_{k,k-2}(a_1, \ldots, a_{k-1}, a_k + 1) + 1,
\end{aligned}$$

which proves (2). The proposed statement is equivalent to (1).

537. There exist y_i, $1 \leq i \leq n$, such that $y_i \geq 0$ for $2 \leq i \leq n$ and

$$x_k = \sum_{i=1}^{k} y_i.$$

Define

$$\sum b_i^2 = \sum a_i^2 = T.$$

Then

$$\sum (x_i - b_i)^2 = \sum x_i^2 + \sum b_i^2 - 2\sum x_i b_i = T + \sum x_i^2$$
$$- 2\sum (y_1 + \ldots + y_i) b_i = T + \sum x_i^2 - 2\sum (b_i + \ldots + b_n) y_i = T + \sum x_i^2$$
$$+ 2\sum (b_1 + \ldots + b_{i-1}) y_i.$$

Since $y_i \geq 0$ for $i \geq 2$, if $b_1 + \ldots + b_{i-1} \geq 0$, then the sum is at least T, value attained when all y_i are 0.

Suppose now that $b_1 + \ldots + b_k = -c < 0$ and take all $y_i = 0$, except $y_{k+1} = Y$. Then we have $T + (n-k)Y^2 - 2cY$ and taking Y sufficiently small, but positive, we can make this expression less than T, which finishes the proof.

538. Answer: for all $n \geq 4$.

If $n = 3$ and a, b, c are integers such that $c = a + b$ and $\frac{1}{a} = \frac{1}{b} + \frac{1}{c}$, then

$$b^2 - ab - a^2 = 0,$$

with discriminant $\Delta = 5a^2$, which is not a perfect square.

For $n \geq 4$, we will prove by induction the existence of solutions. If $n = 4$, then $(1, 2, 3, 6)$ is a solution. If $n = 5$, then $(9, 25, 30, 36, 100)$ is a solution. Assume that $(a_1, \ldots, a_n)$ is a solution for some $n \geq 4$. It is easy to verify that

$$(a_1, \ldots, a_{n-1}, 2a_n, 3a_n, 6a_n)$$

is a solution with $n + 2$ components, which finishes the proof by induction.

539. Write $3P_n(x) + 1 = (3P_{n-1}(x) + 1)(3P_{n-2}(x) + 1)$. From $u_n = u_{n-1}u_{n-2}$, we get

$$u_n = u_{n-2}^2 u_{n-3} = u_{n-3}^3 u_{n-4}^2 = \ldots = u_1^{F_n} u_0^{F_{n-1}},$$

where $(F_n)_{n \geq 0}$ is the Fibonacci sequence, so

$$3P_n(x) + 1 = (3P_1(x) + 1)^{F_n} (3P_0(x) + 1)^{F_{n-1}} = (3x + 7)^{F_n}.$$

Now, the result follows from $F_n | F_{mn}$ for any positive integers m, n.

540. Let $f(x) = \frac{2x}{1+x^2}$. Then $f(f(f(a))) = a$. The fixed points of the function f are $x \in \{-1, 0, 1\}$. We have $f(i\tan\alpha) = i\tan 2\alpha$, so the fixed points of $f \circ f \circ f$ are solutions of the equation $\tan 8\alpha = \tan\alpha$, whence $\alpha = \frac{k\pi}{7}, k \in \mathbb{Z}$. The real solutions are $(-1, -1, -1), (0, 0, 0)$ and $(1, 1, 1)$.

The complex solutions are $\left(i\tan\frac{\pi}{7}, i\tan\frac{2\pi}{7}, i\tan\frac{4\pi}{7}\right)$ and circular permutations.

541. Let A, B, $C \in \mathbb{R}^2$ such that ΔABC is an equilateral triangle with $AB = 1$. Reflect A across BC to get the rhombus $ABA'C$. Then

$$f(A)+f(B)+f(C)=f(A')+f(B)+f(C)=0,$$

hence $f(A)=f(A')$ for any two points with $AA'=\sqrt{3}$.

Take any point A, then there exists a finite sequence of points $A=X_0, X_1, \ldots, X_n = O$ such that $X_iX_{i+1}=\sqrt{3}$, so $f(A)=f(O)$, hence $f(A)=0$.

Construction of the sequence: if $AO=r<\sqrt{3}$, then draw the perpendicular bisector of AO and let X_1 be a point on it such that $AX_1=\sqrt{3}$. The sequence A, X_1, O satisfies the condition. If $r>\sqrt{3}$, then define X_1, …, $X_{\lfloor r\rfloor}$ such that $X_iX_{i+1}=\sqrt{3}$ and A, X_i, O are collinear. As $\{r\}<1<\sqrt{3}$, we can use the first case.

542. Simplifying, we get

$$\frac{a_{n+2}}{a_{n+1}}-1=\frac{\left(\frac{a_2^2}{a_1^2}+\frac{a_2}{a_1}+1\right)\ldots\left(\frac{a_{n+1}^2}{a_n^2}+\frac{a_{n+1}}{a_n}+1\right)}{2^n}.$$

By the AM−GM inequality,

$$\frac{a_{i+1}^2}{a_i^2}+\frac{a_{i+1}}{a_i}+1\geq\frac{3a_{i+1}}{a_i},$$

hence

$$\frac{a_{n+2}}{a_{n+1}}>\frac{3^n a_{n+1}}{2^n}\Longrightarrow\frac{a_{n+1}^2}{a_{n+2}}<\left(\frac{2}{3}\right)^n.$$

Now, we need to show that

$$\frac{1}{2}+\sum_{n\geq 1}\left(\frac{2}{3}\right)^n<3,$$

which is true.

543. Answer: $P(x) = c$; $P(x) = x + 1$.

Assume that P is nonconstant and choose n such that $P(n) \neq 0$. If $m = q - n$, then q divides $P^{(m)}(n) - P^{(n)}(-n)$. The sequence $a_m = P^{(m)}(n)$ is eventually periodic modulo q. Let T be the length of the fundamental period, then $T \leq q$. If $T = q$, then $\{P^{(r)}(n), \ldots, P^{(r+q)}(n)\}$ is a complete residue system mod q.

If $T < q$, then gcd $(T, q) == 1$. For $1 \leq s \leq q$, there is a positive integer k such that $m + qk \equiv s(\bmod T)$. Hence

$$q\left|P^{(m+qk)}(n) - P^{(n)}(m+qk) \Rightarrow q\right|P^{(s)}(n) - P^{(n)}(m).$$

In this case, $T = 1$. If for all large enough q we have $T = q$, then for all large enough q the set $\{P(0), \ldots, P(q-1)\}$ is a complete residue system modulo q. If deg $P > 1$, then consider $P(x + 1) - P(x)$ and use Schur's theorem to arrive at a contradiction.

We conclude that there are infinitely many q with $T = 1$. Observe that in this case,

$$P^{(2q-n+1)}(n) \equiv P^{(2q-n)}(n)(\bmod q)$$

for a fixed n and infinitely many q. Since

$$P^{(2q-n)}(n) - P^{(n)}(-n)$$
$$\times\left|P\left(P^{(2q-n)}(n)\right) - P\left(P^{(n)}(-n)\right) = P^{(2q-n+1)}(n) - P^{(n+1)}(-n),\right.$$

it follows that $q|P^{(2q-n+1)}(n) - P^{(n+1)}(-n)$. Since $P^{(2q-n+1)}(n) \equiv P^{(2q-n)}(n)(\bmod q)$,

it follows that $P^{(n+1)}(-n) \equiv P^{(n)}(-n)(\bmod q)$, that is, for all large $l \geq n$, $P^{(l)}(-n)$ is a solution of $P(x) \equiv x(\bmod q)$. Choose $q > \max(|P^{(l)}(-n)|, |P^{(l+1)}(-n)|)$, we deduce $P^{(l+1)}(-n) = P^{(l)}(-n)$, that is, $P^{(l)}(-n)$ is a root of $P(x) - x$, hence $P(x) - x$ has the roots $P^{(n)}(-n)$, $P^{(n+1)}(-n)$, $P^{(n+2)}(-n), \ldots$

If $P(x) - x$ is not constant, then $|P(x)| > |x|$ for all sufficiently large x, hence

$$|-n| < |P(-n)| < \left|P^{(2)}(-n)\right| < \ldots$$

are all distinct roots of $P(x) - x$, contradiction.

544. If $p = 2$, then deg $P < 1$, therefore P is constant.

If $p > 2$ and $P(1) = 0$, then P has $p >$ deg P roots, so $P = 0$. If $P(1) \neq 0$, then we may suppose that $|P(1)| = |P(2)| = \ldots = |P(p)| = 1$. By the Lagrange Interpolation,

$$\sum_{i=0}^{p-1}(-1)^i\binom{p-1}{i}P(i+1) = 0.$$

Since $\binom{p-1}{i} \equiv (-1)^i(\bmod p)$, we get

$$0 = \sum_{i=0}^{p-1} (-1)^i \binom{p-1}{i} P(i+1) \equiv \sum_{i=0}^{p-1} P(i+1) (\text{mod}\, p).$$

Using $P(i + 1) = \pm 1$, we obtain

$$-p \leq \sum_{i=0}^{p-1} P(i+1) \leq p,$$

hence

$$\sum_{i=0}^{p-1} P(i+1) \in \{-p, 0, p\}.$$

The sum cannot be 0, because p is odd, so it is either p or $-p$, hence all values are 1 or -1. In both cases we conclude that P is constant.

545. Answer: $P(x) = x^{2023}$.

Let $-x_1 \leq \ldots \leq -x_k \leq 0 \leq x_{k+1} \leq \ldots \leq x_{2023} < 1$ be the roots of P, with $x_i \geq 0$ for all i. By the Viète's relation,

$$\sum_{1 \leq i \leq k} x_i = \sum_{k+1 \leq i \leq 2023} x_i.$$

We have $P(1) = (1 + x_1)\ldots(1 + x_k)(1 - x_{k+1})\ldots(1 - x_{2023}) = 1$. Each factor is positive, so we can use the AM–GM inequality:

$$1 \leq \left(1 + \frac{1}{2023}\left(\sum_{1 \leq i \leq k} x_i - \sum_{k+1 \leq i \leq 2=23} x_i\right)\right)^{2023} = 1,$$

whence $1 + x_1 = \ldots = 1 + x_k = 1 - x_{k+1} = \ldots = 1 - x_{2023}$. This implies $x_i + x_j = 0$ for all $i \in \{1, \ldots, k\}, j \in \{k + 1, \ldots, 2023\}$, so all x_i are 0.

546. From

$$P(x) = x^n Q\left(\frac{1}{x}\right) \geq Q(x) = x^n P\left(\frac{1}{x}\right) \geq x^n Q\left(\frac{1}{x}\right),$$

it follows that $P(x) = Q(x)$ for all x.

3.53 2024

547. We have

$$\sum_{\text{cyc}}\frac{a^6}{b^4+c^3}\geq\sum_{\text{cyc}}\frac{a^6}{b^4+3c^4}\geq\frac{\left(\sum\limits_{\text{cyc}}a^4\right)^2}{\sum\limits_{\text{cyc}}\left(a^2b^4+3a^2c^4\right)}\geq\frac{\left(\sum\limits_{\text{cyc}}a^4\right)^2}{\frac{1}{4}\sum\limits_{\text{cyc}}a^2\sum\limits_{\text{cyc}}a^4+\frac{3}{4}\sum\limits_{\text{cyc}}a^2\sum\limits_{\text{cyc}}a^4}$$

$$=\frac{\sum\limits_{\text{cyc}}a^4}{\sum\limits_{\text{cyc}}a^2}\geq\frac{a^2+b^2+c^2+d^2}{4}\geq\left(\frac{a+b+c+d}{4}\right)^2.$$

Equality holds when $a=b=c=d=\frac{1}{3}$.

548. Define $g(x)=x^3+dx+e$ to be the monic polynomial with roots x_1-x_2, x_2-x_3, x_3-x_1. Then

$$d=\sum_{\text{cyc}}(x_1-x_2)(x_2-x_3)=-\sum_{\text{cyc}}x_2^2+\sum_{\text{cyc}}x_1x_2=-a^2+3b,$$

hence $|d|\leq M^2+3M$. Observe that e is the square root of the discriminant of f.

Case 1. $e\neq 0$. Then $|e|\geq 1$, $g'(x)=3x^2+d$. Since $g(0)=e$, the roots must be at least

$$\frac{|e|}{|d|+1}\geq\frac{1}{M^2+3M+1}.$$

Case 2. $e=0$. Then the roots of g are $-\sqrt{-d},0,\sqrt{-d}$. Since $x_1\neq x_2$, we cannot choose the root 0. We have

$$\left|\sqrt{-d}\right|\geq\frac{1}{M^2+3M+1}.$$

549. Answer: $f(x)=c$ for all x; $f(x)=x$ for all x.

If f is nonconstant, then

$$\sum_{k=0}^{2b}f(a+k)\equiv 0\equiv\sum_{k=0}^{2b}f(a+k+1)(\text{mod }(2b+1))\Rightarrow f(a)$$
$$\equiv f(a+2b+1)(\text{mod }(2b+1)).$$

If f is bounded, $f(x) < N$ for all x, where N is a positive integer, then $2N + 1$ divides $f(x + 2N + 1) - f(x)$ and $2N + 3$ divides $f(x + 2N + 3) - f(x)$, therefore

$$f(x) = f(x + 2N + 1) = f(x + 2N + 3).$$

From $\gcd(2N + 1, 2N + 3) = 1$, it follows that f is constant, contradiction. We deduce that f is unbounded. If $f(x) = f(y)$, $x \neq y$, then

$$\sum_{k=0}^{2b} f(x+k) = \sum_{k=0}^{2b} f(y+k) \Rightarrow \sum_{k=0}^{y-x-1} f(x+k) = \sum_{k=2b-y+x+1}^{2b} f(y+k).$$

If $y - x - 1 > 0$, then f is bounded, contradiction. If $y - x - 1 = 0$, then $f(x) = = f(x + 2b + 1)$, contradiction. We deduce that f is injective. For b odd, we have

$$\sum_{k=0}^{2b} f(a+k) \equiv f(f(a)) (\text{mod } b), \sum_{k=0}^{4b} f(a+k)$$
$$\equiv f(f(a)) (\text{mod } b) \Rightarrow \sum_{k=2b+1}^{4b} f(a+k)$$
$$\equiv f(f(a)) (\text{mod } b) \Rightarrow \sum_{k=1}^{2b} f(a+2b+k) \equiv 0$$
$$\equiv f(f(a+2b)) (\text{mod } b) \Rightarrow f(a+2b) \equiv f(f(a+2b)) (\text{mod } b) \Rightarrow f(a)$$
$$\equiv f(f(a)) (\text{mod } b) \Rightarrow f(a) = f(f(a)), \forall a \Rightarrow f(a) = a, \forall a,$$

which is indeed a solution.

550. Apply induction on $m + n$: if $m = 0$ or $n = 0$, then the assertion is obvious. Assume that it holds for $m + n - 1$ and let

$$A - \sum_{i=0}^{m} u_i, B = \sum_{j=0}^{n} b_j.$$

By the induction hypothesis,

$$\sum_{k=1}^{m+n} c_k \geq \frac{m+n}{m(n+1)} \sum_{i=1}^{m} a_i \sum_{j=0}^{n} b_j = \frac{m+n}{m(n+1)} (A - a_0) B,$$

$$\sum_{k=1}^{m+n} c_k \geq \frac{m+n}{(m+1)n} \sum_{i=0}^{m} a_i \sum_{j=1}^{n} b_j = \frac{m+n}{(m+1)n} A(B-b_0).$$

Now, it suffices to show that

$$a_0 b_0 + \max\left\{\frac{m+n}{m(n+1)}(A-a_0)B, \frac{m+n}{(m+1)n}A(B-b_0)\right\} \geq \frac{(m+n+1)AB}{(m+1)(n+1)}.$$

For

$$x = \frac{a_0}{A}, y = \frac{b_0}{B},$$

we have to prove that

$$(m+1)(n+1)xy + (m+n)\max\left\{\frac{m+1}{m}(1-x), \frac{n+1}{n}(1-y)\right\} \geq m+n+1.$$

Denote

$$t = \max\left\{\frac{m+1}{m}(1-x), \frac{n+1}{n}(1-y)\right\},$$

then

$$x \geq 1 - \frac{mt}{m+1}, y \geq 1 - \frac{nt}{n+1}$$

and it remains to show that $(m+1-mt)(n+1-nt)+(m+n)t \geq m+n+1$, or equivalently, $mn(t-1)^2 \geq 0$, which is obvious.

551. Denote

$$P_n = \prod_{k=1}^{n} a_k.$$

Then $nP_n = P_{n-1} + n$. Now, we have to prove that $P_n \leq P_{n-1}$, $n \geq 3$, which means

$$P_{n-1} \geq \frac{n}{n-1}, n \geq 3.$$

For $n = 3$ we have

$$P_2 = \frac{3}{3-1}.$$

Assume that the inequality holds for n, then

$$P_n = \frac{1}{n} P_{n-1} + 1 \geq \frac{1}{n} \cdot \frac{n}{n-1} + 1 = \frac{n}{n-1} > \frac{n+1}{n}.$$

552. We will use the Viète's relations. If p is a prime divisor of

$$a_0 = (-1)^n r_1 r_2 \ldots r_n,$$

where $r_1, r_2, \ldots, r_n$ are the roots, then $p|r_k$ for some k and $p \nmid r_i$ for $i \neq k$, since the roots are pairwise coprime. It follows that

$$p \left| \sum_{i \neq k} \prod_{j \neq i} r_j, p \nmid \prod_{j \neq k} r_j, \right.$$

so

$$p \nmid a_1 = (-1)^{n-1} \sum_{i=1}^{n} \prod_{j \neq i} r_j = (-1)^{n-1} \left(\left(\sum_{i \neq k} \prod_{j \neq i} r_j \right) + \left(\prod_{j \neq k} r_j \right) \right),$$

hence $\gcd(a_0, a_1) = 1$.

553. Observe that

$$\text{LHS} = -\sum_{i=1}^{1012} x_{2i-1}^2 x_{2i} + \sum_{i=1}^{1012} \left(\left(x_{2i}^2 - x_{2i-1}^2 \right) \left(\sum_{i < j \leq 1012} (x_{2j} - x_{2j-1}) \right) \right).$$

Using $x_{2i}^2 \geq x_{2i-1}^2$ and $x_{2j} \geq x_{2j-1}$, it suffices to prove that

$$\sum_{i=1}^{1012} x_{2i-1}^2 x_{2i} \leq 1012.$$

We have

$$3x_{2i-1}^2 x_{2i} \leq 2x_{2i-1}^3 + x_{2i}^3 \leq 3 \cdot \frac{x_{2i-1}^3 + x_{2i}^3}{2},$$

so we need the inequality

$$\sum_{i=1}^{1012} \frac{x_{2i-1}^3 + x_{2i}^3}{2} \le 1012,$$

which is just the given condition.

Equality holds when $x_{2i-1} = x_{2i}$ for all i.

554. If $|x| > 1$, then

$$1 \ge |x+y+z| + |xy+yz+zx| + |xyz| > \left|\frac{x+y+z}{x}\right| + \left|\frac{xy+yz+zx}{x^2}\right| + \left|\frac{xyz}{x^3}\right|$$
$$= \left|1 + \frac{y}{x} + \frac{z}{x}\right| + \left|-\frac{y}{x} - \frac{z}{x} - \frac{yz}{x^2}\right| + \left|\frac{yz}{x^2}\right| \ge \left|1 + \frac{y}{x} + \frac{z}{x} - \frac{y}{x} - \frac{z}{x} - \frac{yz}{x^2} + \frac{yz}{x^2}\right| = 1,$$

contradiction. Hence $|x| \le 1$ and similarly, $|y| \le 1$, $|z| \le 1$.

555. Answer: $P = mx^2 + nx + p$, $m, n, p \in \mathbb{R}$, $m > 0$, $p \le 4m$.

The given inequality is

$$m\left(a^2+a\right)^2 + n\left(a^2+a\right) + p \ge a\left[m(a+1)^2 + n(a+1) + p\right], \forall a \ge 1 \Rightarrow ma^4$$
$$+ ma^3 - ma^2 - (m+p)a + p \ge 0, \forall a \ge 1,$$

whence $m > 0$. Denote $t = \frac{p}{m}$, then

$$a^4 + a^3 - a^2 - (t+1)a + t \ge 0, \forall a \ge 1 \Rightarrow (a-1)\left(a^3 + 2a^2 + a - t\right) \ge 0, \forall a$$
$$\ge 1 \Rightarrow a^3 + 2a^2 + a - t \ge 0, \forall a \ge 1.$$

The function f: $(0, \infty) \to \mathbb{R}$, $f(a) = a^3 + 2a^2 + a - t$ is increasing, so it suffices to have $f(1) \ge 0$, so $t \le 4$, whence $p \le 4m$.

556. Answer: $f(x) = 0$ for all x; $f(x) = x$ for all x;

$$f(x) = \begin{cases} 0, x \equiv 0 \pmod 2 \\ 2a, x \equiv 1 \pmod 2 \end{cases}, a \in \mathbb{Z};$$

$$f(x) = \begin{cases} 0, x \equiv 0, 2 \pmod 4 \\ 4b+2, x \equiv 1 \pmod 4 \\ 4c+2, x \equiv 3 \pmod 4 \end{cases}, b, c \in \mathbb{Z}.$$

If f is injective, then for $x = 2f(y)$ we have $f(2f(y) - 2y) = 0$, so $f(y) - y$ is constant. We can see that f is the identity in this case.

If $f(u) = f(v)$ for integers $u > v$, then we find after substituting $y = u$ and $y = v$ that f is periodic with period $2(u - v)$, so the image of f is finite and has a maximal element M. If $f(k) = M$ and $f(f(y_0)) > 0$ for some y_0, then $f(k + 2y_0 - f(y_0)) > M$, contradiction. In conclusion, $f(f(y)) \le 0$ for all y. Considering the minimal element of the image, we arrive at $f(f(y)) \ge 0$ for all y, so $f \circ f = 0$. This implies $f(0) = 0$ and the functional equation becomes

$$f(x - f(y)) = f(x - 2y).$$

Let $P = \{c \in \mathbb{Z} | f(x) = f(x + c), \forall x \in \mathbb{Z}\}$ be the set of periods of f and let k be minimal positive element of P. Observe that

1. $f(y - 2y) \in P$ for any y, in particular $2f(y) \in P$ for all y;
2. P is closed under linear combinations, so every element of P is divisible by k.

Then $f(1) - 2 \in P$, $2f(1) \in P$, $4 = 2f(1) - 2(f(1) - 2) \in P$, so $k|4$. Considering all cases and the requirement $k|f(y) - 2y$, we find the above functions.

557. The AM–GM inequality implies

$$\text{LHS} \leq \frac{\sum_{cyc} ab(b + 1 - a)}{2} = \frac{\sum_{cyc} ab(2b + c)}{2} = ab^2 + bc^2 + ca^2 + \frac{3}{2}abc$$
$$< \frac{1}{4} \Leftrightarrow 4\left(ab^2 + bc^2 + ca^2\right) + 6abc < (a + b + c)^3. \quad (1)$$

By the Schur's inequality for $r = 1$,

$$a^3 + b^3 + c^3 + 3abc \geq \sum_{\text{sym}} ab^2.$$

The inequality (1) is equivalent to $4(a^2b + b^2c + c^2a) > 6abc$, which is true by the AM–GM inequality.

558. *a*) Answer: yes.
Choose

$$n = 1012, a_0 = \frac{1 + i}{\sqrt{2}}, a_{1012} = \frac{1 - i}{\sqrt{2}}, a_1 = \ldots = a_{1011} = 0.$$

b) Answer: no.
If $a_0 a_n \neq 0$, then $\deg PQ = 2n = 2024$, whence $n = 1012$. Moreover,

$$Q(x) = x^{1012} P\left(\frac{1}{x}\right),$$

hence

$$x^{1012} P(x) P\left(\frac{1}{x}\right) = x^{2024} + 1.$$

The coefficients of terms of degree 1012 in the both sides of the equality are equal:

$$a_0^2 + a_1^2 + \ldots + a_{1012}^2 = 0,$$

therefore $a_0 = a_1 = \ldots = a_{1012} = 0$, contradiction.

559. Answer: $P(x) = (x + a)^n$, with a integer and n nonnegative integer.

Assume that $P(x)$ is a nonconstant solution. We can show by induction that for any integers $b_1, b_2, \ldots, b_k$, there exists an integer c such that

$$P(b_1)P(b_2)\ldots P(b_k) = P(c).$$

It follows that for any positive integer k and integer b, there is an integer c such that

$$(P(b))^k = P(c).$$

We will show that if $P(x)$ is not of the form $P(x) = (x + a)^n$, then there exists a constant $C > 0$ such that for any integer c with $|c| > C$, we have that $P(c)$ is never a nth power. Let

$$P(x) = x^n + a_{n-1}x^{n-1} + \ldots + a_0, \quad Q(x) = \left(x + \frac{a_{n-1}}{n}\right)^n.$$

For any $\varepsilon > 0$, there exists C_ε such that $|x| > C_\varepsilon$ implies

$$|Q(x - \varepsilon)| < |P(x)| < |Q(x + \varepsilon)|.$$

560. Answer: $f(x) = \frac{\sqrt{5}-1}{2}$ for all $x; f(x) = \frac{x}{x+1}$ for all x.

Denote $\varphi = \frac{\sqrt{5}-1}{2}$. For $x = y = \frac{1}{\varphi}$, we get $f\left(\frac{1}{\varphi}\right) = \varphi$. It is clear that $f(x) < 1$ for all x. For $y = \frac{1}{x+1}$, we get

$$f(x+1) = \frac{1}{1 + f\left(\frac{1}{x}\right)},$$

so $f(x) > \frac{1}{2}$ for all $x > 1, f(x) < \frac{2}{3}$ for all $x < 2, f(x) > \frac{3}{5}$ for all $x > \frac{2}{3}$, etc. Iterating and passing to the limit, we get that $f(x) \geq \varphi$ for all $x \geq \frac{1}{\varphi}$ and $f(x) \leq \varphi$ for all $x \leq \frac{1}{\varphi}$. Also, f is continuous at $x = \frac{1}{\varphi}$. Choosing y fix and a sequence $x_n \to \varphi$, we find that f is continuous at $\varphi y + y + \varphi$, hence on (φ, ∞) and applying once f, we find that f is continuous on $(0, \infty)$.

For $x \leq \varphi$, we have $f(xy + x + y) \leq f(\varphi y + y + \varphi)$, so f is monotonic on $(\varphi, \varphi + 1]$, similarly on $[\varphi + 1, \varphi + 2]$, etc., hence on $[\varphi, \infty)$ and applying once f we get that f is monotonic on $(0, \infty)$. In particular, f has limits at $x \to 0$ and at $x \to \infty$, so f can be extended by continuity at $x = 0$. Then

$$\lim_{x\to\infty} f(x) := M \le 1,\ \lim_{x\to 0}\left[f(xy+x+y) + f\left(\frac{1}{x}\right)f\left(\frac{1}{y}\right)\right] = f(y) + Mf\left(\frac{1}{y}\right) = 1.$$

If $M < 1$, then f is constant. If $M = 1$, then

$$f(x) + f\left(\frac{1}{x}\right) = 1.$$

From this and the above relation for $f(x+1)$, it is easy to get $f(x) = \frac{x}{x+1}, \forall x \in \mathbb{Q}_+$ and by continuity, $f(x) = \frac{x}{x+1}, \forall x \in (0, \infty)$.

561. If all terms of the sequence are negative, then we have a contradiction. Let a_k be the smallest nonnegative term, then $a_n \ge 0$ for $n \ge k$ and it is easy to see that $k = 3$. In this case, $a_3 \ge 0 > a_1 + a_2$. We will omit this situation and suppose that all terms are nonnegative.

Consider the given condition as a quadratic function of a_n:

$$a_n^2 - 2a_{n-1}a_n + a_{n-1}^2 - 2a_{n-1}a_{n-2} \ge 0,$$

with discriminant $\Delta = 8a_{n-1}a_{n-2}$ and roots $a_{n-1} \pm \sqrt{2a_{n-1}a_{n-2}}$. Then

$$a_n \ge a_{n-1} + \sqrt{2a_{n-1}a_{n-2}}, n \ge 3.$$

We will prove the inequality by induction. Denote $S_k = a_1 + \ldots + a_k$ and assume that the inequality is true for $n-1$ and $n-2$. Then

$$a_n \ge a_{n-1} + \sqrt{2a_{n-1}S_{n-2}} = a_{n-1} + \sqrt{2a_{n-2}^2 + 2a_{n-2}S_{n-3}} \ge$$

$$\ge a_{n-1} + \sqrt{2a_{n-2}^2 + 2S_{n-3}^2} \ge a_{n-1} + (a_{n-2} + S_{n-3}) = S_{n-1}.$$

562. Answer: $Q = \pm P^k$, k positive integer.

We can suppose that the leading terms of P, Q have positive coefficients, otherwise we will replace P, Q by $-P$, $-Q$. Now, we need the following result.

Theorem. If a, b are integers greater than 1 and $a^n - 1$ divides $b^n - 1$ for every positive integer n, then b is a power of a.

Proof. Lemma. Let $x_1, x_2, \ldots, x_k \ge 1$ be distinct rational numbers, $c_1, c_2, \ldots, c_k$ nonzero real numbers and (a_n) a sequence of integers such that

$$\lim_{n\to\infty}\left(a_n - \sum_{i=1}^{k} c_i x_i^n\right) = 0.$$

Then $x_1, x_2, \ldots, x_n$ are integers.

Proof of the lemma. Induction on k. Let $x_k = \frac{p}{q}$, with p, q positive integers, $p \geq q$ and $\gcd(p,q) = 1$. Denote

$$b_n = a_n - \sum_{i=1}^{k} c_i x_i^n.$$

Then

$$\lim_{n\to\infty}(qb_{n+1} - pb_n) = \lim_{n\to\infty}\left(qa_{n+1} - pa_n - \sum_{i=1}^{k-1} c_i(qx_i - p)x_i^n\right) = 0.$$

For $k = 1$, we have $a_n \neq 0$ and $qa_{n+1} - pa_n = 0$ for n sufficiently large, hence $q = 1$.

For $k > 1$, denote $c_i' = (qx_i - p)c_i, 1 \leq i \leq k-1, a_n' = qa_{n+1} - pa_n$. Then $c_i' \neq 0$, since x_i are distinct, and we can apply the induction hypothesis.

Now, we will show how the theorem follows from the lemma. There is a positive integer k such that $a^k \leq b < a^{k+1}$. Let

$$x_i = \frac{b}{a^i}, c_i = 1, 1 \leq i \leq k, a_n = \frac{b^n - 1}{a^n - 1}, n \geq 1.$$

Then

$$\lim_{n\to\infty} r_n := \lim_{n\to\infty}\left(a_n - \sum_{i=1}^{k} c_i x_i^n\right) = \lim_{n\to\infty}\left(\frac{b^n-1}{a^n-1} - \sum_{i=1}^{k}\frac{b^n}{a^{in}}\right) = \lim_{n\to\infty}\frac{b^n - a^{kn}}{(a^n-1)a^{kn}} = 0.$$

By the lemma, all x_i are integers, hence r_n is an integer, therefore $r_n = 0$ for n sufficiently large. But then $b^n = a^{kn}$ for n sufficiently large, whence $b = a^k$.

Back to the problem, applying the theorem, we conclude that $Q = P^k$ for some fixed positive integer k and for infinitely many integer values of the variable, hence $Q = P^k$ as polynomials.

References

1. Barbeau, E. J., Polynomials, Springer, 1989
2. Becheanu, M., *Problems from Mathematical Olympiads 1987–1994* (in Romanian), GIL, 1995
3. *Camp Selection Problems 2010–2018,* New Zealand Mathematical Olympiad Committee, https://www.mathsolympiad.org.nz
4. Djukić, D., Janković, V., Matić, I., Petrović, N., *The IMO Compendium, Second Edition,* Springer, 2011
5. *Estonian Math Competitions 1998–2021,* University of Tartu Youth Academy, 2021
6. Gunderson, D. S., Handbook of Mathematical Induction, CRC Press, 2011
7. Herman, J., Kučera, R., Šimša, J., Equations and Inequalities, Springer, 2000
8. India IMO Training Camp, https://olympiads.hbcse.in
9. Kin Y. Li, *Math Problem Book I,* Hong Kong Mathematical Society, 2001
10. *Lősungen zur IMO Selektion,* https://mathematical.olympiad.ch
11. Manfrino, R. B., Ortega, J. A. G., Delgado, R. V., Inequalities. A Mathematical Olympiad Approach, Birkhäuser, 2009
12. *Mathematical Olympiad in China (2007–2014),* World Scientific
13. *Olimpiadas Matemáticas Españolas 1963–2004,* Real Sociedad Matemática Española, 2004
14. *Romanian Mathematical Competitions 1996–2008,* Societatea de Ştiinţe Matematice din România
15. Sedrakyan, H., Sedrakyan, N., Algebraic Inequalities, Springer, 2018
16. Small, C. G., Functional Equations and How to Solve Them, Springer, 2007
17. *Team Selection Tests,* https://artofproblemsolving.com
18. Uţă, M., Catzaiti, A., Maftei, I. V., (eds.), *Team Selection Tests for the IMO 1975–1983* (in Romanian), Piatra Neamţ, 1984
19. Venkatachala, B. J., Inequalities. An Approach Through Problems, Second Edition, Springer, 2017
20. *Vietnamese Mathematical Competitions 2017–2019,* Hanoi, 2019

C. Mănescu-Avram, *Selection Tests in Algebra for Mathematical Olympiads*, Problem Books in Mathematics, https://doi.org/10.1007/978-3-032-01032-2

Index

C. Mănescu-Avram, *Selection Tests in Algebra for Mathematical Olympiads*, Problem Books in Mathematics, https://doi.org/10.1007/978-3-032-01032-2